Communications
in Computer and Information Science 2096

Rationale

The CCIS series is devoted to the publication of proceedings of computer science conferences. Its aim is to efficiently disseminate original research results in informatics in printed and electronic form. While the focus is on publication of peer-reviewed full papers presenting mature work, inclusion of reviewed short papers reporting on work in progress is welcome, too. Besides globally relevant meetings with internationally representative program committees guaranteeing a strict peer-reviewing and paper selection process, conferences run by societies or of high regional or national relevance are also considered for publication.

Topics

The topical scope of CCIS spans the entire spectrum of informatics ranging from foundational topics in the theory of computing to information and communications science and technology and a broad variety of interdisciplinary application fields.

Information for Volume Editors and Authors

Publication in CCIS is free of charge. No royalties are paid, however, we offer registered conference participants temporary free access to the online version of the conference proceedings on SpringerLink (http://link.springer.com) by means of an http referrer from the conference website and/or a number of complimentary printed copies, as specified in the official acceptance email of the event.

CCIS proceedings can be published in time for distribution at conferences or as postproceedings, and delivered in the form of printed books and/or electronically as USBs and/or e-content licenses for accessing proceedings at SpringerLink. Furthermore, CCIS proceedings are included in the CCIS electronic book series hosted in the SpringerLink digital library at http://link.springer.com/bookseries/7899. Conferences publishing in CCIS are allowed to use Online Conference Service (OCS) for managing the whole proceedings lifecycle (from submission and reviewing to preparing for publication) free of charge.

Publication process

The language of publication is exclusively English. Authors publishing in CCIS have to sign the Springer CCIS copyright transfer form, however, they are free to use their material published in CCIS for substantially changed, more elaborate subsequent publications elsewhere. For the preparation of the camera-ready papers/files, authors have to strictly adhere to the Springer CCIS Authors' Instructions and are strongly encouraged to use the CCIS LaTeX style files or templates.

Abstracting/Indexing

CCIS is abstracted/indexed in DBLP, Google Scholar, EI-Compendex, Mathematical Reviews, SCImago, Scopus. CCIS volumes are also submitted for the inclusion in ISI Proceedings.

How to start

To start the evaluation of your proposal for inclusion in the CCIS series, please send an e-mail to ccis@springer.com.

Abbas M. Al-Bakry · Mouayad A. Sahib ·
Safaa O. Al-Mamory · Jaafar A. Aldhaibani ·
Ali N. Al-Shuwaili · Haitham S. Hasan ·
Rula A. Hamid · Ali K. Idrees
Editors

New Trends in Information and Communications Technology Applications

7th National Conference, NTICT 2023
Baghdad, Iraq, December 20–21, 2023
Proceedings

 Springer

Editors
Abbas M. Al-Bakry ⓘ
University of Information Technology
and Communications
Baghdad, Iraq

Safaa O. Al-Mamory ⓘ
University of Babylon
Babylon, Iraq

Ali N. Al-Shuwaili ⓘ
University of Information Technology
and Communications
Baghdad, Iraq

Rula A. Hamid ⓘ
University of Information Technology
and Communications
Baghdad, Iraq

Mouayad A. Sahib ⓘ
University of Information Technology
and Communications
Baghdad, Iraq

Jaafar A. Aldhaibani ⓘ
University of Information Technology
and Communications
Baghdad, Iraq

Haitham S. Hasan ⓘ
University of Information Technology
and Communications
Baghdad, Iraq

Ali K. Idrees ⓘ
University of Babylon
Babylon, Iraq

ISSN 1865-0929   ISSN 1865-0937 (electronic)
Communications in Computer and Information Science
ISBN 978-3-031-62813-9   ISBN 978-3-031-62814-6 (eBook)
https://doi.org/10.1007/978-3-031-62814-6

This Springer imprint is published by the registered company Springer Nature Switzerland AG
The registered company address is: Gewerbestrasse 11, 6330 Cham, Switzerland

If disposing of this product, please recycle the paper.

Preface

The 7th National Conference on New Trends in Information and Communications Technology Applications (NTICT 2023) was hosted and organized by the University of Information Technology and Communications, Baghdad, Iraq. It was held in Baghdad on December 20–21, 2023. NTICT 2023 was a specialized conference that aimed to present the latest research related to information and communications technology applications. NTICT was the first conference in Iraq to have its proceedings published in Communications in Computer and Information Science (CCIS) by Springer. NTICT 2023 aimed to provide a meeting for an advanced discussion of evolving applications in artificial intelligence, machine learning and computer networks. The conference brought both young researchers and senior experts together to share novel findings and practical experiences in their fields. The call for papers resulted in a total of 103 submissions from which only 92 articles passed the prescreening stage. In the second stage, each submission was assigned to at least three international reviewers for double-blind reviews. After receiving the review comments, the Program Committee decided to accept only 28 papers based on a highly selective review process with an acceptance rate of 30%. The accepted papers are categorized in two fields; 21 papers in AI and machine learning and 7 papers in computer networks. The conference organizing committee would like to thank all who contributed to the success of this conference, in particular the members of the Program Committee and the respected international reviewers for their scientific efforts in carefully reviewing the contributions and selecting high-quality papers. The reviewers' efforts to submit the review reports within the specified period were greatly appreciated. Their comments were very helpful in the selection process. Furthermore, we would like to convey our gratitude to the keynote speakers for their excellent presentations. A word of thanks is extended to all authors who submitted their papers and for letting us evaluate their work. The submitted papers were managed using Microsoft's Conference Management Toolkit. Thanks to the Springer team for publishing these proceedings. We hope that all participants enjoyed a successful conference, made a lot of new contacts, engaged in fruitful discussions, and had a pleasant stay in Baghdad.

December 2023

Abbas M. Al-Bakry
Mouayad A. Sahib
Safaa O. Al-Mamory
Jaafar A. Aldhaibani
Ali N. Al-Shuwaili
Haitham S. Hasan
Rula A. Hamid
Ali K. Idrees

Organization

General Chair

Abbas M. Al-Bakry — President of UOITC, Iraq

Program Chairs

Mouayad A. Sahib	UOITC, Iraq
Safaa O. Al-mamory	UOITC, Iraq
Jaafar A. Aldhaibani	UOITC, Iraq
Ali N. Al-Shuwaili	UOITC, Iraq
Haitham S. Hasan	UOITC, Iraq
Rula A. Hamid	UOITC, Iraq
Ali K. Idrees	University of Babylon, Iraq

Steering Committee

Loay Edwar George	UOITC, Iraq
Mahdi N. Jasim	UOITC, Iraq
Inaam Rikan Hassan	UOITC, Iraq
Ahmed Sabah Ahmed	UOITC, Iraq
Nagham Hamid Abdul-Mahdi	UOITC, Iraq

International Scientific Committee

Abdelnaser Omran	Bright Star University, Libya
Adham R. Azeez	UOITC, Iraq
Ahmad Abdulfattah	Ninevah University, Iraq
Ahmad Salim	Middle Technical University, Iraq
Ahmed Al-Azawei	University of Babylon, Iraq
Ahmed Ali	Al-Iraqia University, Iraq
Ahmed Fanfakh	University of Babylon, Iraq
Ahmed Hussein	UOITC, Iraq
Ahmed Jader	Al-Furat Al-Awsat Technical University, Iraq
Ahmed Hussein	University of Babylon, Iraq

Aini Syuhada	Md Zain Universiti Malaysia Perlis, Malaysia
Alaa Abdulateef	Universiti Utara Malaysia, Malaysia
Alaa Faieq	UOITC, Iraq
Alaa Yaseen Taqa	University of Mosul, Iraq
Aladdin Alsharify	University of Babylon, Iraq
Alejandro Zunino	National Univ. of Central BA, Argentina
Ali Al-Shuwaili	UOITC, Iraq
Ali Kareem	University of Technology, Iraq
Ali Alnooh	UOITC, Iraq
Ali Kadhum Idrees	University of Babylon, Iraq
Amera Melhum	University of Duhok, Iraq
Anas Al-Shabandar	UOITC, Iraq
Assad H. Al-Ghrairi	Al-Nahrain University, Iraq
Atheer Al-Chalabi	UOITC, Iraq
Athraa Jani	Al-Nahrain University, Iraq
Bahaa Al-Musawi	University of Kufa, Iraq
Balqees Talal Hasan	Ninevah University, Iraq
Bayadir Razaq	Southern Technical University, Iraq
Buthaina Abed	UOITC, Iraq
Emad Mohammed	Northern Technical University, Iraq
Esraa Alwan	University of Babylon, Iraq
Farah Alani	University of Anbar, Iraq
Firas Abedi	Al-Furat Al-Awsat Technical University, Iraq
Ghada Alkateb	UOITC, Iraq
Haitham Hasan	UOITC, Iraq
Hamidah Ibrahim	Universiti Putra Malaysia, Malaysia
Haneen Ahmed	University of Baghdad, Iraq
Harith Al-Badrani	Ninevah University, Iraq
Hasan Al-Khaffaf	University of Duhok, Iraq
Hassan Harb	American University of the Middle East, Kuwait
Haydar Al-Tamimi	University of Technology, Iraq
Hiba Salim	University of Baghdad, Iraq
Hind Ghazi	UOITC, Iraq
Ibtisam Aljazaery	University of Babylon, Iraq
Idress Husien	University of Kirkuk, Iraq
Iman Abduljaleel	University of Basrah, Iraq
Intisar Al-Mejibli	UOITC, Iraq
Jaafar Aldhaibani	UOITC, Iraq
Joseph Azar	University Bourgogne Franche-Comté, France
Kadhim Swadi	Esraa University, Iraq
Khitam Salman	University of Technology, Iraq
Layla H. Abood	University of Technology, Iraq

Litan Daniela	Hyperion University, Romania
Lubna Alkahla	Ninevah University, Iraq
Maha Ibrahim	UOITC, Iraq
Moceheb Shuwandy	Tikrit University, Iraq
Mohammad Alhisnawi	University of Babylon, Iraq
Mohammad Rasheed	UOITC, Iraq
Mohammed Ali	UOITC, Iraq
Mohammed Al-Neama	University of Mosul, Iraq
Muhammad Raheel Mohyuddin	National University of Science & Technology, Oman
Mustafa Kadhm	Mustansiriyah University, Iraq
Muthana Mahdi	Mustansiriyah University, Iraq
Nada Rasheed	Al-Karkh University of Science, Iraq
Nadhir Abdulkhaleq	UOITC, Iraq
Nadia Albakri	Al-Nahrain University, Iraq
Nor Azliana Jamaludin	Universiti Pertahanan Nasional Malaysia, Malaysia
Omar Abdulrahman	University of Anbar, Iraq
Qabeela Thabit	Minstry of Education, Iraq
Rabab Abbas	University of Technology, Iraq
Rula Hamid	UOITC, Iraq
Ruslan Al-Nuaimi	Al-Nahrain University, Iraq
Saba Hamada	UOITC, Iraq
Saba Tuama	UOITC, Iraq
Safaa Al-mamory	UOITC, Iraq
Saif Mahmood	Middle Technical University, Iraq
Samar Taha	UOITC, Iraq
Sanaa Kadhim	UOITC, Iraq
Sarah Abdullah	UOITC, Iraq
Sarmad Ibrahim	Mustansiriyah University, Iraq
Sarmad Hadi	Al-Nahrain University, Iraq
Sawsen Mahmood	Mustansiriyah University, Iraq
Shaheen Abdulkareem	University of Duhok, Iraq
Shaimaa Al-Abaidy	University of Baghdad, Iraq
Shayma Nourildean	University of Technology, Iraq
Suhad Ali	University of Babylon, Iraq
Sumaya Hamad	University of Anbar, Iraq
Tara Yahiya	Université Paris-Saclay, France
Thaker Nayl	UOITC, Iraq
Vijay Kumar	Visvodaya Institute of Technology and Science, India
Wathiq Al-Yaseen	Al-Furat Al-Awsat Technical University, Iraq

Wisal Abdulsalam	University of Baghdad, Iraq
Yaseen Ismael	University of Mosul, Iraq
Yasmin Mohialden	Mustansiriyah University, Iraq
Yossra Ali	University of Technology, Iraq
Yousif Hamad	University of Kirkuk, Iraq
Yousra Fadil	University of Diyala, Iraq
Zeyad Younus	University of Mosul, Iraq
Ziad AlAbbasi	Middle Technical University, Iraq

Secretary Committee

Suhaib S. Al-Shammari	UOITC, Iraq
Atheer M. Al-Chalabi	UOITC, Iraq
Shoug R. Noori	UOITC, Iraq
Ahmed Jassim Jabur	UOITC, Iraq

Contents

Artificial Intelligence and Machine Learning

Nanoscale Communication Redefined: Exploring Bio-Inspired Molecular Systems

Athraa Juhi Jani[1(✉)] and Jafar J. Jani[2]

[1] Al-Nahrain University, Kadhmiya, Baghdad, Iraq
athraajuhi@nahrainuniv.edu.iq
[2] Ministry of Labour and Social Affairs, Baghdad, Iraq

Abstract. This paper explores a novel communication model employing a channel with multiple interconnected nodes between the sender and recipient nanomachines. The channel mimics a tissue-like structure, where each cell represents a node, and communication between adjacent cells is based on calcium signalling, a well-established biological mechanism. The nodes' behaviour within the channel is inspired by the Abelian sandpile model (ASM). The suggested model in this paper has been evaluated using the verification tool PRISM model checker. Various experiments had been conducted for the verification of acknowledgement reception. The experiments were performed on channels of various sizes to assess the system's performance and behaviour. The results highlight that the reception of acknowledgments is influenced by the size of the channel between the sender and recipient nanomachines.

Keywords: Bio-inspired Communication · Abelian sandpile model · Molecular Communication · Calcium Signalling · PRISM Verification

1 Introduction

Future network applications encounter numerous challenges that need to be addressed. As networks continue to expand in size and complexity, they encounter a multitude of challenges, such as the dynamic characteristic of diverse architectures, limited resources, and various other hurdles. However, nature has already tackled and overcome these challenges through millions of years of evolution. This gives an inspiration to draw from biological mechanisms and integrate them into the planning and execution of network designs and implementations [1,2]. Biological systems possess several inherently appealing features, including robustness and the ability to withstand failures, adaptability to changing environments, and the capability to exhibit intricate behaviours while relying on a limited set of fundamental principles [1]. Drawing inspiration from biological systems, the nodes within the network are conceptualized as group of entities or individuals capable of interacting and performing specific tasks. As

A. M. Al-Bakry et al. (Eds.): NTICT 2023, CCIS 2096, pp. 3–17, 2024.
https://doi.org/10.1007/978-3-031-62814-6_1

a result, accomplishing intricate objectives becomes possible through the cooperative behaviour demonstrated by this community of agents [3,4]. Drawn from biological inspiration, the communication in nano scale emerges as an extremely encouraging approach for facilitating information exchange among devices at the nano-scale [5].

Research Background: Nanotechnology is a field with a great potential for breakthroughs, encompasses the manipulation of substances at the atomic and molecular levels to fabricate structures, devices, systems, and materials. It facilitates the creation of devices spanning a range of sizes, from one to hundreds of nanometers. Within this domain, nanonetworks have emerged as a novel research branch, combining nanotechnology with digital communication [6,7]. Within any nano-system, the nanomachine is a fundamental block, serving as an essential functional unit. Nanomachines consist of components designed to execute specific tasks, including communication, data sorting, sensing, computation, and actuation, all at the nano scale. When these nanomachines communicate with one another, they form a nanonetwork, amplifying the capabilities and potential applications beyond those of individual nanomachines. This interaction enables the execution of complex tasks in a distributed manner, allowing for enhanced functionality and expanding the possibilities within the nano-system. [6,8]. Molecular communication stands out as a highly promising paradigm that have been used for facilitating interaction among nanomachines. Drawing inspiration from biological systems, this approach emulates the observed communication patterns in nature, where molecules serve as the main intermediary of information exchange. Thus, within nanonetworks, molecules are facilitating the transmission of information between nanomachines by functioning as carriers of communication. The biochemical responses induced by these informational molecules on the receiving end, play a crucial role in decoding and processing the communicated information. Molecular communication offers a compelling approach by emulating these natural processes, for enabling effective communication and coordination among nanomachines [9,10].

Related Work: Various types of molecular communication systems draw inspiration from the communication mechanisms observed in living cells [11]. The exploration of cellular communication processes have been driven by the remarkable characteristics of cells, such as their size (typically around 10 μm) and mass (approximately 1 nanogram). This has led to studies focused on understanding these mechanisms cells [12], with the aim of applying them to the implementation of nanonetworks in biological scenarios [13]. In molecular communication methods, information is conveyed using the quantity of molecules. In such way that the transmitter nanomachine uses molecules to represent the encoded information, it initiates their propagation throughout the medium of communication towards the recipient nanomachine(s) [14]. The authors in [15] presented a novel assessment of the concept of "age of information" in the channels of molecular systems. They had focused on the widely applicable and straightforward discrete-time channel model to describe these molecular channels. One of the employed molecular communication techniques is calcium signalling. In molecular communication based on calcium signalling, the information is represented using calcium ions to facilitate communication from the sender nanomachine to the recipient nanomachine(s) [16]. Within biological systems, calcium signalling serves as a means of communication between neighboring cells through direct access. In cellular tissues, the intercellular membrane contains gap junctions, allowing the passage of molecules and ions, including calcium ions, from the sender cell to adjacent cells [6]. The authors in [17] have proposed and analyzed a communication system employing calcium signalling as its foundation. Additionally in [18], an investigation comparing calcium signalling in three different cell types was performed. In [19] an extensive review of the mechanisms utilized for the advancement of molecular communication systems using calcium signaling, had been presented. It covers various aspects such as information molecule encoding, modulation techniques, propagation mechanisms, and decoding methods. An analytical approximation for a 2-Rx single-input multiple-output (SIMO) topology has been presented in [20]. Where the authors proposed this approximation based on analytical expressions derived from molecular single-input single-output (SISO) systems. The propagation of molecules through the medium has been modeled as a continuous Wiener process with independent step increments. The authors defined an absorbing recipient, in such a way that once a molecule is absorbed by one Rx, it cannot be absorbed by another Rx. For each Rx, the channel response was analytically derived from the molecular SISO response, assuming no other Rx is present. In [21] the authors explored the capability of nanonodes in distinguishing specific diffusion characteristics, which could provide valuable insights into the dynamics of the networks based on molecular communication. The primary emphasis of their research was to assess the channel performance in molecular communication systems. Thus, an algorithm was proposed to determine the maximum distance diffused molecules can effectively propagate. In addition, they have investigated the nanomachines' ability to distinguish the diffusion patterns of one nanomachine located at a distance d_0 from another nanomachine within the same nanonetwork.

The Aim of this Paper. In this paper, we investigate a communication model that employs a channel composed of multiple locations (nodes) situated between two nanomachines a sender and a recipient. This channel has the potential to be conceptualized as a tissue-like structure consisting of cells that are interconnected. The initial cell symbolizes the sender, while the final cell represents the recipient, on the other hand, the intermediary cells represent the channel. The communication technique employed between neighbouring cells in this model is calcium signalling, which is well-established in biology [22]. However, the mechanism utilized by the nodes in the channel is inspired by the Abelian sandpile model (ASM) [23]. To evaluate the proposed model and analyze the communication channel's performance, using the model checker tool PRISM, the verification of acknowledgments' reception in a one-dimensional channel was conducted. The verification experiments were performed on channels of different sizes, allowing us to assess the performance and behaviour of the system.

2   Model

The considered model comprises of a single sender nanomachine and a single recipient nanomachine. The channel facilitating communication between the sender and recipient nanomachines comprises a series of interconnected locations. The model's main objective is ensuring that the receiver nanomachine effectively detects the information molecules emitted by the transmitter nanomachine. This is achieved by using a verification tool to validate the reception of acknowledgments [24]. We can visualize the proposed system as a tissue comprising neighboring cells, through which the information molecules released by the sender nanomachine travel to make their way towards the recipient nanomachine, similar to calcium signalling communication concept [22]. Yet, the governing rules of the channel nodes draw inspiration from the ASM [23,25], The fundamental concept of ASM revolves around various locations, each associated with values symbolizing grains on a sandpile slope. With time, these locations accumulate sand grains, gradually forming a pile. As the pile grows, reaching a certain predefined value (a threshold), the location collapses, and some sand grains move to neighboring locations. This process raises the slope of these adjacent locations [26]. However, the distinction that in our implementation, molecules replace the sand particles.

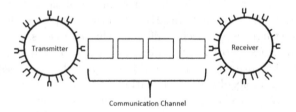

Fig. 1. System Configuration

The system's environment comprises of n locations denoted as S_i, arranged in a linear configuration. Where i belongs to the set $1, 2, \ldots, n$, represents the individual nodes that make up the channel. The nodes, represented by S_i, are positioned at a distance $dist$ separating the sender nanomachine S_T and the recipient nanomachine S_R within a liquid medium. At first, the molecular concentration is randomly distributed among the nodes of the channel S_i, the sender S_T, and the recipient S_R. In the context of the ASM, overtime, more sand grains possibly added up to the pile. In the model we present, the recipient nanomachine S_R is capable of detecting and gathering molecules from the environment, with the condition that the sensed molecules are at least α, where $\alpha \in 0, 1, 2, 3, \ldots, L$. In this model, we will assume the possibility that the environment may contain molecules that could be considered as noise. However, for the sake of simplicity, we will neglect the presence of these noise molecules. Incorporating noise into the model will be considered for future work [24].

In Fig. 2, the vertical arrows represent the thresholds present in the sender, recipient, and intermediate nodes. Within the channel nodes S_i, whenever the molecular concentration exceeds the threshold L, it will be uniformly distributed among its neighboring nodes on both sides, as the bold arrows illustrate in Fig. 2. For instance, if the molecular concentration in node S_2 exceeds the threshold, the excess molecules will be passed on to nodes S_1 and S_3. Similarly, the surplus molecules in node S_4 will be conveyed to nodes S_3 and S_R. Nevertheless, for node S_R, any molecular concentration above the threshold L will be directed to node S_4. Meanwhile, sender nanomachine S_T operates with a distinct threshold Y, which surpasses L. Therefore, S_T will gather excess molecules from its neighboring nodes until it reaches a concentration of Y molecules. Once the concentration reaches Y, the sender nanomachine τ emits this concentration to its neighbouring node. It is important to note that All nodes in the network, including S_T and S_R, have buffer sizes of E, where $E > Y$, and $Y > L$. This allows the nodes to store molecules in their buffers up to their specified capacities [24].

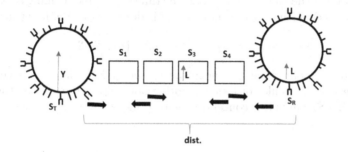

Fig. 2. Model Representation

The model operates on a mechanism based on time slots, with a time slot duration represented as τ. The value of τ is is determined by the geometric properties of the network and serves as a system parameter. Hence, it is feasible to compute τ using the formula $\tau = q\frac{dist^2}{L}$, In this context, q is a constant that can be assigned the value of 1, *dist* denotes the distance between the sender nanomachine S_T and the recipient nanomachine S_R, and L is a threshold value. In every time slot, both the nodes in S_i and the recipient nanomachine S_R check their molecular concentrations with respect to the threshold L. On the other hand, the sender nanomachine, S_T, considers the threshold Y when comparing its own molecular concentration.

S_i **Rules:** When the molecular concentration in a node S_i surpasses the threshold L, the surplus is evenly distributed among the adjacent nodes. This behaviour can be described through the following steps outlined in Algorithm 1. However, it is important to begin by assuming that at time slot τ, $S_i(\tau)$ represents the molecular concentration level in node i. e_i represents the surplus molecules originating from a specific node:

$$e_i = S_i(\tau) - h \tag{1}$$

8   A. J. Jani and J. J. Jani

Algorithm 1. Rules of Channel Nodes

---
if $S_i(\tau - 1) > L$ then
   $S_{i-1}(\tau) \leftarrow S_{i-1}(\tau - 1) + max\{e, 0\}/2$
   $S_{i+1}(\tau) \leftarrow S_{i+1}(\tau - 1) + max\{e, 0\}/2$
   $S_i(\tau) = S_i(\tau - 1) - e_i$
end if

The model verification experiment utilized floor and ceiling functions to address situations where the surplus molecular concentration e_i resulted in an odd number. By applying this method, the distribution of molecules among adjacent nodes remained balanced and consistent.

Rules of Sender Nanomachine: The sender (transmitter) nanomachine S_T has the capability to continue accepting excess molecular concentrations from its neighboring node, i.e. S_1. Assuming that $S_T(\tau)$ represents the molecular concentration of the transmitter during the time slot τ, and defining e_T as $S_1(\tau - 1) - h$, the mechanism of the nanomachine is described as the following:

$$S_T(\tau) = S_T(\tau - 1) + max\{e_T, 0\}/2 \tag{2}$$

Thus, sender nanomachine S_T endeavors to gather Y molecules concentration from the excess molecules of its neighboring node S_1. Subsequently, the sender nanomachine S_T emits this concentration of Y molecules to S_1, as demonstrated in Algorithm 2:

Algorithm 2. Sender nanomachine Rules

---
if $S_T(\tau - 1) > X$ then
   $S_T(\tau) \leftarrow S_T(\tau - 1) - X$
   $S_1(\tau) \leftarrow S_1(\tau - 1) + X$
end if

The sender nanomachine S_T employs individual mechanisms that distinguish it from the rest of the channel nodes. After S_T emits a concentration of Y molecules to S_1, it no longer receives any excess molecules from its the adjacent node S_1. This pause in activity is designed to facilitate the passage of the emitted concentration of Y molecules through the intermediate channel nodes S_i across the distance denoted as $dist$, as well as to account for the time needed for the nanomachine S_R to confirm receipt. Considering that the acknowledgment molecules must also pass through the intermediate channel nodes S_i covering the distance denoted as $dist$ between the recipient nanomachine S_R, and the sender nanomachine S_T, the overall waiting period for S_T can be computed in the following manner:

$$\text{Cumulative waiting duration of } S_T = 2dist \tag{3}$$

The value of the parameter *dist*, representing the distance between the sender nanomachine S_T and the recipient nanomachine S_R, is not known. Hence, it becomes necessary for the sender nanomachine to make an estimation of the value of *dist*. To do this, the sender nanomachine releases a predetermined quantity of molecules, referred to as Y. Nevertheless, it's important to note that the recipient nanomachine, S_R, might not receive the complete quantity of molecules transmitted by S_T due to the passage of Y through the channel nodes S_i. However, it's probable that S_R can detect an elevation in the average amount of molecules it receives from its neighbor S_n, which happens to be the final node in the channel. To simplify matters, it's assumed that this increase is no less than L. Consequently, the sender nanomachine, S_T, can interpret the potential rise in molecule concentration received by S_R from S_n as being equal to L. The determination of the distance between the sender nanomachine S_T and the recipient nanomachine S_R can be established in the following manner:

$$dist \leq \log\left(\frac{Y}{L}\right) \tag{4}$$

The sender nanomachine S_T will pause for a duration of at least $2dist$ before it resumes its task of receiving an excess molecular concentration from S_1. Nevertheless, there is a notable distinction this time: the sender nanomachine S_T endeavours to gather a molecular concentration of $(Y + 1)$ and subsequently transmits it to S_1. Following this, S_T undergoes an additional pause of $2dist$ in duration. However, in this scenario, the value of *dist* needs to be constrained such that *dist* is less than or equal to $dist \leq \log\left(\frac{Y+1}{L}\right)$. In each phase, the sender nanomachine S_T aims to raise the accumulated molecular concentration level until it reaches the target of collecting E molecules. The collection of E molecules signifies the full capacity of each node, signifying that the overall molecular concentration in the network has increased as a result of the molecules acquired by S_R from the environment over the preceding period. Subsequently, the sender nanomachine S_T expels this accumulated reservoir of E molecular concentration into the environment, marking the commencement of the communication process once more.

Rules of Recipient Nanomachine: As previously mentioned, the recipient nanomachine S_R possesses the ability to detect and gather molecules from the surrounding environment. This capability serves to safeguard against the depletion of molecular concentration within the system. The molecules present in the environment propagate through spontaneous diffusion, as discussed in [27], resulting in variable arrival times at the recipient nanomachine S_R, as noted in [28]. Within this model, we assume that during each time slot τ, the recipient nanomachine S_R is capable of sensing at least α molecules, with α being constrained to a maximum value of L. In simpler terms, α can vary across the spectrum of values in the range $\{0, 1, 2, 3, \ldots, L\}$. The recipient nanomachine S_R operates using the same procedures as the channel nodes S_i when it comes to handling excess molecules. However, when S_R detects that the sender nanomachine S_T has transmitted its concentration (i.e., S_T sent Y molecules), it initiates a sequence of processes to release all the surplus concentration to its neighbor, S_n (the terminal node within the channel). Assuming that $S_R(\tau)$ signifies the molecular concentration at the recipient at time slot τ, and introducing the variable e_R as $S_R(\tau - 1) - L$, this procedure can be outlined as demonstrated in Algorithm 3.

Algorithm 3. Recipient Nanomachine Rules

if $S_R(\tau - 1) > L$ then
  $S_n(\tau) \leftarrow S_n(\tau - 1) + max\{e_R, 0\}/2$
end if

The recipient nanomachine S_R assesses the quantity of molecules it receives from its neighbour S_n by comparing the current received molecular concentration to the previous value. The recipient nanomachine S_R employs a mechanism where it emits all its molecular concentration to S_n if the current received concentration is at least L units higher than the previous one. Let's denote e_n as $e_n = S_n(\tau - 1) - L$.

This behaviour of the recipient nanomachine can be expressed as the following:

$$S_R(\tau) = S_R(\tau - 1) + max\{e_n, 0\}/2 \tag{5}$$

The recipient nanomachine S_R initially regards the first received surplus molecular concentration as max_R. Subsequently, when S_R receives another excess concentration from S_n, it compares this new concentration to the current max_R; if the new concentration surpasses the current max_R, it is designated as the new max_R. Subsequently, S_R assesses the incremental rise in the received concentration from S_n by calculating the difference between the current max_R and the previous max_R. If this difference surpasses a threshold L, then the recipient nanomachine S_R infers that the transmitting nanomachine S_T has released its accumulated concentration to the channel nodes S_i. As a result, it proceeds to transmit an 'acknowledgment' to S_T, accomplished by transmitting its entire molecular concentration to its neighbouring node, S_n. To enable the emitted concentration from S_R to disperse through the channel and potentially reach the transmitter nanomachine S_T, the recipient nanomachine S_R refrains from receiving any additional surplus concentration from its neighbour for a minimum duration of $\log \frac{S_R(\tau)}{L}$. However, the recipient nanomachine S_R remains receptive to molecules from the environment.

This process can be elucidated as follows: S_R initially regards the first amount of molecular concentration received from its neighbor as max_R, as indicated in Eq. (6).

$$max_R = (S_n(\tau) - L)/2 \tag{6}$$

Next, the procedure involves comparing the received molecular concentration to assess the incremental increase in the concentration level received from the neighboring node S_n. This process is outlined in Algorithm 4:

Algorithm 4. S_R checks for the transmission of Y molecules by S_T

---

if $(S_n(\tau + 1) - L)/2 > max_R$ **then**
    $max_R \leftarrow (S_n(\tau + 1) - L)/2)$
    if $((S_n(\tau + 1) - L)/2) - ((S_n(\tau) - L)/2) \geq L$ **then**
        $S_n(\tau + 1) = S_R(\tau)$
    end if
end if

The notation $S_n(\tau + 1) = S_R(\tau)$ signifies that the molecular concentration of S_R at time slot τ will be transferred to its neighbour, S_n, in the subsequent time slot $\tau + 1$.

The transmitter nanomachine S_T follows analogous procedures, which involve comparing the received molecular concentration from the neighbouring node (once the waiting time has elapsed) to identify the acknowledgment from the recipient nanomachine S_R.

3   Verifying the Model Using PRISM Tool

PRISM is a robust tool employed for formal system modelling and analysis. PRISM creates a mathematical model to represent system behaviour, and then utilizes formally specified quantitative properties in a temporal logic language for analysis [29]. One of the key advantages of PRISM is its capability to analyse not only the likelihood of a system demonstrating a particular behaviour but also to assess different quantitative measures related to the system's behaviour [30,31]. Model checking comprises two primary phases: model construction and model verification. In the initial stage of model construction, the model is transformed into a PRISM language representation. In contrast, model checking involves the examination of the constructed model to verify a specific property and obtain the corresponding result of that property [32]. Hence, for model construction and analysis using PRISM, it is essential to express the model in a state-based PRISM language [29].

The fundamental elements of the PRISM language consist of *modules* and *variables*. A PRISM model is constructed using multiple modules, that can interact with each other. Local *variables* are defined and used within these modules to perform specific processes and actions. The values of these variables determine the state of each module at any given time. Thus, the global state of the entire model is formed by the combination of individual local states from all modules. [31]. However, the probabilities of the transitions used in this model are deterministic, this implies that if the specified conditions are met, a particular event will take place. As an example, let's examine the PRISM command associated with the second node, S_2, in the channel.

[receive]   (active=1)&(u1<u)&(x1=1) -> p: (u1'=9) & (x1'=2) +   (1-p):   (u1'=0) & (x1'=2);
‾‾‾‾‾‾   ‾‾‾‾‾‾‾‾‾‾‾‾‾‾‾‾‾‾‾‾‾‾‾‾   ‾‾   ‾‾‾‾‾‾‾‾‾‾‾‾‾‾‾‾   ‾‾‾‾   ‾‾‾‾‾‾‾‾‾‾‾‾‾‾‾‾
action   guard   probability   update   probability   update

In this channel module command, the end conditions are as follows: if the molecular concentration of S_2 is higher than threshold L, and the molecular concentrations in nodes S_1 and S_3 are less than the capacity E, and if the molecular concentration of S_2 remains larger or equal to zero after emitting the excess concentration, and if the molecular concentrations in nodes S_1 and S_3 remain less than E after receiving the excess concentration from S_2, then this command is enabled, and the following events can occur. These events involve deducting the excess concentration from the molecular concentration of S_2, and distributing half of it to each of the nodes S_1 and S_3. This is just one example of the commands used to represent the model in the PRISM language. Similar commands are used in the sender and recipient modules, following the conditions explained in Sect. 2. In addition to deterministic transitions, the modules utilize flags and labels to indicate specific conditions. For example, flags play a crucial role in informing S_R when S_T emits Y molecular concentration. Likewise, flags can serve to notify S_T when S_R has acknowledged the receipt of information, indicating that S_T can now begin receiving molecular concentration from its neighbouring node. Moreover, flags play a central role in the process of analysing the model, particularly during property verification. The analysis of the constructed model using PRISM involves specifying properties that are related to the model and can be evaluated within the tool [31]. The property specification language utilized in PRISM relies on probabilistic temporal logic [33]. Determining a specific set of states within the model is crucial when defining properties for it. For example, to verify whether "an algorithm eventually terminates successfully with probability 1," it is necessary to identify the states in the model that signify the successful termination of the algorithm [31]. Constructing an expression in the PRISM language involves referencing variables and constants from the model.

This expression evaluates a Boolean value based on the set of states it corresponds to. When the set of states evaluates as true, the expression is considered "satisfied" in those states [32]. Within the property specification language of PRISM, one of the key operators is the **P** operator. This operator facilitates reasoning about the probabilities of event occurrences [31]. Within the process of properties verification in PRISM, various path properties are utilized in conjunction with the **P** operator. A path property can be defined as a formula that determines whether a specific path in the model evaluates to true or false [29]. Several temporal operators are utilized inside the **P** operator, including the 'Eventual path property' denoted as **F** *prop*. This property is satisfied for a particular path when the condition *prop* eventually becomes true at some stage along that path [31]. The formulation and verification processes of the model were performed on multiple instances of the system using PRISM language.

The initial experiments involved a model with two channel nodes, a sender nanomachine, and a recipient nanomachine. The thresholds were set as follows: $E = 20$, $L = 5$, and $Y = 10$, starting with molecular concentrations of $\{9\}$ in S_R, $\{7\}$ in the S_R, and $\{5, 4\}$ in the nodes between S_T and S_R. Subsequently, the verification process was repeated on models with five, ten, and twenty channel nodes, all sharing the same threshold values and initial molecular concentrations for the sender and recipient nanomachines. However, the initial molecular concentrations for the channel nodes varied, with values of $\{4, 3, 8, 7, 4\}$ for the five-node model, $\{7, 4, 9, 4, 7, 3, 10, 3, 8, 4\}$ for the ten-node model, and $\{3, 8, 10, 5, 7, 10, 5, 11, 4, 9, 13, 6, 4, 3, 8, 7, 4, 9, 4, 7\}$ for the twenty-node model. The last two experiments maintained the same threshold values and initial molecular concentrations for the sender and recipient nanomachines as the first two experiments. Table 1 summarizes the results of evaluating theses proposed models.

Table 1. Results following Building the PRISM Models

Channel Nodes Number	2	5	10	20
Iterations of Reachability	19	48	68	95
Model Construction Time	0.122	11.93	425.412	6373.23
States Number	1136	125399	10296733	207448902
Transitions Number	2530	222490	40140786	516044219

The outcomes presented in Table 1 were obtained using the PRISM model checker after ensuring the correct construction of the model. In PRISM, each module comprises two essential components: variables and commands. Though, the potential states that a module can adopt are determined by the variables, while the commands govern how the state evolves over time. Transitions represent changes between specific states, and they occur by assigning alternative values to the variables within the module. It's worth noting that a linear increase in the system's size may lead to an exponential growth in the model's complexity. The verified properties are associated with the reception of molecular concentration via S_R when S_T releases its Y molecular concentration, and the reception of acknowledgments by S_T after S_R releases all the molecules it contains. To express these properties, specific flags are used to indicate these events. The flag $sent_Y$ is related to S_T, initialized as *false*, and becomes *true* when S_T secretes Y molecular concentration. Moreover, the variables, $maxR_{prv}$ and $maxR_{new}$, are introduced to signify the past and present levels of received molecular concentrations at S_R. The property can be verified using the following expression:

$$P =?[F(S_R = 0)] \tag{7}$$

This expression checks if there exists a path (**F**) where S_R eventually reaches a state where its molecular concentration becomes zero. The condition for this event to occur is that the difference between the current and previous received concentrations ($maxR_{new} - maxR_{prv}$) is either equal to or exceeds L, and simultaneously, S_T has already transmitted its Y molecular concentration ($sent_Y$ is *true*). By evaluating this property, it can be confirmed that S_R releases its entire molecular concentration under these specified circumstances. When Formula (7) is evaluated over the model, it calculates the likelihood that the molecular concentration of S_R will eventually reach zero. Figure 3 illustrates the verification results of this property.

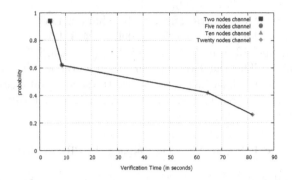

Fig. 3. Results of Verifying S_R Detecting S_T Emitting Y Molecular Concentration

In Fig. 3, the x-axis corresponds to the time duration required to complete the verification process of Property 7, whereas on the y-axis, we illustrate the probability of this property eventually becoming true. The size of the verified networks is represented by the data points in the figure. The plot illustrates an inverse relationship between the probability of S_R detecting S_T emitting molecular concentration and the quantity of intermediary nodes between them. Whereas number of nodes linking S_T and S_R increases, the likelihood of molecules reaching S_R from S_T decreases significantly. Moreover, it demonstrates that the required duration for the checking of the model increases as the number of nodes in the channel grows. This observation aligns with the fact that larger systems entail more complex computations, leading to longer verification times. The small square in the figure corresponds to the verification process involving a channel with only two nodes, whereas the circle, triangle, and diamond shapes represent experiments with channel sizes of five, ten, and twenty nodes, respectively. When $S_R = 0$, it triggers the flag $sent_{all}$ to change from *false* (its initial state) to *true*. Together with the parameters $maxT_{prv}$ and $maxT_{new}$, which monitor the molecular concentration received by S_T in both the previous and current instances, the flag *ack* can be utilized to determine if S_T acknowledges that S_R has sent an acknowledgement, denoted by $ack = true$. This verification is done through the conditions specified in the property expression.

$$\mathbf{P} =?[\mathbf{F}(ack = true)] \tag{8}$$

Formula (8) denotes the probability of the sender nanomachine S_T eventually receiving an acknowledgement, indicated by the *ack* flag becoming *true*. The verification process has been applied to the property in Formula (8), and various experiments with different network sizes have been conducted. The results of these experiments are displayed in Fig. 4, where the number of nodes between S_T and S_R in the verification experiments were 2, 5, 10, and 20, respectively.

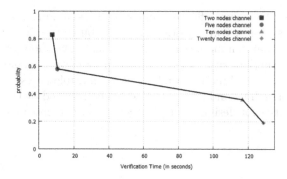

Fig. 4. Probability of *ack* flag becoming *true* over time for different system sizes (2 to 20 channel nodes).

Figure 4 illustrates the possibility of an acknowledgement reception by S_T from S_R. In this figure, the needed time for the verification of property 8 is represented, along with the likelihood of *ack* flag ultimately changing to *true*. Each point on the graph corresponds to the network's size, i.e., the number of nodes in the channel between S_T and S_R. The figure demonstrates how the acknowledgment reception is affected by the network's size, revealing that the probability of acknowledgment reception decreases as the number of nodes between S_T and S_R rises. Nevertheless, the probabilities in Fig. 4, in contrast to those in Fig. 3, indicate a general trend of lower acknowledgment reception probability compared to the likelihood of S_R detecting molecular concentration emitted by S_T. This could be attributed to S_T frequently emitting a larger number of molecules than those transmitted by S_R as an acknowledgment.

4 Conclusions

In this paper, we have explored communication model that utilizes a channel composed of multiple nodes positioned between the sender and recipient nanomachines. This channel is akin to a tissue-like structure, with interconnected cells representing the intermediate nodes, the first cell representing the transmitter, and the last cell representing the recipient. For communication between adjacent cells in this model, we adopted calcium signalling, a well-established biological technique [22]. Interestingly, the nodes within the channel employ a process motivated by the ASM [23], adding a unique dimension to the communication process. To comprehensively assess the performance and behaviour of the proposed model by utilizing the PRISM tool to verify the receiving an acknowledgment in a single-dimensional pathway. This verification process was conducted on channels of varying sizes, enabling us to gain valuable insights into the system's dynamics. The results have revealed a noteworthy observation: the reception of acknowledgments is significantly influenced by the size of the channel between the sender and recipient nanomachines. This underscores the importance of careful channel design and sizing to ensure effective and reliable communication in molecular communication systems. Our investigation contributes to advancing the understanding of molecular communication and lays the foundation for future studies exploring more complex situations, such as investigating channels that allow communication in both directions and facilitate multiple access. The incorporation of calcium signalling and the novel mechanism inspired by the ASM open up exciting possibilities for designing efficient and robust nanoscale communication networks.

References

1. Dressler, F., Akan, O.B.: A survey on bio-inspired networking. Comput. Netw. **54**(6), 881–900 (2010)
2. Jani, A.J.: Estimating nodes number in a nanonetwork using two algorithms. In: Nagar, A.K., Singh Jat, D., Mishra, D.K., Joshi, A. (eds.) Intelligent Sustainable Systems. LNNS, vol. 578, pp. 645–652. Springer, Singapore (2023). https://doi.org/10.1007/978-981-19-7660-5_58
3. Di Lorenzo, P., Barbarossa, S., Sayed, A.H.: Bio-inspired decentralized radio access based on swarming mechanisms over adaptive networks. IEEE Trans. Signal Process. **61**(12), 3183–3197 (2013)
4. Jani, A.J.: Anti-quorum sensing nanonetwork. Indian J. Public Health Res. Dev. **9**(12), 1108–1114 (2018)
5. Knyazev, S., Tarakanov, S., Kuznetsov, V., Porozov, Y., Koucheryavy, Y., Stepanov, E.: Coarse-grained model of protein interaction for bio-inspired nano-communication. In: Ultra Modern Telecommunications and Control Systems and Workshops (ICUMT), 2014 6th International Congress on, pp. 260–262. IEEE (2014)
6. Akyildiz, I.F., Brunetti, F., Blázquez, C.: Nanonetworks: a new communication paradigm. Comput. Netw. **52**(12), 2260–2279 (2008)
7. Jani, A.J.: Consensus problem with the existence of an adversary nanomachine. In: Rocha, Á., Adeli, H., Reis, L.P., Costanzo, S. (eds.) WorldCIST'19 2019. AISC, vol. 931, pp. 407–419. Springer, Cham (2019). https://doi.org/10.1007/978-3-030-16184-2_39
8. Jani, A.J.: Challenges and distinctions in nanonetworks design. In: 2019 2nd International Conference on Engineering Technology and its Applications (IICETA), pp. 219–224. IEEE (2019)
9. Hiyama, S., et al.: Molecular communication. J.-Inst. Electron. Inf. Commun. Eng. **89**(2), 162 (2006)
10. Jani, A.J.: Pattern of diffusion recognition in a molecular communication model. In: Khalaf, M.I., Al-Jumeily, D., Lisitsa, A. (eds.) ACRIT 2019. CCIS, vol. 1174, pp. 349–363. Springer, Cham (2020). https://doi.org/10.1007/978-3-030-38752-5_28
11. Nakano, T., Eckford, A.W., Haraguchi, T.: Molecular communication. Cambridge University Press, Cambridge (2013)
12. Gregori, M., Akyildiz, I.F.: A new nanonetwork architecture using flagellated bacteria and catalytic nanomotors. IEEE J. Sel. Areas Commun. **28**(4), 612–619 (2010)
13. Martı, I.L.: Exploring the scalability limits of communication networks at the nanoscale. Master's thesis, Universitat Politècnica de Catalunya (2011)

14. Farsad, N., Yilmaz, H.B., Eckford, A., Chae, C.B., Guo, W.: A comprehensive survey of recent advancements in molecular communication. IEEE Commun. Surv. Tutorials **18**(3), 1887–1919 (2016)
15. Gómez, J.T., Pitke, K., Stratmann, L., Dressler, F.: Age of information in molecular communication channels. Digit. Sig. Process. **124**, 103–108 (2022)
16. Nakano, T., et al.: Microplatform for intercellular communication. In: Nano/Micro Engineered and Molecular Systems, 2008. NEMS 2008. 3rd IEEE International Conference on, pp. 476–479. IEEE (2008)
17. Barros, M.T., Balasubramaniam, S., Jennings, B., Koucheryavy, Y.: Transmission protocols for calcium-signaling-based molecular communications in deformable cellular tissue. IEEE Trans. Nanotechnol. **13**(4), 779–788 (2014)
18. Barros, M.T., Balasubramaniam, S., Jennings, B.: Comparative end-to-end analysis of ca 2+-signaling-based molecular communication in biological tissues. IEEE Trans. Commun. **63**(12), 5128–5142 (2015)
19. Barros, M.T.: Ca2+-signaling-based molecular communication systems: design and future research directions. In: Nano Communication Networks, pp. 103–113 (2017)
20. Yaylali, G., Akdeniz, B.C., Tugcu, T., Pusane, A.E.: Channel modeling for multi-receiver molecular communication systems. IEEE Trans. Commun. **71**, 4499–4512 (2023)
21. Juhi, A., Kowalski, D.R., Lisitsa, A.: Performance analysis of molecular communication model. In: 2016 IEEE 16th International Conference on Nanotechnology (IEEE-NANO), pp. 826–829 (2016). https://doi.org/10.1109/NANO.2016.7751543
22. Clapham, D.E.: Calcium signaling. Cell **80**(2), 259–268 (1995)
23. Járai, A.A.: Sandpile models. arXiv preprint arXiv:1401.0354, pp. 1–66 (2014)
24. Al-Krizi, A.J.: Communication Models and Protocols for Diffusion Based Networks. The University of Liverpool (United Kingdom) (2018)
25. Paoletti, G.: The abelian sandpile model. In: Deterministic Abelian Sandpile Models and Patterns, pp. 9–35. Springer, Cham (2014). https://doi.org/10.1007/978-3-319-01204-9_2
26. Bak, P., Tang, C., Wiesenfeld, K.: Self-organized criticality: an explanation of the 1/f noise. Phys. Rev. Lett. **59**(4), 381 (1987)
27. Atakan, B.: Molecular communications and nanonetworks: from nature to practical systems. Springer Science & Business Media (2014)
28. Moore, M.J., Suda, T., Oiwa, K.: Molecular communication: modeling noise effects on information rate. IEEE Trans. Nanobiosci. **8**(2), 169–180 (2009)
29. Kwiatkowska, M., Norman, G., Parker, D.: Advances and challenges of probabilistic model checking. In: Communication, Control, and Computing (Allerton), 2010 48th Annual Allerton Conference on, pp. 1691–1698. IEEE (2010)
30. Bernardo, M., Degano, P., Zavattaro, G.: Formal Methods for Computational Systems Biology: 8th International School on Formal Methods for the Design of Computer, Communication, and Software Systems, SFM 2008 Bertinoro, Italy, 2-7 June 2008, vol. 5016. Springer, Cham (2008)
31. Parker, D.A.: Probabilistic model checking. University Lecture (2011)
32. Kwiatkowska, M., Norman, G., Parker, D.: PRISM: probabilistic symbolic model checker. In: Field, T., Harrison, P.G., Bradley, J., Harder, U. (eds.) TOOLS 2002. LNCS, vol. 2324, pp. 200–204. Springer, Heidelberg (2002). https://doi.org/10.1007/3-540-46029-2_13
33. KWIATKOWSKA, M.: Probabilistic model checking. In: Modeling and Verification of Parallel Processes. Summer school, pp. 189–204 (2001)

Hybrid Edge Detection and Singular Value Decomposition for Image Background Removal

Zahraa Faisal[1]($\boxtimes$) (iD), Esraa H. Abdul Ameer[2] (iD), and Nidhal K. El Abbadi[3] (iD)

[1] Computer Science Department, Education College, University of Kufa, Najaf, Iraq
zahraaf.shouman@uokufa.edu.iq
[2] Computer Science Department, College of Education for Girls, University of Kufa, Najaf, Iraq
[3] Al-Mustaqbal Center for AI Applications, Al-Mustaqbal University, Babylon, Iraq

Abstract. Image background removal is a crucial technique for enhancing the visual impact of images or altering their composition, finding applications in various fields such as photography and computer vision. This process can be executed manually through conventional image editing software or automated using advanced image processing algorithms. In this context, we present a novel algorithm that combines edge detection and singular value decomposition (SVD) to precisely segment the primary connected object within RGB images.

The proposed methodology initiates with pre-processing and edge detection, leveraging an innovative filter amalgamating Markov and Laplace filters. Subsequently, the image undergoes block division, and features are extracted through the application of SVD transformation. To ensure optimal threshold determination, a unique approach is employed, resulting in the generation of a binary image. In the final stage, morphological operations are implemented to rectify fragmented object sections, eliminate small artifacts, and fill in gaps. The binary image, when multiplied by the original image, yields a meticulously segmented color object.

This paper's distinctive contributions include the introduction of a novel threshold determination approach and the utilization of SVD for image background removal. Comparative assessments against alternative strategies consistently affirm the efficacy of our proposed technique, with accuracy measurements reaching up to 99%. The experimental results underscore the robustness and superiority of our approach, establishing it as a valuable addition to the repertoire of image background removal methodologies.

Keywords: Morphology · Segmentation · Optimal threshold · SVD · Image Processing · Markov · Laplacian

1 Introduction

An image serves as a visual representation, rich in valuable information. In the realm of image processing technology, a fundamental endeavor is the extraction of data from images while preserving their inherent qualities [1].

A. M. Al-Bakry et al. (Eds.): NTICT 2023, CCIS 2096, pp. 18–34, 2024.
https://doi.org/10.1007/978-3-031-62814-6_2

Within the domain of digital image processing, background removal emerges as a pivotal technique. It empowers the differentiation of regions within an image into two categories: those that are extraneous and those that hold interest. This technique assumes a vital role across various facets of image processing and computer vision. It frequently serves as a crucial precursor, allowing the elimination of unwanted background elements before subsequent analyses and manipulations. An illustrative application involves object segmentation within photographs, where the precision of isolating desired objects heavily relies on effective background removal [2].

However, the utility of background removal extends well beyond individual images, encompassing a series of images, including videos and multi-viewpoint images. This remarkable versatility grants the ability to extract foreground elements from videos, enabling their isolation and manipulation devoid of their original background context. For instance, the application of background removal proves invaluable in the separation of foreground objects within videos, thus enhancing their visibility and facilitating subsequent analysis or editing procedures [3].

The removal of image backgrounds carries substantial potential for enhancing computational efficiency by reducing the need to extract and match numerous features. Presently, many background removal methods rely on manual masking, a task performed by the user [4]. Nevertheless, when the background remains static relative to the camera, as is the case in measurement systems employing rotation stages, this characteristic can be harnessed to automate the background removal process effectively [5]. Furthermore, the elimination of static background elements proves an advantage for reconstruction algorithms, as their presence can impede algorithmic performance. The removal of these features contributes to the stability of the reconstruction process. Acknowledging these benefits, commercial photogrammetry software packages now incorporate the capability to integrate masks within their reconstruction algorithms [6].

Background removal, a pivotal facet of image editing, has gained paramount importance across various industries, including graphic design, e-commerce, marketing, video monitoring, intelligent transportation, sports video analysis, industrial vision, and numerous others. The precise separation of subjects from their backgrounds bestows greater flexibility for image manipulation and integration into diverse contexts. While an array of techniques and tools exist for background removal, recent advancements have introduced innovative references that promise heightened accuracy and efficiency in this endeavor [7, 8].

Developing an effective background model that can aptly describe background changes in both space and time domains poses a considerable challenge. The actual variations in background can be exceedingly intricate, encompassing factors such as variations in brightness, rapid changes in lighting conditions, the presence of shadows, swaying foliage, water ripples, and intermittent motion, among others [9].

The subsequent sections of this paper are structured as follows: Section Two delves into related works, while Section Three provides an in-depth explanation of the Singular Value Decomposition (SVD). Section Four presents the tools employed for performance measurement. Section Five elucidates the intricacies of the proposed methodology. Results and discussions unfold in Section Six. Finally, Section Seven offers the conclusion of this study.

2 Literature Review

(**J Eastwood, et. al., 2023**) presents an autonomous method for removing background interference in a photogrammetric dataset. By utilizing masked images directly in reconstruction, processing times and memory usage are reduced. The method increases point density on the object surface while minimizing superfluous background points. Comparative analysis shows that background removal decreases the standard deviation of point-to-mesh distance by up to 30 μm. The approach is robust across different artifacts, offering improved efficiency and measurement results for photogrammetry coordinate systems [6].

(**Bici, M., et. al., 2023**) presents an efficient approach for background removal using a CNN-based method. A comparison is made between CNN and manual assessment in terms of accuracy and automation, focusing on cultural heritage targets. The U-NetMobilenetV2 architecture is employed, enabling fast convergence and high efficiency with small datasets. Over 700 RGB images are utilized to train the CNN and distinguish statue pixels from the background. The CNN's performance is evaluated using the Dice coefficient. Results demonstrate improved automation with over 50% reduction in processing time, despite a few errors caused by lighting conditions. The approach proves effective for photogrammetric reconstruction and is compared to a 3D scanner model for evaluation [10].

(**Chang-Chieh Cheng, 2021**) presents a novel method for background removal in single images. The proposed approach utilizes Shannon entropy to measure the texture complexity of both the background and foreground regions. A normalized entropy filter is employed to calculate the entropy value for each pixel. By distinguishing the entropy distributions of the background and foreground, effective pixel classification is achieved. To enhance performance, an image pyramid is constructed, enabling the labeling of most background pixels in a lower-resolution image, thereby reducing computational costs associated with entropy calculation in the original-resolution image. Additionally, connected component labeling is applied for denoising, ensuring the preservation of the primary subject area [11].

(**Chunyan Huang, Xiaorui Li, 2021**) To create an FOA-OTSU segmentation method, this research integrates the fruitfly optimization algorithm (FOA) with OTSU segmentation. According to the simulations, the method outperforms the conventional OTSU algorithm in terms of convergence speed and processing time without compromising segmentation accuracy [12].

(**JIAXIN ZHANG, et. al., 2021**) addresses the challenge of object removal and façade inpainting in urban cityscapes from two perspectives. Firstly, an image-based method is proposed for automatic object removal and façade inpainting using semantic segmentation to detect various classes such as pedestrians, riders, vegetation, and cars. Generative adversarial networks (GANs) are employed to fill the detected regions with appropriate background textures and patching information derived from street-level imagery. Secondly, a workflow is introduced to automatically filter unoccluded building façades from street view images, and a dataset is tailored specifically for the GAN-based image inpainting model, comprising original and mask images. Additionally, several full-reference image quality assessment (IQA) metrics are utilized to evaluate the quality of the generated images. The validation results demonstrate the feasibility and effectiveness of the proposed method, producing visually realistic and semantically consistent synthetic images [13].

3 Singular Value Decomposition (SVD)

SVD is a mathematical algorithm that emerged and altered a way of thinking. This approach begins in tiny fields and quickly spreads to a variety of different fields [14].

SVD is a robust and reliable orthogonal matrix decomposition method. Due to SVD's conceptual and stability reasons, it has become more and more popular in the signal processing area. SVD is an attractive algebraic transform for image processing. SVD has prominent properties in imaging. Although some SVD properties are fully utilized in image processing, others still need more investigation and contribution [15].

To extract the image's geometrical and algebraic features, the SVD factorization matrix technique is employed. SVD has been widely used in the field of object identification because it possesses characteristics that are described as stability; scaling, and rotation invariance.

When a matrix is factored into three matrices using the SVD technique, some of the advantageous and significant characteristics of the original matrix are displayed. Three deconstructed matrices exist for any matrix (A) [16].

$$A = USV^T \tag{1}$$

where

$U \in R_{mxm}$ is an m $\times$ m orthogonal matrix (called the left singular vectors).

$S \in R_{mxn}$ is an m $\times$ n diagonal matrix of the singular values.

$V \in R_{nxn}$ is an n $\times$ n orthogonal matrix (called the right singular vectors).

The SVD can be applied to matrices $A \in R_{mxn}$, where m $\geq$ n. In the case that m = n, diagonal elements of S have only positive non-zero values.

In the case that m > n, diagonal elements of S are both non-zero positive descending values, and zero values [17].

Where $Q1 \geq Q2 \geq \ldots \geq Qn > 0, Qn + 1 = Qn + 2 = \ldots Qm = 0$, Q is the singular value of matrix S [17]. Matrix S is used for object detection in an image in this proposal.

4 Performance Measuring Tools

The performance of each segmented method is measured by some of the tools used for this purpose, almost these tools compare the segmentation between two images (the image segmented manually (A), and image (B) segmented by an auto technique) the tools used in this proposal are:

4.1 Hammoude Distance (HM)

Used to compare the segmented parts of the proposed method and the segmented parts of the manually segmented data, this distance metric is utilized to calculate the errors. The HM algorithm compares pixels within two borders pixel-by-pixel [18]. The HM computes by Eq. 2.

$$HM = \frac{N\left(A \cup^B\right) - N\left(A \cap^B\right)}{N\left(A \cup B\right)} \qquad (2)$$

4.2 True Detection Rate (TDR)

TDR metrics track the rate of pixels that are categorized as objects by both manual and suggested segmentation [19]. As in Eq. 3:

$$TDR = \frac{N\left(A \cap^B\right)}{N\left(B\right)} * 100\% \qquad (3)$$

4.3 Jaccard Similarity (J(A, B))

Is a measure of the similar rate between (A, B) based on the object present in both (A, B) [20]. The J(A, B) computes by Eq. 4.

$$J(A, \ B) = \frac{\left|A \cap^B\right|}{\left|A \cup B\right|} * 100\% \qquad (4)$$

For all the above equations, (N) represents the number of pixels; A is the segmented object manually and B is the segmented object obtained by the suggested method.

5 Methodology

In algorithm 1, the proposed approach for object detection in the image is summarized. The block diagram of the proposed method is shown in Fig. 1.

Algorithm (1): The proposed methodology's steps
Input: color image.
Output: binary image.

Step 1: Pre-processing (includes convert image to grayscale image; de-noising; and image resizing)
Step 2: Apply the hybrid filter generated by combining Markov basis elements (Z) with Laplace filter (L) for edge detection

$$\text{hybrid filter} = (Z\,(x,y) + L\,(x,y))*(3/8)$$

Step 3: Divide image to many nonoverlapping blocks.
Step 4: Using SVD transformation, extract the feature for each block..
Step 5: Select the diagonal matrix (S) of the singular values only from resulted of three decomposed matrices (USV^T).
Step 6: Create a new image by using the diagonal matrix, replace each image block.
Step 7: Convert the image into binary image based on threshold determined by suggested method.
Step 8: Using the mathematical morphology (such as dilation) to fill the small holes; removing the small objects; and it utilized to link the object's narrow break sections.

Fig. 1.The flowchart for the suggested technique

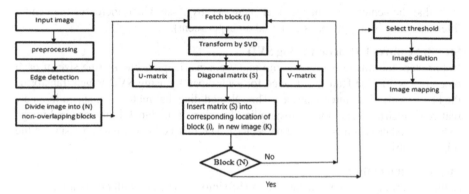

Fig. 1. The flowchart for the suggested technique

The suggested approach takes the following actions:

Step 1: Pre-processing

Three stages for preprocessing are suggested in this proposal:

1) Converting the input color image into a grayscale image.
2) Resize the image to (180 x 180).
3) Image denoising by using the median filter.

Step 2: Edges Detection

In this step, we suggested a new method for edge detection which is more suitable for this proposal. Increasing the edge content and smoothing the color change are accomplished with the help of this filter. Figure 2 depicts the creation of a hybrid filter by merging Laplace and Markov basis elements (Z, L). Note that, this filter does not change the image into a binary image, but it keeps the image as a gray image with a highlight on the edge of the image.

hybrid filter = (Z (x, y) + L (x, y)) * (3/8)

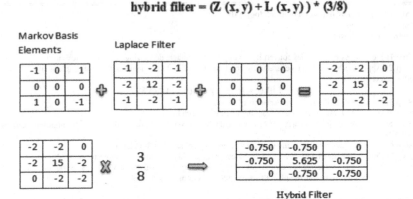

Fig. 2. Steps of creating a hybrid filter.

Note that the center value of the filter increased by three which increases the central filter weight (it is an optional step based on experiment).

Step 3: Features Extraction Using SVD
In this step, the image is divided into many nonoverlapping blocks (the size of each block is 3 × 3). The feature extracted for each block by using SVD which transforms the block matrix into three matrices. The selected diagonal matrix is mapped in a new real value matrix (SS) at the corresponding location of the block in the original image (while the other two matrices are neglected). The matrix (SS) is the same size as the original matrix.

Step 4: Segmentation
In this step, we aim to separate the matrix (SS) into two regions with binary numbers (0, 1), where the result is a binary image. This is done by using the threshold, and because the value of matrix results from SVD transformation is a real value, the classical algorithms for determining the threshold may not work fine, for that, we suggested a new method for determining the threshold as shown in algorithm 2.

Algorithm 2: Determine the optimal threshold

Input: matrix (SS).
Output: binary image.

Step 1: Select an initial threshold T0 = 0

Step 2: Generate two counters G1=0, and G2=0

Step 3: For all pixels in SS

   If I(x,y) > T0 then

G1 = G1 + I(x,y)

   else   G2 = G2 + I(x,y)

Step 4: Compute

$$T1 = \frac{\frac{G1}{N1} + \frac{G2}{N2}}{2}$$

Where N1 and N2 is the number of pixels in G1 and G2 consequantly.

Step 5: If T1 $\neq$ T0 then

           T0 = T1 ,   Go to step 3

Step 6: T1 is the optimal threshold.

Step 7: end

Step 5: Enhance a Binary Image

The next step is to focus on utilizing mathematical morphology to remove the tiny items and fill the tiny holes, using dilation.

Step 6: Mapping to the Color Image

The final step in this proposal is to build a new color image that removes the background of the image and segments the object. This is done by multiplying the binary image by the original color image, each color channel (Red, Green, and blue) multiplied by the binary image, and finally concatenating the three channels.

6 The Results and Discussion

First, show the results of each step of the proposed method:

The input image is a color image and this image is converted into a gray image; resize, and denoise as shown in Fig. 3.

Fig. 3. Image Pre-processing. (a) Original image; (b) Image after pre-processing.

After preprocessing, the edge detection process is applied by using the suggested filter, as shown in Fig. 4, in this step, the filter will highlight the edge and make the image sharper.

Fig. 4. Edges Detection of the Image. (A) Image after pre-processing; (B) Image after applying Suggested Hybrid Filter

As we mention in the methodology the image after edge detection, will be transformed by using the SVD transformation to produce a real numbers matrix. This real number matrix is converted into a binary image based on determining the optimal threshold by the suggested threshold algorithm. The result binary image is shown in Fig. 5.

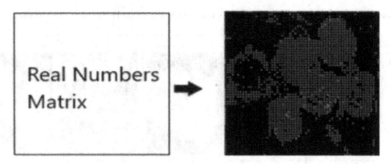

Fig. 5. Convert the real matrix into a binary image.

The resulting image from Fig. 5 has a lot of small objects and holes. To remove these holes and connect the small objects, we suggested using a mathematical morphology (dilation) for this purpose. The result of this step is shown in Fig. 6.

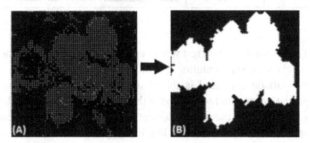

Fig. 6. Enhancing a binary image. (A) After applying the optimal threshold. (B) After applying mathematical morphology

The final step is to map the binary image result from Fig. 6 to the original color image. The result is shown in Fig. 7.

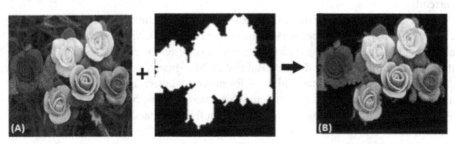

Fig. 7. The final result. (a) Original image; (b) After mapping.

To evaluate how well the suggested strategy worked, we used more than 150 images, where the object is only determined and eliminates all the other structures in the image. A sample of the images used in the test with background removal is listed in Table 1.

Table 1. Sample from the testing images, and segment objects by the proposed method.

The second and most important test was to compare the performance of this proposal with some of the classical segmentation techniques:

The first comparison test used is the Otsu approach, which uses a threshold to divide the image histogram into two groups with extremely little within-class variability. Figure 8 shows the ability of the suggested method to detect the prominent and smooth edges in the image efficiently unlike the Otsu technique.

The second technique we compared with is the k-means clustering method. K-means is an unsupervised learning algorithm that classifies a given set of data with the number of segments or clusters. The comparing results are shown in Fig. 9. The proposed method gives nice outlines of the cat object, which identify all cat bodies; although the image background contains the same color as the object; the edges are more clear. While the k-means technique perceives many edges, they are too spotty and wide to identify objects correctly.

Fuzzy c-mean (FCM) (a fuzzy logic-based clustering technique) is another technique compared. Where data can belong to two or more clusters. FCM is sometimes called soft clustering because there are no clear boundaries between the clusters. Figure 10 shows the results of implementing the proposal and FCM segmentation.

The proposed method has a high performance by detecting the object edges more accurately, where the object is completely separated from the background of the image. While an FCM algorithm fails to grab the object from the background.

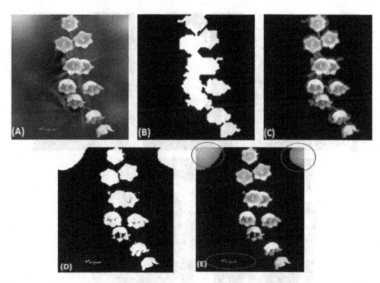

Fig. 8. A comparison of the proposed method's background removal and OTSU technique. A. Original image. B. binary image by the proposed method. (C) segmented object by the proposed method. D. binary image of OTSU technique. E. segmented object by Otsu technique.

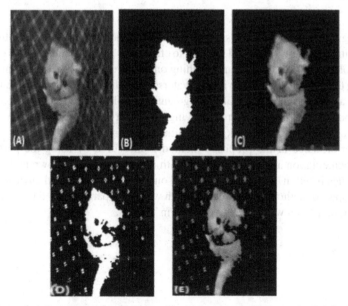

Fig. 9. Comparing the proposed technique with the k-means technique. A. Original image. B. binary image of the proposed method. (C) segment object by the proposed method. D. binary image of K-mean technique. E. segment object by K_mean.

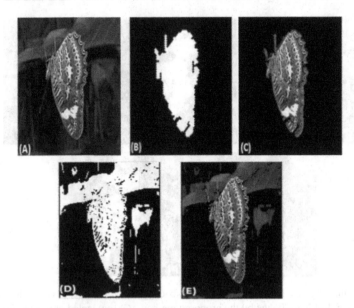

Fig. 10. Comparing the FCM algorithm and the proposed method. A. Original image; B. binary image of the proposed method. (C) segmented objects by the proposed method. D. binary image of FCM technique. E. segmented objects by FCM.

The region-based technique is used by the Watershed transformation, which looks for similarities between pixels and regions, to effectively segment images. It is especially useful for objects whose borders are touching one another. Figure 11 shows the result of comparing the proposed method with the Watershed method. The proposed method gives more accurate background removal and segmentation, while the watershed algorithm did not give good or reliable results in identifying the object in the image, as it divided parts of the object and parts of the background at the same time as a final result.

The performance of the proposed method is measured by using the performance tools for segmentation and also compared with other methods as shown in Table 2. All the techniques listed in Table 2 are measured on the same images and environment.

The proposed method has proven its worth when compared with other techniques in terms of strengths and weaknesses as shown in Table 3.

Fig. 11. Comparing the Watershed segmentation method and the proposed method. A. Original image; B. binary image of the proposed method. (C) a segmented object by the proposed method. D. binary image of watershed technique. E. a segmented object by a watershed.

Table 2. Outcomes of test images for different techniques

Techniques	HM	TDR %	J (A, B) %
K_mean	0.39	93	60
Fuzzy_ C_mean	0.52	51	48
OTSU	0.44	90	55
Watershed	0.78	30	21
Proposed	0.05	99	94

Table 3. Strengths and weaknesses of various techniques

Techniques	Strengths	Weaknesses
Edge detection [21]	Useful for photographs with sharper contrast between objects, such as roads, buildings, or water features	Inappropriate for incorrect detection or excessive edges; In complex, internally heterogeneous objects like forests, things are frequently tough. The similarity criterion that is selected has a significant impact on the outcomes
Otsu's Technique [22]	Automated Threshold Determination, Global Thresholding, Statistical Optimality, and applicable to various types of grayscale images	Sensitive to image characteristics, limited to two classes, global thresholding limitation, and sensitive to noise
Thresholding Method Based on Local Entropy [22]	Adaptive Thresholding, Robust to Image Characteristics, Multilevel Thresholding, and Reduced Sensitivity to Noise	Even though two photos with the same image histogram result in the same threshold, this is incorrect in practice
K-mean [23]	Efficient Clustering, Simple and Intuitive, Versatile, Scalability, Convergence	The value of the cluster, i.e. K, needs to be defined, and the segmentation is poor and slow (especially with huge images)
Watershed Technique [24]	Segmentation of Complex Structures, Automatic, Edge Preservation, Applicability to Various Image Types, and Multiresolution Analysis	Intricate calculation of gradients
Proposed Method	Edges preservation, automatic, Applicability to Various Image Types, applying SVD to extract image features, and determining an ideal threshold for object extraction, in addition to the fast implementation and accuracy of results, reduce sensitivity to noise	–

7 Conclusion

In conclusion, our presented algorithm for image background removal emerges as a highly efficient and precise solution, significantly elevating the quality of image subjects and enabling versatile composition modifications. The seamless integration of edge detection and singular value decomposition (SVD) showcases the algorithm's adeptness in accurately segmenting the primary connected object within RGB images. Notably, our approach introduces an innovative filter that skillfully amalgamates Markov and Laplace filters.

The novel threshold determination method further enhances the algorithm's precision, ensuring optimal threshold selection. The subsequent application of morphological operations serves to rectify fragmented object sections, eliminate inconsequential elements, and fill voids, culminating in a meticulously segmented binary image.

The substantial contributions of this paper extend beyond the algorithm itself. The introduction of a new edge detection technique, a refined threshold determination method, and the utilization of SVD collectively signify significant strides in the realm of automatic image processing algorithms. The achieved remarkable accuracy, reaching up to 99%, not only validates the algorithm's efficacy but also positions it as a valuable tool for broader applications in image editing and computer vision. Consequently, this research lays a solid foundation for advancing image processing capabilities, with practical implications spanning a wide range of domains.

References

1. Desai, B., Kushwaha, U., Jha, S., NMIMS, M.: Image filtering techniques algorithms and applications. Appl. GIS **7**(11), 101 (2020)
2. Chen, T., Zhu, Z., Hu, S.M., Cohen-Or, D., Shamir, A.: Extracting 3D objects from photographs using 3-sweep. Commun. ACM **59**(12), 121–129 (2016)
3. Kumar, S., Yadav, J.S.: Video object extraction and its tracking using background subtraction in complex environments. Persp. Sci. **8**, 317–322 (2016). https://doi.org/10.1016/j.pisc.2016.04.064
4. Woloszyk, K., Bielski, P.M., Garbatov, Y., Mikulski, T.: Photogrammetry image-based approach for imperfect structure modeling and FE analysis. Ocean Eng. **223**, 108665 (2021)
5. Sathirasethawong, C., Sun, C., Lambert, A., Tahtali, M.: Foreground object image masking via EPI and edge detection for photogrammetry with static background. In: Advances in Visual Computing: 14th International Symposium on Visual Computing, ISVC 2019, Lake Tahoe, NV, USA, 7–9 Oct 2019, Proceedings, Part II 14, pp. 345–357. Springer International Publishing (2019)
6. Eastwood, J., Leach, R.K., Piano, S.: Autonomous image background removal for accurate and efficient close-range photogrammetry. Meas. Sci. Technol. **34**(3), 035404 (2022). https://doi.org/10.1088/1361-6501/aca497
7. Zhang, X., Zhang, S., He, X., Wang, J.: A deep learning-based approach for object segmentation and background removal. IEEE Access **7**, 49143–49154 (2019)
8. Hsu, Y.H., Lin, C.H.: Deep interactive image segmentation with human in the loop. In: Proceedings of the IEEE/CVF Conference on Computer Vision and Pattern Recognition (CVPR), pp. 11979–11988 (2021)

9. Zheng, W., Wang, K., Wang, F.Y.: A novel background subtraction algorithm based on parallel vision and Bayesian GANs. Neurocomputing **394**, 178–200 (2020). https://doi.org/10.1016/j.neucom.2019.04.088

10. Bici, M., Gherardini, F., de Los Angeles Guachi-Guachi, L., Guachi, R., Campana, F.: Convolutional neural network for background removal in close range photogrammetry: application on cultural heritage artefacts. In: International Joint Conference on Mechanics, Design Engineering & Advanced Manufacturing, pp. 780–792. Springer International Publishing, Cham (2022)

11. Cheng, C.C.: Single-image background removal with entropy filtering. In: VISIGRAPP (4: VISAPP), pp. 431–438 (2021). https://doi.org/10.5220/0010301204310438

12. Huang, C., Li, X., Wen, Y.: AN OTSU image segmentation based on the fruitfly optimization algorithm. Alex. Eng. J. **60**(1), 183–188 (2021)

13. Zhang, J., Fukuda, T., Yabuki, N.: Automatic object removal with obstructed façades completion using semantic segmentation and generative adversarial inpainting. IEEE Access **9**, 117486–117495 (2021). https://doi.org/10.1109/ACCESS.2021.3106124

14. El Abbadi, N.K., Mohamad, A., Abdul-Hameed, M.: Image Encryption based on singular value decomposition. J. Comput. Sci. **10**(7), 1222 (2014)

15. El abbadi, N., AL-Rammahi, A., Redha, D., Abdul-Hameed, M.: Image compression based on SVD and MPQ-BTC. J. Comp. Sci. **10** (2014). https://doi.org/10.3844/jcssp.2014.2095.2104

16. El abbadi, N., Mohamad, A., Abdul-Hameed, M.: Image encryption based on singular value decomposition. J. Comp. Sci. **10** (10). 1222–1230 (2014). https://doi.org/10.3844/jcssp.2014.1222.1230

17. Konda, T., Nakamura, Y.: A new algorithm for singular value decomposition and its parallelization. Parallel Comput. **35**(6), 331–344 (2009)

18. Abbadi, N.K., Faisal, Z.: Detection and analysis of skin cancer from skin lesions. Int. J. Appl. Eng. Res. **12**(19), 9046–9052 (2017)

19. Faisal, Z., Abbadi, N.: New segmentation method for skin cancer lesions. J. Eng. Appl. Sci. **12**(21), 5598–5602 (2017)

20. Yan, Z., Wu, Q., Ren, M., Liu, J., Liu, S., Qiu, S.: Locally private Jaccard similarity estimation. Concurr. Comput. Pract. Exp. **31**(24), e4889 (2019)

21. Blaschke, T., Kelly, M., Merschdorf, H.: Object-based image analysis: Evolution, history, state of the art, and future vision (2015)

22. Sethy, P.K., Negi, B., Behera, S.K., Barpanda, N.K., Rath, A.K.: An image processing approach for detection, quantification, and identification of plant leaf diseases. Int. J. Eng. Technol. **9**(2) (2017). https://doi.org/10.21817/ijet/2017/v9i2/170902059

23. Hassanat, A.B., et al.: Victory sign biometric for terrorists identification: preliminary results. In: 2017 8th International Conference on Information and Communication Systems (ICICS), pp. 182–187. IEEE. (2017)

24. Barida, B.A.A.H., Chinagolum, I., Shedrack, M.: An enhanced application of artificial neural network algorithm in image segmentation process. J. Environ. Sci. Comp. Sci. Eng. Technol. **8**, 28–37 (2019)

Multi-objective Residential Load Scheduling Approach Based on Pelican Optimization Algorithm

Hiba Haider Taha[1,2]([⊠]) and Haider Tarish Haider[2]

[1] State Company for Steel Industries, Ministry of Industry and Minerals, Baghdad, Iraq
{hibahaider,haiderth}@uomustansiriyah.edu.iq
[2] Department of Computer Engineering, Mustansiriyah University, Baghdad, Iraq

Abstract. Controlling the available power generation to support current residential demand, and managing the local loads is a crucial task. In this paper, a multi-objective pelican optimization algorithm (MOPOA) is proposed for optimal load control of smart grids. The MOPOA algorithm is used to minimize the energy cost for customers and reduce the peak load for utility companies. The proposed algorithm is evaluated using a case study of a smart grid with 20 household appliances. MOPOA algorithm provides all Pareto front solutions that achieve a trade-off between the optimal cost of the customer's load, and the peak load saving. The dominance count concept is used to select the compromise solutions ensuring that the diversity of the population is maintained. The proposed method provides saving more than 42.67% for energy bills. Moreover, the power suppliers have gained benefits by lowering the peak energy demand by 25.8%.

Keywords: Demand response · pelican optimization algorithm · time of use · multi-objectives optimization · dominance count

1 Introduction

For the last decade, global power consumption has significantly increased. To deal with rising demand and changing government policies, it is vital to consider whether the current level of consumption can be sustained in the long run [1]. In comparison to 2015, global electricity consumption will continue to rise by 30% in 2040 [2]. The energy demand during peak hours is the most important issue for utilities. In that instance, the generators must be run at full capacity, which could cause additional strain on the grid [3]. The usual approach to face these difficulties is to expand the available power capacity (e.g., by initiating new conventional coal-fired power stations). Although increasing supply capacity by constantly building traditional power plants is unsustainable due to the high capital expense [4]. The important purpose of the transformation from the traditional power grid to the smart grid is to fulfil future energy requirements [5]. Having improved technologies for measuring network characteristics and advanced metering infrastructure (AMI) for two-way communication of the network's essential data, make continuous growing for smart grid system [6]. Consumer friendliness, hack-proof self-healing,

© The Author(s), under exclusive license to Springer Nature Switzerland AG 2024
A. M. Al-Bakry et al. (Eds.): NTICT 2023, CCIS 2096, pp. 35–47, 2024.
https://doi.org/10.1007/978-3-031-62814-6_3

assault resistance, the capacity to accommodate all types of generating and storage alternatives, energy market-based efficient operation, high power quality, and optimum assets are the major features of a smart grid [7]. As a result, managing the available generation capacity provides an essential approach in the new operating environment [8]. The demand side management (DSM) is an effective solution that contributes to distribution networks in various aspects, such as the transformation towards smart grid liberalization of the electricity market, enhancing the effects of control management, reducing the cost of infrastructure construction, and increasing the feasibility of decentralized energy resources [9]. DSM is composed of demand response (DR), load management solutions that include the planning, integration, and monitoring of pre-assigned regular operations depending on consumer consumption patterns [10] by encouraging users to use less electricity during peak hours or to move their energy consumption to off-peak hours [11]. DR programs are classified into two groups: first, price-driven programs seek to reduce daily power expenses at different price levels throughout the day. Second, incentive-driven and event-driven programs consider a direct load control signal from DR aggregators at critical peak or emergency times [12].

Over the past decade, scheduling and managing household loads have been studied from various perspectives. To provide a more comprehensive analysis of the related work, we have categorized the existing scheduling models for household appliances into two main groups: single-objective and multi-objective. The first group is Single-objective scheduling models that focus on optimizing a single objective, such as energy cost or peak load reduction. The authors in [13] have provided an efficient scheduling model for household appliances. The model incorporates power costs, incentives, and consumer discomfort under the TOU electricity rate. Furthermore, the impact of the inconvenience weighting factor on overall expenses is addressed. The simulation results show the effectiveness of the proposed model, which can reduce 34.71% of consumer's total costs. It also illustrates that the total costs will be raised with the increase in the inconvenience weighting factor. Thus, consumers will choose whether to participate in DR programs according to their preferences. In [14] a new approach was demonstrated for residential load shifting in demand response events. It uses genetic algorithm (GA) to find the optimal scheduling for shifting loads to reduce the customer's electricity bill. The executed GA reduced bills by up to 15%. In [15] a mixed-integer linear programming (MILP) model was presented to formulate the DR scheduling for smart residential communities considering heterogeneous energy consumption among residents by introducing the willingness to pay (WTP) to quantify the heterogeneity. The simulation findings indicate that the algorithm described in this study effectively reduces the amount of transmitted data by around 30% when compared to the demand response scheduling algorithm that does not include the concept of WTP. The algorithm has the potential to provide a reduction in community energy costs of around 10% when compared to a demand response algorithm that uses game theory.

The second group is multi-objective scheduling models that consider multiple objectives simultaneously, such as energy cost, peak load reduction, and user comfort. These models are more complex to solve than single-objective models, but they can provide a more holistic view of the scheduling problem and lead to more efficient solutions. In [16], a dynamic residential load scheduling (DRLS) system was proposed to use an adaptive consumption level pricing scheme (ACLPS) to encourage customers to manage their energy consumption within a certain level. The ACLPS is based on the concept of time-of-use (TOU) pricing, but it is adaptive in the sense that the pricing levels are adjusted based on the current load demand. This allows the DRLS system to achieve a balance between the customer's comfort and utility. The simulation results show that the DRLS system is able to reduce the customer's energy bill by up to 53% and the peak load by around 35% while still taking into account the customer's comfort. They also show that the DRLS system is more effective than a fixed TOU pricing scheme. In [17] an analytical optimization method (AM) for scheduling the electrical energy of smart homes in a smart grid and the scenario based on TOU pricing rates were proposed. The algorithm saves 44% on energy prices and 35% on peak demand. The AM provides more efficient results to solve optimization problems than GA and particle swarm optimization (PSO) for the minimum time of execution. In [18] an optimum load scheduling model of household appliances for smart home energy management is developed. The model considers various pricing schemes, such as ACLPs and TOU. The proposed approach is based on the Henry gas solubility optimization method (HGSOM) and the VIKOR multi-criteria decision-making method. The results save more than 80% and 76% of end-user energy costs for TOU and ACLPs schemes respectively. Meanwhile, the end-user peak demand reduced by 55%. In [19] a self-adapting differential evolution (SaMODE) algorithm for multi-objective residential load management was addressed. The SaMODE algorithm is a modified version of the differential evolution algorithm using TOU tariffs. The main strengths of the SaMODE algorithm are the ability to find good solutions for multi-objective problems and its ability to adapt to different problem settings. The main weaknesses of the SaMODE algorithm are its computational complexity and sensitivity to the initial population. The self-adapting multi-objective differential evolution (SaMODE) algorithm reduces electricity bills and peak demand while providing consumer convenience. The results show that users benefit by lowering their energy bills by up to 53.34%, while power suppliers' profit by lowering peak demand by 43.04%. In [20] a hybrid heuristic algorithm to solve the multi-objective optimal scheduling problem was suggested for household appliances for demand side management to minimize the electricity cost and maximize user satisfaction. The hybrid algorithm is based on multi-objective particle swarm optimization (MOPSO) and uses the genetic operator of the non-inferior solution sorting genetic algorithm (NSGA-II). Results show that it can find high-quality solutions to the problem.

Based on the aforementioned research works in the current state of the arts, load management of residential customers is still a critical and hot topic. The present research addresses the household load management system for scheduling user profile of household appliances in smart grid, with the goal of minimizing peak load in the power demand and lowering consumer electricity costs. Pelican optimization algorithm (POA) is a relatively new and promising algorithm that has not been as widely studied as other algorithms, such as GA and PSO. POA is a population-based algorithm, which has ability to avoid getting stuck in local optima. It is effective in solving a variety of problems, including engineering design, scheduling, and financial optimization. These objectives are stated using POA to produce every potential solution, and the compromise solutions are chosen utilizing multi-objective approaches and the dominance count concept to rank the given solutions. Furthermore, TOU is utilized to compare the outcomes in relation to various pricing strategies. As it was stated previously, TOU is a pricing scheme that charges different rates for electricity depending on the time of day.

2 Problem Formulation

This research presents two objectives (cost and peak load), each customer has a set of loads/appliances powered by the electricity grid. Each residence has a smart meter that is linked to a centralized utility [19], the smart meter collects all necessary information about customer loads and transmits it to the centralized scheduler through a neighborhood area network (NAN). The internet of things (IOT) can be used to connect household appliances with smart meter for scheduling processes via home area network (HAN). The formulation of objective functions is presented in the following subsections.

2.1 Energy Cost

For each time slot $ts = [1, 2...., T]$, the utility provides the price of electricity C_{ts}. Optimal load management for end-user appliances may be achieved by lowering energy costs and consuming less power during peak hours. Energy cost (ENC) is calculated using the following formula [21]:

$$ENC = \sum_{a=1}^{A} \sum_{ts=1}^{T} C_{ts} P_a NF_{a,ts}^o * \Delta t \tag{1}$$

where A is the total number of home appliances and a stands for every single item. Where P_a represents the appliance's a power consumption (kW). $NF_{a,ts}^o$ is the optimal ON/OFF status of ath appliance at ts^{th} the time slot. Δts stands for the sample time.

2.2 Peak Load Energy

Customers gain from lower energy prices, while utility companies benefit from lowering peak load energy. The peak power load energy has to be:

$$PL = \sum_{a=1}^{A} P_a NF_{a,ts}^o ; \ \forall ts \in [1, T] \tag{2}$$

PL indicates the amount of power used at a time slot ts.

3 Methodology

To address the objective function efficiently, it is advisable to utilize the optimization technique referred to as POA. This approach enables the attainment of an optimal load scheduling solution, simultaneously minimizing customer costs and peak load.

3.1 Pelican Optimization Algorithm (POA)

POA, the pelican serves as a member of the population, often engaging in collaborative efforts during prey pursuit [22]. Derived from the natural behaviors of pelicans, the POA incorporates both exploitation and exploration phases to seek the optimal solution. The mathematical formulation of POA involves two phases: the first phase involves moving towards prey (exploration), while the second phase entails winging on the water surface (exploitation) [23].

3.1.1 Initialization

The POA is a population-based algorithm in which each member of the pelican population is considered a candidate solution. The optimization step begins with creating an initial population POP consisting of np individual vectors, also known as solutions. Each vector in the set has DV decision variables normally distributed in the search area. The search area is constrained by the lower and upper boundaries, defined as $X_{J,L}$ and $X_{J,H}$, respectively. The initial individual vector (solution) is initiated by [23]:

$$X_{J,I,G} = X_{J,L} + rand * (X_{J,H} - X_{J,L}) \tag{3}$$

Here, rand represents a random number within the range [0, 1]. The index $G = [1,..$ Gmax] denotes the generation to which a vector belongs, and the index I refers to the individual vector $I = [1, 2, np]$. Vector parameters are indexed with $J = [1, 2, ...DV]$.

3.1.2 Moving Towards Food Source (Exploration Phase)

In this phase, the pelican engages in locating and swiftly descending to consume its prey. The prey's random distribution within the search region enhances the pelican's exploration capability. Equation (4) encapsulates the representation of the pelican's location and a mathematical simulation of its location update throughout each iteration [23].

$$X_{J,I}^{P1} = \begin{cases} X_{J,I} + rand * (P_j - I * X_{R,J}) & F_P < F_I \\ X_{J,I} + rand * (X_{J,I} - P_J) & else \end{cases} \tag{4}$$

Here, $X_{J,I}^{P1}$ represents the updated state of the Ith pelican in the Jth dimension, as determined by phase one. P_J is the location of prey in the Jth dimension, and F_P is its objective function value. Parameter's value is chosen at random to be either 1 or 2.

In the proposed POA, a pelican's new location is accepted only if the objective function's value improves in that position. This form of updating, termed effective updating prevents the algorithm from migrating to suboptimal regions.

$$X_I = \begin{cases} X_I^{P1}, & F_I^{P1} < F_I \\ X_I & else \end{cases} \tag{5}$$

Variable X_I^{P1} represents the updated status of the Ith pelican, whereas F_I^{P1} represents its objective function value derived from phase one.

3.1.3 Winging on the Water Surface (Exploitation Phase)

During pelicans reach the water's surface, they spread their wings to entice fish into shallower waters, where they can be caught more easily. Equation (6) is a possible mathematical representation of this behavior [23]:

$$X_{J,I}^{P2} = X_{J,I} + R * \left(1 - \frac{G}{Gmax}\right) * (2 * rand - 1) * X_{J,I} \tag{6}$$

Here, $X_{J,I}^{P2}$ represents the updated status of the Ith pelican in the Jth dimension, considering phase two, R is a constant with a value of 0.2 and Gmax represents the maximum number of iterations. In order to accept or reject the new pelican position at this stage, which is represented by Eq. (7), effective updating has also been applied.

$$X_I = \begin{cases} X_I^{P2} F_I^{P2} < F_I \\ X_I \, else \end{cases} \tag{7}$$

Variable X_I^{P2} represents the updated status of the Ith pelican, whereas F_I^{P2} represents its objective function value derived from second phase.

3.2 Dominance-Based Approaches

A multi-objective optimization problem involves several objective functions that must be simultaneously minimized or maximized. Similar to a single-objective optimization problem, a set of constraints must be satisfied by every potential solution, including all optimal solutions, in a multi-objective optimization scenario [24]. Multiple objectives in a problem lead to Pareto-optimal solutions, which cannot be considered better than one another without additional information. Dominance-based ranking strategies, such as dominance rank, dominance depth, and dominance count, are commonly employed. This paper utilizes dominance count for ranking solutions in each generation based on their dominance relationships. The dominance count is a concept in multi-objective optimization. It is used to measure the quality of a solution in terms of how many other solutions are worse than it for all objectives [25]. For example, solution A is said to dominate solution B if A is better than B in all criteria without being worse in any criterion, as shown in Fig. 1. A solution with a high dominance count is better than many other solutions in the population.

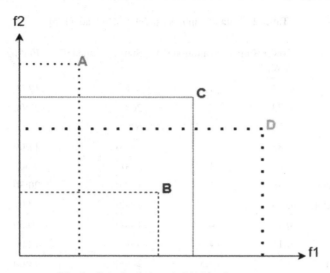

Fig. 1. Set of solutions in bi-objectives space

4 Case Study

This study employs the end-user energy usage profile data. The twenty appliances are chosen in the optimal scheduling model for a typical residential home. As shown in Table 1, a TOU tariff has been set up by an electricity provider [26]. The peak times are from 11:00 to 14:00 and from 18:00 to 23:00. Valley times occur between 00:00 and 07:00. There are flat periods of time during the day that were taken to cover the remainder of the time. Table 2 includes information on household appliances, including the power rating and starting and ending times for each device (*ST* and *ET*, respectively), and the time duration needed to complete normal operation.

Table 1. The TOU tariff [26]

Load	Time ranges	Power price ($/kWh)
Peak	11:00–14:00	0.84
	18:00–23:00	
Valley	00:00–07:00	0.31
	8:00–10:00	
Flat	15:00–17:00	0.61

Table 2. Data of appliances (deferrable loads) [26]

Appliance	Power Rating (kW)	Duration (h)	Starting Time (ST)	Ending Time (ET)
Dishwasher	0.73	1	8:00	12:00
Dishwasher	0.73	1	20:00	23:00
Dishwasher	0.73	1	6:00	7:30
Rice cooker	0.8	1	15:00	17:00
Rice cooker	0.8	1	6:00	9:30
Washing machine	0.38	2	18:00	20:30
Washing machine	0.38	2	00:00	9:00
Humidifier	0.15	4	14:00	20:30
Humidifier	0.15	4	9:00	12:00
Laundry drier	1.26	1.5	20:00	23:00
Laundry drier	1.26	1.5	4:00	8:00
Water heater	1.85	2	16:00	20:00
Water heater	1.85	2	21:00	24:00
Electric kettle	1.5	0.5	6:00	7:30
Electric kettle	1.5	0.5	20:00	23:30
Electric kettle	1.5	0.5	9:00	12:30
Electric oven	1.3	1	15:00	17:00
Electric oven	1.3	1	19:00	22:00
Air conditioner	2.2	4	11:00	15:00
Air conditioner	2.2	4	17:00	23:00

5 Experimental Results

The proposed algorithm is executed using the MATLAB software package. The operation time considered in this paper is set to one day; by specifying 30 min as a sampling time, the entire day might thus be split into 48 slots. The smart home can shift the loads from peak time of high price rate to the off-peak time of lower price rate. This feature enhances the adaptability of the smart home in managing load schedules, enabling it to identify a cost-effective. In the optimization algorithm, the number of decision variables (DV) that must be optimized using the POA method is set to twenty, the same as the number of household devices. POA was implemented on the objective functions for populations of 50 members. The increase in population has resulted in an increase in the algorithm's exploratory capacity and the identification of more optimal regions. The maximum number of iterations is 100. This is sufficient for the current optimization problem-solving process to obtain optimum solutions with the lowest cost and peak. POA optimization strategies contribute to providing optimal sets within the Pareto front, aiming to redistribute load and avoid peak times to lower average energy costs. The Pareto front, obtained from the POA algorithm, represents the set of all non-dominated solutions to the optimization problem, with each point comprising a new solution's cost and peak values. The count dominance method serves as a mechanism for selecting the optimal solution on the Pareto front, where the solution with the highest count is optimal. Figure 2 show load management after implementing the POA algorithm to optimize customer load management for 30 min. The x-axis represents the time of day, and the y-axis represents power consumption. The peak load consumption after scheduling is 4.45 kW, while for baseline end-users, was about 6 kW (25.8% saving). However, the total cost was reduced to 29.78 $, while for baseline end-users, it was about 51.95 $ (about 42.67% cost savings).

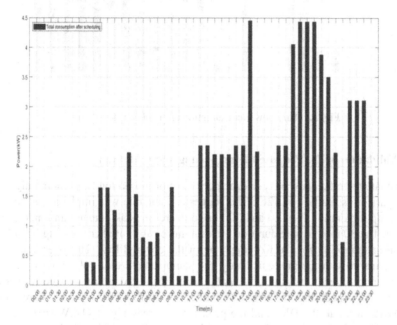

Fig. 2. Daily power consumption after scheduling (30 min)

Achieving a reduction in electrical consumption in high-demand regions is a desirable result, as it matches our objective of cost reduction. When the level of consumption is decreased, there is a corresponding reduction in costs, as the price of power in peak regions is significantly higher compared to other regions.

Figure 3 illustrates load management after implementing the POA algorithm to optimize customer load management for 60 min. The peak load consumption is reduced to 4.41 kW, resulting in a 26.5% saving compared to baseline end-users. Moreover, the total cost is reduced to $30.75, reflecting a substantial 40.81% cost savings compared to baseline end-users.

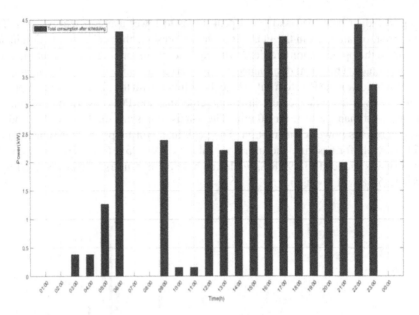

Fig. 3. Daily power consumption after scheduling (60 min)

5.1 Validation of the Proposed Load Management Method

To evaluate the efficiency and performance of the proposed load scheduling algorithm, a comparison is made with [26]. The suggested schedule was implemented, utilizing both sampling durations of 60 and 30 min to decrease peak demand and increase cost savings. The 30 min sample duration yields a more substantial cost savings of 42.67%, indicating suitability for the working time of the covered load. However, for fairness in comparison with the previous study, the same sample time (60 min) is utilized, a TOU pricing scheme and consumer data consumption. According to the results after the proposed scheduling, the total cost reduces to 30.75 $ (the total costs are reduced by 40.81% through the DR), and the peak load decreases to 4.41 kW (the total peak is reduced by 26.5%). Table 3 compares the results from [26] with the results of the proposed method for sample time (60 min) and (30 min), showcasing the improvements achieved through MOPOA-based scheduling. Before scheduling, the reference study reported a total energy consumption of 80.28 kWh, a peak load of 6 kW, and a total cost of $51.95. Following the implementation of MOPOA-based scheduling with a sample time of 60 min, there is a notable reduction in total cost to $30.75 for about 26.5% savings and a decrease in peak load to 4.41 kW. Further refinements are observed with a sample time (30) and duration, resulting in a total cost of $29.78 for about 25.8% savings and decrease in peak load to 4.45 kW.

According to the above results, this method not only aligns with the economic interests of end-users but also contributes to the overall resilience and efficiency of the power system and increases power system stability by minimizing swings in load demand.

Table 3. Comparison results with [26]

Before Scheduling				After Scheduling				
References	Total energy (kWh)	Peak load (kW)	Total cost ($)	Total energy (kWh)	Peak load (kW)	Total cost ($)	Cost %	Peak %
[26]	80.28	6	51.95	80.28	5.5	38.59	25.71	8.3
Proposed method (60 min)	80.28	6	51.95	80.28	4.41	30.75	40.81	26.5
Proposed method (30 min)	80.28	6	51.95	80.28	4.45	29.78	42.67	25.8

6 Conclusion

The proposed multi-objective residential load scheduling approach that employing the POA optimization algorithm helps to provide optimal load scheduling for helping end-users save money on their energy bills while helping the utility company to reduce peak load demand. The POA algorithm is a promising approach for load scheduling. The proposed method achieves remarkable reductions in energy costs (42.67%) and peak load (25.8%), demonstrating the effectiveness of the proposed solution. It provides dual benefits for end-users (cost savings on energy bills) and utility companies (improved peak load management).To ensure a fair and comparable evaluation of the proposed load scheduling algorithm, we conducted experiments using both 60-min and 30-min sample durations. While the previous study employed a 60-min sample time, we recognized that this choice could potentially limit the algorithm's ability to capture the dynamic nature of energy consumption patterns and optimize load scheduling effectively. Therefore, we included the 30-min sample time to investigate whether a shorter interval could improve scheduling outcomes. This dual approach allowed us to assess the performance of our proposed algorithm under both sample durations, providing a more comprehensive evaluation and enabling a more accurate comparison with the previous study. The results showed that the POA algorithm was able to achieve a cost reduction of 40.81% and a peak reduction of 26.5% with a sample time of 60 min, and a cost reduction of 42.67% and a peak reduction of 25.8% with a sample time of 30 min. The decrease in sample time from (60 min) to (30 min) further improved the solutions.

In future work, we plan to explore the use of other methods for selecting solutions from the Pareto front.

Acknowledgments. The authors would like to thank Mustansiriyah University (www.uomustans iriyah.edu.iq) Baghdad-Iraq for its support in the present work.

References

1. Logenthiran, D.S., Phyu, E.: Particle swarm optimization for demand side management in smart grid. In: Proceedings of 2015 IEEE Innov. Smart Grid Technol. – Asia, ISGT ASIA 2015 (2016). https://doi.org/10.1109/ISGT-Asia.2015.7386973
2. Abdel-Mawgoud, H., Kamel, S., Khasanov, M., Khurshaid, T.: A strategy for PV and BESS allocation considering uncertainty based on a modified Henry gas solubility optimizer. Electr. Power Syst. Res. **191** (2021). https://doi.org/10.1016/j.epsr.2020.106886
3. Rajarajeswari, R., Suchitra, D., Vijay Krishna, J., Das Gupta, J.: Priority based scheduling of residential users devices in smart grid including waitingtime. Int. J. Recent Technol. Eng. **8**(1, 4), 438–442 (2019)
4. Haider, H.T., See, O.H., Elmenreich, W.: Residential demand response scheme based on adaptive consumption level pricing. Energy **113**, 301–308 (2016). https://doi.org/10.1016/j.energy.2016.07.052
5. Niharika, V.M.: Day-ahead demand side management using symbiotic organisms search algorithm. IET Gener. Transm. Distrib. 12(14), 3487–3494 (2018). https://doi.org/10.1049/iet-gtd.2018.0106
6. Kakran, S., Chanana, S.: Energy scheduling of residential appliances by a pigeon-inspired algorithm under a load shaping demand response program. Int. J. Electr. Eng. Informatics **11**(1), 18–34 (2019). https://doi.org/10.15676/ijeei.2019.11.1.2
7. Logenthiran, T., Srinivasan, D., Shun, T.Z.: Demand side management in smart grid using heuristic optimization. IEEE Trans. Smart Grid **3**(3), 1244–1252 (2012). https://doi.org/10.1109/TSG.2012.2195686
8. Radhakrishnan, A., Selvan, M.P.: Load scheduling for smart energy management in residential buildings with renewable sources. In: 2014 18th Natl. Power Syst. Conf. NPSC 2014 (2015). https://doi.org/10.1109/NPSC.2014.7103825
9. Logenthiran, T., Srinivasan, D., Shun, T.Z.: Multi-agent system for demand side management in smart grid. Proc. Int. Conf. Power Electron. Drive Syst. **954**(December), 424–429 (2011). https://doi.org/10.1109/PEDS.2011.6147283
10. Panda, S., et al.: Residential demand side management model, optimization and future perspective: a review. Energy Rep. **8**, 3727–3766 (2022). https://doi.org/10.1016/j.egyr.2022.02.300
11. Shaban, A., Maher, H., Elbayoumi, M., Abdelhady, S.: A cuckoo load scheduling optimization approach for smart energy management. Energy Rep. **7**, 4705–4721 (2021). https://doi.org/10.1016/J.EGYR.2021.06.099
12. Setlhaolo, D., Xia, X., Zhang, J.: Optimal scheduling of household appliances for demand response. Electr. Power Syst. Res. **116**, 24–28 (2014). https://doi.org/10.1016/j.epsr.2014.04.012
13. Lu, X., Zhou, K., Chan, F.T.S., Yang, S.: Optimal scheduling of household appliances for smart home energy management considering demand response. Nat. Hazards **88**(3), 1639–1653 (2017). https://doi.org/10.1007/s11069-017-2937-9
14. Mota, B., Faria, P., Vale, Z.: Residential load shifting in demand response events for bill reduction using a genetic algorithm. Energy **260**(July), 124978 (2022). https://doi.org/10.1016/j.energy.2022.124978
15. Fan, X.M., Li, X.H., Ding, Y.M., He, J., Zhao, M.: Demand response scheduling algorithm for smart residential communities considering heterogeneous energy consumption. Energy Build. **279** (2023). https://doi.org/10.1016/j.enbuild.2022.112691
16. Haider, H.T., See, O.H., Elmenreich, W.: Dynamic residential load scheduling based on adaptive consumption level pricing scheme. Electr. Power Syst. Res. **133**, 27–35 (2016). https://doi.org/10.1016/j.epsr.2015.12.007

17. Sultan, M.J., Tawfeeq, M.A., Haider, H.T.: Residential load scheduling based analytical optimization method. IOP Conf. Ser. Mater. Sci. Eng. **1076**(1), 012005 (2021). https://doi.org/10.1088/1757-899x/1076/1/012005
18. Haider, H.T., Muhsen, D.H., Al-Nidawi, Y.M., Khatib, T., See, O.H.: A novel approach for multi-objective cost-peak optimization for demand response of a residential area in smart grids. Energy **254**, 124360 (2022). https://doi.org/10.1016/j.energy.2022.124360
19. Ibrahim Mansoor, M., Haider, H.T.: A multi-objective residential load management based on self-adapting differential evolution. Renew. Energy Focus **38**(September), 44–56 (2021). https://doi.org/10.1016/j.ref.2021.05.004
20. Liu, Y., Li, H., Zhu, J., Lin, Y., Lei, W.: Multi-objective optimal scheduling of household appliances for demand side management using a hybrid heuristic algorithm. Energy **262**, 125460 (2023). https://doi.org/10.1016/j.energy.2022.125460
21. Sultan, M.J., Tawfeeq, M.A., Haider, H.T.: Residential load control system based analytical optimization method for real residential data consumption. J. Phys. Conf. Ser. **1**, 2021 (1973). https://doi.org/10.1088/1742-6596/1973/1/012018
22. Xiong, Q., She, J., Xiong, J.: A new pelican optimization algorithm for the parameter identification of memristive chaotic system. Symmetry (Basel) **15**(6), 1279 (2023). https://doi.org/10.3390/sym15061279
23. Trojovský, P., Dehghani, M.: Pelican optimization algorithm: a novel nature-inspired algorithm for engineering applications. Sensors **22**(3) (2022). https://doi.org/10.3390/s22030855
24. Talbi, E.-G.: Metaheuristics: From Design to Implementation. Wiley (2009)
25. Reyes Fernández de Bulnes, D.: Multi-objective optimization approach based on minimum population search algorithm, vol. 7, pp. 1–19 (2019)
26. Rong, J., Liu, W., Jiang, F., Cheng, Y., Li, H., Peng, J.: Privacy-aware optimal load scheduling for energy management system of smart home. Sustain. Energy Grids Netw. **34** (2023). https://doi.org/10.1016/j.segan.2023.101039

Incorporating Dilation Convolution into Mask Region Convolution Neural Network for Advanced Fruit Classification and Freshness Evaluation

Rafah Adnan Shandookh$^{(\boxtimes)}$ ⓘ, Tariq M. Salman ⓘ, and Abbas H. Miry ⓘ

Electrical Engineering Department, Collage of Engineering, Al Mustansiriyah University, Baghdad, Iraq
rafah88@uomustansiriyah.edu.iq

Abstract. Fruit supply and storage chains play a significant role in the food economies of all countries that export and consume these crops. Reducing the losses caused by the damage to these crops and organizing their marketing, relying on artificial intelligence methods, has begun to receive increasing attention from researchers. To ensure food safety and meet consumer expectations. This paper investigates the integration of dilation convolution into the Mask R-CNN framework to improve the accuracy of produce classification and detection of freshness. This paper seeks to address the challenges associated with traditional methods of fruit quality assessment, such as visual inspection and manual grading, which are time-consuming and subjective. To ensure the robustness of our approach, we collected a diverse set of complex images under various lighting conditions, and we implemented preprocessing techniques, such as histogram equalization and color correction, to enhance image quality. Using instance segmentation allowed us to distinguish and classify overlapped objects within each image. The practical results demonstrate that integrating dilation convolution improves the fruit classification, with an accuracy of 97.8% over the baseline Mask R-CNN. Real-time images and video were applied using the trained model, proving its practical utility in dynamic environments. It demonstrates the potential for direct object detection and classification by proposing a valuable tool to enhance productivity and ensure product quality.

Keywords: Deep learning · Pre-Processing · Dilated Convolution · Mask Regin Convolution Neural Network (R-CNN) · Real-Time image and video stream

1 Introduction

Fruits are an essential part of agriculture because they provide food and income for farmers. They also play a role in the maintenance of the health and diversity of ecosystems. For humans, fruits are an essential source of nutrients, including vitamins and minerals, and can help prevent several diseases [1]. Fruit rotting can happen for many reasons,

A. M. Al-Bakry et al. (Eds.): NTICT 2023, CCIS 2096, pp. 48–62, 2024.
https://doi.org/10.1007/978-3-031-62814-6_4

including during harvest, transit, storage, etc. The implementation of artificial intelligence in agriculture solved many challenges. It helped to diminish many disadvantages of traditional farming and classification by enabling machines to understand and analyze visual data with unprecedented accuracy using Deep Learning (DL) and Computer Vision (CV) [2]. These technologies have found extensive applications in fields such as image classification and object detection, where their ability to process large amounts of data and extract meaningful patterns has led to significant advancements [3].

Automated fruit classification is challenging in computer vision(CV) due to the variability in size, shape, color, and texture of fruits and the occlusions and clutter often occurring in natural scenes [4]. In recent years, deep learning-based approaches have shown promising results in many fields, such as medicine, self-driving, face recognition, agriculture, and fruit classification. Among the most successful methods are convolutional neural networks (CNNs) and their variants, such as Mask R-CNN, which can simultaneously detect and segment objects in images [5].

This study aims to evaluate the impact of dilation convolution on Mask R-CNN, not only in terms of improved object detection and classification but also in its application to real-time video data. This research shows what is achievable with DL and CV techniques, highlighting their potential in real-world, time-critical applications.

After training, the model can be utilized to classify new images of fruits based on their types and freshness levels. To classify the input image, the model takes the image as input and produces a probability distribution over the different freshness labels. The model then assigns the label with the highest probability to the fruit in the image, indicating its type and freshness status. The following is a list of the study's contributions:

- Design an automatic fruit detection and classification system using deep learning and computer vision techniques.
- Collect a dataset of multiple fruits (Apple, Banana, and Orange), fresh and rotten, with complicated backgrounds and different lighting levels.
- Apply pre-processing steps to the images.
- Train the data using mask R-CNN with dilation convolution.

This work first summarizes previous overview papers on fruit and vegetable classification using ML and DL algorithms. Then, it will explain the materials and methods for detecting and classifying fruits, describe Mask R-CNN architecture, and the type of backbone architectures used in the study, including information on the analytical findings and experiment results. Finally, we will discuss the results and conclusions introduced.

2 Related Works

Several significant studies have successfully developed automated fruit recognition and classification systems using computer vision and deep learning techniques, such as:

(Lee & Shin, 2020) [6] the Mask R-CNN architecture was used to develop a precise and efficient potato detection and segmentation algorithm. A collection of potato images was collected and manually labeled for object detection and segmentation before being used to train the model using transfer learning, with the pre-trained model on the COCO dataset fine-tuned on the potato dataset. According to the test results of 69 randomly chosen images, the potato's average detection precision, recall, and F1 scores were 90.8%, 93.0%, and 91.9%, respectively.

(Roy et al., 2021) [7] created a system to distinguish between fresh and rotten apples. The dataset came from the website kaggle.com. The binary masks were created using data pre-processing. To increase accuracy, the upgraded U-Net was constructed utilizing the U-Net frame. With the GPU Tesla K80 in Google Colab, the model was trained. The upgraded U-Net model's validation accuracy was 97.54% compared to the U-Net model's 95.36% validation accuracy.

(Tian et al., 2021) [8] developed a novel multi-scale Dense classification network to diagnose 11 types of apple fruit and leaf images, including healthy and diseased samples. To address the issue of limited images for certain diseases, the authors use the Cycle-GAN algorithm to generate additional images for anthracnose and ring rot. They then employ DenseNet and Multi-scale connections to create two models, Multi-scale Dense Inception-V4 and Multi-scale Dense Inception-Resnet-V2, which achieve high accuracy in image classification. The models outperform the DenseNet-121 network and achieve a state-of-the-art accuracy of 94.31% and 94.74%, respectively.

(Dhiman et al., 2022) [9] proposed a novel deep-learning model for detecting the disease severity level in citrus fruits at four levels: low, medium, high, and healthy. A convolutional neural network (CNN) with a transfer learning approach was used to classify the severity level of citrus canker disease. Test accuracy achieved on randomly selected images for healthy, low-level, high-level, and medium levels of disease was 96%, 99%, 98%, and 97%.

(Minagawa & Kim, 2022) [10] develop an automated system for predicting the harvest time of tomatoes using Mask R-CNN. A dataset of tomato images was collected at different stages of maturity and used to train the Mask R-CNN model to detect and segment the tomatoes in the images. Then, the color and texture features of the tomato images were used to predict their maturity level and estimate the harvest time. The results showed that the proposed system achieved high accuracy in predicting the tomatoes' maturity level and harvest time, with an overall accuracy of 92%. Also, it was able to distinguish between ripe and unripe tomatoes.

(J. Lu et al., 2022) [11] enhanced the Mask R-CNN for citrus green fruit extraction, achieving an enhanced mask accuracy of 95.3% using the Conditional Batch Normalization Network CB-Net architecture.

(Cong et al., 2023) [12] proposed an improved version of the Mask R-CNN algorithm to improve the accuracy and efficiency of sweet pepper detection in greenhouse environments. UNet3 + was used to improve the mask head and segmentation quality. They also conducted experiments to evaluate the performance of the proposed algorithm and compared it with other state-of-the-art methods. The results showed that the proposed algorithm achieved high accuracy and efficiency in sweet pepper detection and segmentation, with a processing time of 0.2 s per image.

Previous studies highlight the importance of transfer-learning and pre-processing in improving the results of agricultural image analysis and the progress in this field. Innovative methods such as Mask R-CNN, up-to-date U-Net, and multi-scale dense models demonstrate the versatility of deep learning in crop monitoring. Many factors affect the training results and the performance of the masker and lead to their differences, such as image pre-processing and segmentation methods, in addition to the training parameters.

3   Material and Methods

In this section, we describe the steps taken to analyze the input images and conduct training to perform classification and detection tasks using the Mask R-CNN architecture, Fig. 1 illustrates the steps of our work. Our method comprises steps, including preprocessing the input images, model training, and deploying the Mask R-CNN framework to achieve accurate classification and detection results.

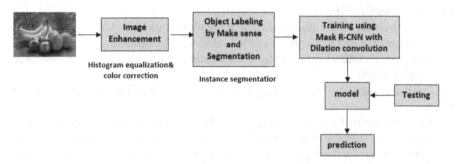

Fig. 1. Block diagram of the training model

3.1   Dataset

This paper presents a dataset of fruits collected from Kaggle (https://www.kaggle.com/datasets/muhriddinmuxiddinov/fruits-and-vegetables-dataset) and Google, consisting of six classes: fresh apple, rotten apple, fresh banana, rotten banana, fresh orange, and rotten orange. The dataset contains a total of (1014) images with varying lighting conditions, backgrounds, and poses.

Some images have more than one type of fruit, so the training fruits were 1420, validation fruits 150, and testing fruits 273, divided into six classes as shown in Table 1.

Table 1. Represent training, validation, and testing data.

	Fresh Apple	Rotten Apple	Fresh Banana	Rotten Banana	Fresh Orange	Rotten Orange
Training	246	207	203	267	262	235
Validation	25	25	25	25	25	25
Testing	53	37	54	49	49	31

Before feeding the input images into the modified Mask R-CNN architecture, a series of pre-processing steps, like histogram equalization and color correction, were applied to enhance the image quality and address the challenges posed by varying illumination across the dataset.

First, the input images were converted into the grayscale form, and then the image's histogram was calculated using histogram equalization to analyze the distribution of pixel intensities. It helps improve the visibility of features and textures, making detecting and classifying fruits based on shape and texture easier [13].

Second, color balance adjustment was applied to all images in the dataset to standardize their color appearance and improve the overall color distribution and contrast [14]. Doing this ensures that object features are more distinguishable regardless of illumination conditions and that the neural network models trained on the dataset are robust to color variations [15].

Third, the image was converted to L*a*b *, and contrast-limited adaptive histogram Equalization(CLAHE) was applied to the L-channel (lightness) of the L*a*b* image to enhance local contrast in the lightness information of an image, leading to improved visibility of details.

Finally, convert the L*a*b* images to RGB color space and save the enhanced images.

The next step is the labeling process. The Make Sense program was used to label each image with instance masks for each object in the image. We used polygon tools in Make Sense to create these annotations, as they allow for precise delineation of object boundaries and are particularly well-suited for instance segmentation tasks.

Once we had labeled all the images with instance masks, we exported the annotations in the COCO JSON format, widely used in the computer vision community for instance segmentation tasks. This allowed us to load the dataset into popular machine learning frameworks quickly and use it to train instance segmentation models [16].

3.2 Dilated Convolution (DC)

Dilated convolution, known as Atrous Convolution, is a type of convolutional operation that allows for an extended receptive field without increasing the number of parameters or the computational complexity of the network [17]. In standard convolution, the filter slides over the input data with a particular stride, and at each position, it computes the dot product between the filter weights and the input values within its receptive field. The receptive field represents the input data region contributing to the computation at a given position. On the other hand, Dilation convolution modifies the receptive field. Instead of sliding the filter with a stride of 1, dilation convolution skips some positions in between depending on the dilation rate. This is done by inserting gaps between the filter values, effectively increasing the size of the receptive field. Equations (1) shows the standard convolution [18]:

$$(I * W)_{(t)} = \sum_{p+q=t} I(p)W(q) \tag{1}$$

While Eqs. (2) shows Dilated convolution [18]:

$$(I * lW)_{(t)} = \sum_{p+lq=t} I(p)W(q) \tag{2}$$

where I is a discrete signal, W is a convolution kernel, and l is the dilated convolution, so when $l = 1$ it's a standard convolution, and when $l > 1$ it's a dilated convolution.

3.3 Mask R-CNN Architecture

Mask R-CNN is a deep learning algorithm introduced by Facebook AI Research (FAIR) in 2017 [19]. It builds upon the Faster R-CNN architecture [20] by adding a parallel branch to the network that outputs a binary mask for each object instance to identify the pixel-level location of each object instance in an image and the bounding box coordinates and class labels [21].

The main components of Mask R-CNN are the Backbone network, Region Proposal Network (RPN), Region of Interest Align layer (RoIAlign), Classification and Regression Heads, and Mask branch. Combining these components allows Mask R-CNN to detect objects, classify them, refine the bounding box locations, and generate precise instance-level segmentation masks. Figure 2 shows mask architecture.

After the Faster R-CNN detects the target, FCN is used for mask prediction, border regression, and classification. Combining the two makes Mask R-CNN an excellent tool for object detection and segmentation [22].

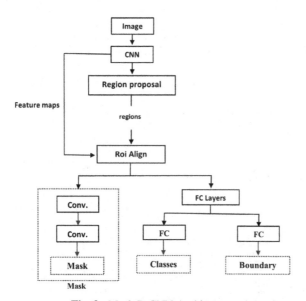

Fig. 2. Mask R-CNN Architecture

3.4 Proposed Method

Google Colaboratory, an online platform that provides a Jupyter Notebook environment with Python support, was used to write, debug, and implement the code.

The backbone network in Mask R-CNN is Residual Network (ResNet101), a convolutional neural network (CNN) architecture used as a feature extractor to analyze an input image and extract features useful for object detection and instance segmentation. The backbone network typically consists of multiple convolutional layers divided into five stages. Low-level features such as textures, edges, and other features are extracted from the input image at the initial stage. In contrast, high-level features and spatial data are extracted at the Final Stages.

Dilated Convolution was used on ResNet101 to improve the network's receptive field without significantly increasing the number of parameters or computational cost, which is done by modifying the configuration of the network and replacing the standard kernel (3×3) convolutional layers in the last two stages with dilated convolutional layers that have a dilation rate of 2, as seen in Fig. 3. Then, fine-tuned the entire Mask R-CNN architecture on our fruit images dataset using stochastic gradient descent (SGD) with a learning rate of 0.0001, the momentum of 0.9, batch size equal to 2, images per Batch equal to 2, and the number of epochs is 10,000 epochs.

```
(res5): Sequential(
  (0): BottleneckBlock(
    (shortcut): Conv2d(
      1024, 2048, kernel_size=(1, 1), stride=(1, 1), bias=False
      (norm): FrozenBatchNorm2d(num_features=2048, eps=1e-05)
    )
    (conv1): Conv2d(
      1024, 512, kernel_size=(1, 1), stride=(1, 1), bias=False
      (norm): FrozenBatchNorm2d(num_features=512, eps=1e-05)
    )
    (conv2): Conv2d(
      512, 512, kernel_size=(3, 3), stride=(1, 1), padding=(2, 2), dilation=(2, 2), bias=False
      (norm): FrozenBatchNorm2d(num_features=512, eps=1e-05)
    )
    (conv3): Conv2d(
      512, 2048, kernel_size=(1, 1), stride=(1, 1), bias=False
      (norm): FrozenBatchNorm2d(num_features=2048, eps=1e-05)
    )
  )
  (1): BottleneckBlock(
    (conv1): Conv2d(
      2048, 512, kernel_size=(1, 1), stride=(1, 1), bias=False
      (norm): FrozenBatchNorm2d(num_features=512, eps=1e-05)
    )
    (conv2): Conv2d(
      512, 512, kernel_size=(3, 3), stride=(1, 1), padding=(2, 2), dilation=(2, 2), bias=False
      (norm): FrozenBatchNorm2d(num_features=512, eps=1e-05)
    )
    (conv3): Conv2d(
      512, 2048, kernel_size=(1, 1), stride=(1, 1), bias=False
      (norm): FrozenBatchNorm2d(num_features=2048, eps=1e-05)
    )
  (2): BottleneckBlock(
    (conv1): Conv2d(
      2048, 512, kernel_size=(1, 1), stride=(1, 1), bias=False
      (norm): FrozenBatchNorm2d(num_features=512, eps=1e-05)
    )
    (conv2): Conv2d(
      512, 512, kernel_size=(3, 3), stride=(1, 1), padding=(2, 2), dilation=(2, 2), bias=False
      (norm): FrozenBatchNorm2d(num_features=512, eps=1e-05)
    )
    (conv3): Conv2d(
      512, 2048, kernel_size=(1, 1), stride=(1, 1), bias=False
      (norm): FrozenBatchNorm2d(num_features=2048, eps=1e-05)
    )
```

Fig. 3. Screenshot of Applying Dilation Convolution

4 Results and Discussion

In this section, we present the results of our experiments that involved training Mask R-CNN with and without the integration of dilation convolution. The primary objective was to assess the impact of dilation convolution on Mask R-CNN's performance in various aspects.

In our study, we used TensorBoard, a powerful visualization tool provided by TensorFlow, to facilitate the training and evaluation of our deep learning models. We utilized TensorBoard to monitor key training metrics such as loss functions, learning rate schedules, and performance indicators. This allowed us to understand the model comprehensively and helped us to make the necessary adjustments.

Figure 4 shows the effect of applying dilation convolution on Region Proposal Network (RPN) Fig. 4.a, Region of Interest Align (RoIAlign) Fig. 4.b, and mask accuracy Fig. 4.c.

Incorporating dilated convolution into the ResNet101 backbone of Mask R-CNN yields noticeable improvements in the RPN and RoIAlign. The introduced dilation convolution enhances the feature extraction capabilities of the backbone, leading to more effective region proposals and refined object alignments. As illustrated in the curves, comparative analysis against the non-dilated counterpart significantly enhances both the object detection and classification processes and lowers classification losses with better mask accuracy.

A confusion matrix was used to gain a more comprehensive understanding of the model's performance and effectiveness by indicating the proportion of correct and incorrect predictions made by the model in comparison to the actual test results using metrics including precision, recall, F1 score, and accuracy where: [23, 24]

$$\text{Accuracy} = (\text{TP} + \text{TN})/(\text{TP} + \text{TN} + \text{FP} + \text{FN}) \tag{3}$$

$$\text{Precision} = \text{TP}/(\text{TP} + \text{FP}) \tag{4}$$

$$\text{Recall} = \text{TP}/(\text{TP} + \text{FN}) \tag{5}$$

$$\text{F1 Score} = 2 * (\text{Precision} * \text{Recall})/(\text{Precision} + \text{Recall}) \tag{6}$$

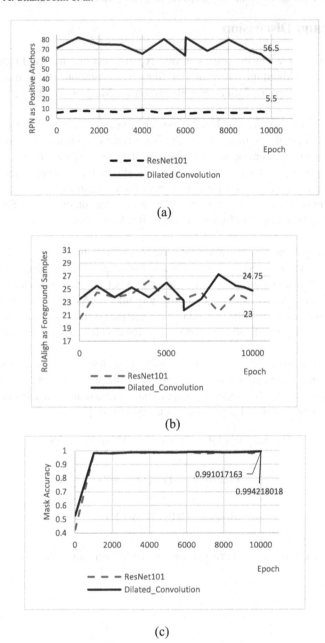

(a)

(b)

(c)

Fig. 4. Effect of Using Dilated Convolution on (a) RPN(positive anchors), (b) RoIAligh (Foreground samples), and (c) Mask accuracy

Figure 5 shows the confusion matrix with and without dilated convolution to compare the classification performance for each class.

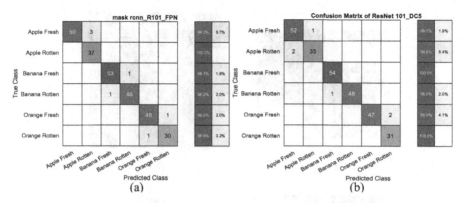

Fig. 5. Confusion Matrix (a) Without Dilation Convolution, and (b) With Dilation Convolution

We noticed that dilated convolution increased the accuracy of some classes, especially fresh fruits. Suggests that the utilization of dilated convolution contributes positively to the model's ability to accurately classify various classes, with particularly notable improvements in specific instances. Results demonstrate that dilation convolution with Mask R-CNN outperformed using only ResNet101 regarding several metrics, as seen in Table 2.

Table 2. Performance of each model

Metrics	Mask R-CNN without DC	Mask R-CNN with DC
Precision	97.22%	97.6%
Recall	97.53%	97.8%
F1 score	97.34%	97.7%
Accuracy	97.5%	97.8%

Incorporating dilation convolution (DC) into Mask R-CNN improved the prediction time from 0.4 s for ResNet101 to 0.28 s for dilation convolution. However, it is essential to note that this enhancement came at the cost of a longer training time, requiring 2 h and 40 min 49 s for 10,000 epochs compared to 1 h and 45 min 3 s for the baseline ResNet-101 model. Figure 6 shows the outcome of our experiment.

A comparison between our improved model and modern similar technologies is made, such as YOLOv7 [25], Faster R-CNN [26], and VGG-based CNN [27], which can be seen in Table 3.

Our Improved Mask R-CNN stands out clearly in comparison to other methods. With a precision of 97.8%, it surpasses YOLOv7, Faster R-CNN, VGG-based CNN, and the conventional Mask R-CNN, showcasing its superior accuracy in correctly identifying positive predictions. Furthermore, the recall of 97.8% indicates an exceptional ability to capture a more significant proportion of actual positives, outperforming other methods. The F1 Score, reaching 97.7%, is higher than all the compared methods, reflecting a

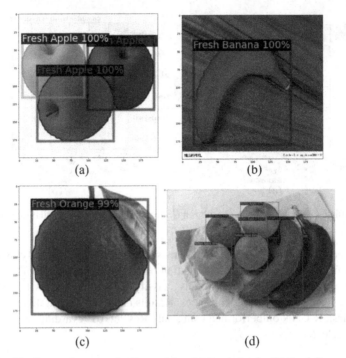

Fig. 6. Classification performance for the model as (a) Fresh Apple, (b) Fresh Banana, (c) Fresh Orange, and (d) Multiple Fruits

Table 3. Performance Comparison with Other Methods

Methods	Precision	Recall	F1 Score
YOLOv7	93%	89%	87.8%
Faster R-CNN	75.39%	90.37%	82.21%
VGG-based CNN	96.5%	96.5%	96%
Mask R-CNN	97.22%	97.53%	97.34%
Improved Mask R-CNN	97.6%	97.8%	97.7%

balanced performance between precision and recall. It highlights our method's robust capability to achieve accurate positive predictions and comprehensive coverage of actual positives, solidifying its effectiveness in object detection.

The impressive prediction speed achieved by our model is an essential factor in enabling real-time object detection and instance segmentation. Also, using the capabilities of dilation convolution and our optimized model, our system can seamlessly analyze video frames as they are captured, providing instantaneous insights and object recognition. A USB-connected camera is used to capture both images and live video streams and enables real-time analysis. The resolution is set to 1920 × 1080. A stand with adjustable

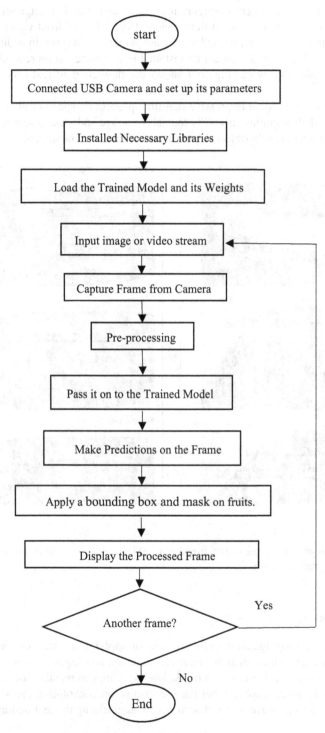

Fig. 7. Flow chart of the Real-Time process

dimensions was employed to provide flexibility in our fruit detection setup. This stand was utilized in three different configurations, each with varying heights((maximum, medium, and minimum)and from different views (top, side, and front view). This setup is complemented by using our trained model and associated weights. In addition, essential libraries, such as OpenCV, must be installed to complete the prerequisites for our real-time analysis framework. Figure 7 shows the flow chart of real-time image and video stream.

Our system achieved an impressive real-time processing speed of about 0.36 s, as seen in Fig. 8, demonstrating its ability to analyze image and video streams. This level of efficiency ensures timely object detection and instance segmentation.

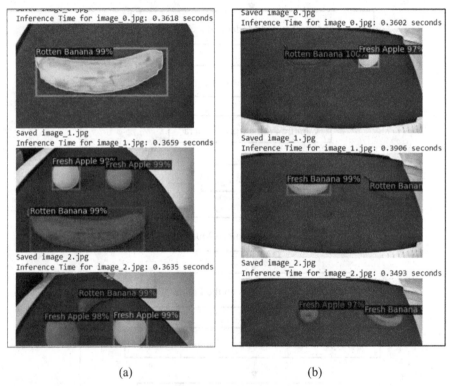

(a)   (b)

Fig. 8. Example of working the system in real-time (a) Front view with minimum height, and (b) Side view with medium height

5   Conclusion

In this paper, we investigated the effectiveness of applying a dilation convolution on Mask R-CNN architecture for fruit image classification and segmentation, intending to distinguish between fresh and rotten fruits. The experimental results show an accuracy of about 97.8%, which explains that the use of dilation convolution can significantly impact the model's performance and accuracy by augmenting the feature map.

Notably, the pre-processing step played a substantial role in enhancing result quality, and the labeling process, facilitated by polygon tools, notably increased mask accuracy, contributing to successful training.

Furthermore, the model's efficacy was underscored when applied to real-time images and videos with a prediction time of about 0.36%. The model was tested from various heights and perspectives, revealing optimal results. Notably, the highest accuracy was observed when the camera was positioned at the top view. This real-time application showcases the model's robustness and efficiency in addressing dynamic scenarios. With an impressive processing speed, our model facilitates seamless analysis of real-time streams, catering to applications where both speed and precision are imperative. This work emphasizes the potential of advanced convolutional techniques and underscores the critical role of meticulous pre-processing and accurate labeling in achieving optimal model performance and training success.

Acknowledgments. The author is grateful to Mustansiriyah University (www.uomustansiriyah. edu.iq) in Baghdad –Iraq for its assistance and encouragement with the current study.

References

1. Kanupuru, P., Uma Reddy, N.V.: A deep learning approach to detect spoiled fruits. WSEAS Trans. Comput. Res. **10**(July), 74–87 (2022). https://doi.org/10.37394/232018.2022.10.10
2. Faisal, M., Albogamy, F., Elgibreen, H., Algabri, M., Alqershi, F.A.: Deep learning and computer vision for estimating date fruits type, maturity level, and weight. IEEE Access **8**, 206770–206782 (2020). https://doi.org/10.1109/ACCESS.2020.3037948
3. Palakodati, S.S.S., Chirra, V.R.R., Yakobu, D., Bulla, S.: Fresh and Rotten Fruits Classification Using CNN and Transfer Learning. Rev. d'Intelligence Artif. **34**(5), 617–622 (2020). https://doi.org/10.18280/ria.340512
4. Ireri, D., Belal, E., Okinda, C., Makange, N., Ji, C.: A computer vision system for defect discrimination and grading in tomatoes using machine learning and image processing. Artificial Intelligence in Agriculture **2**, 28–37 (2019). https://doi.org/10.1016/j.aiia.2019.06.001
5. O'Mahony, N., et al.: Deep learning vs. traditional computer vision. In: Advances in Computer Vision: Proceedings of the 2019 Computer Vision Conference (CVC), Volume 1 1, pp. 128–144. Springer International Publishing (2020). https://doi.org/10.1007/978-3-030-17795-9_10
6. Lee, H.S., Shin, B.S.: Potato detection and segmentation based on mask R-CNN. J. Biosys. Eng. **45**, 233–238 (2020). https://doi.org/10.1007/s42853-020-00063-w
7. Roy, K., Chaudhuri, S.S., Pramanik, S.: Deep learning based real-time Industrial framework for rotten and fresh fruit detection using semantic segmentation. Microsyst. Technol. **27**, 3365–3375 (2021). https://doi.org/10.1007/s00542-020-05123-x
8. Tian, Y., Li, E., Liang, Z., Tan, M., He, X.: Diagnosis of typical apple diseases: a deep learning method based on multi-scale dense classification network. Front. Plant Sci. **12**, 698474 (2021). https://doi.org/10.3389/fpls.2021.698474
9. Dhiman, P., et al.: A novel deep learning model for detection of severity level of the disease in citrus fruits. Electronics **11**(3), 495 (2022). https://doi.org/10.3390/electronics11030495
10. Minagawa, D., Kim, J.: Prediction of harvest time of tomato using mask R-CNN. Agri. Engineering **4**(2), 356–366 (2022). https://doi.org/10.3390/agriengineering4020024

11. Lu, J., et al.: Citrus green fruit detection via improved feature network extraction. Frontiers in Plant Science **13**, 946154 (2022). https://doi.org/10.3389/fpls.2022.946154

12. Cong, P., Li, S., Zhou, J., Lv, K., Feng, H.: Research on instance segmentation algorithm of greenhouse sweet pepper detection based on improved mask RCNN. Agronomy **13**(1), 196 (2023). https://doi.org/10.3390/agronomy13010196

13. Singh, G., Mittal, A.: Various image enhancement techniques-a critical review. Int. J. Inno. Sci. Res. **10**(2), 267–274 (2014)

14. Qi, Y., et al.: A comprehensive overview of image enhancement techniques. Archives of Computational Methods in Engineering, 1–25 (2021). https://doi.org/10.1007/s11831-021-09587-6

15. Maini, R., Aggarwal, H.: A comprehensive review of image enhancement techniques. arXiv Prepr. arXiv1003.4053 (2010). https://doi.org/10.48550/arXiv.1003.4053

16. Tian, D., Han, Y., Wang, B., Guan, T., Gu, H., Wei, W.: Review of object instance segmentation based on deep learning. J. Electron. Imaging **31**(4), 041205 (2022). https://doi.org/10.1117/1.jei.31.4.041205

17. Lin, G., Wu, Q., Qiu, L., Huang, X.: Image super-resolution using a dilated convolutional neural network. Neurocomputing **275**, 1219–1230 (2018). https://doi.org/10.1016/j.neucom.2017.09.062

18. Zhang, Z., Wang, X., Jung, C.: DCSR: Dilated convolutions for single image super-resolution. IEEE Trans. Image Process. **28**(4), 1625–1635 (2018). https://doi.org/10.1109/TIP.2018.2877483

19. He, K., Gkioxari, G., Dollár, P., Girshick, R.: Mask r-cnn. In: Proceedings of the IEEE international conference on computer vision, pp. 2961–2969 (2017). https://doi.org/10.1109/TPAMI.2018.2844175

20. Hafiz, A.M., Bhat, G.M.: A survey on instance segmentation: state of the art. Int. J. Multimed. Info. Retr. **9**, 171–189 (2020). https://doi.org/10.1007/s13735-020-00195-x

21. Amo-Boateng, M., Sey, N.E.N., Amproche, A.A., Domfeh, M.K.: Instance segmentation scheme for roofs in rural areas based on Mask R-CNN. The Egypt. J. Remote Sens. Space Sci. **25**(2), 569–577 (2022). https://doi.org/10.1016/j.ejrs.2022.03.017

22. Nie, X., Duan, M., Ding, H., Hu, B., Wong, E.K.: Attention mask R-CNN for ship detection and segmentation from remote sensing images. Ieee Access **8**, 9325–9334 (2020). https://doi.org/10.1109/ACCESS.2020.2964540

23. Kumar, S.D., Esakkirajan, S., Bama, S., Keerthiveena, B.: A microcontroller based machine vision approach for tomato grading and sorting using SVM classifier. Microprocess. Microsyst. **76**, 103090 (2020). https://doi.org/10.1016/j.micpro.2020.103090

24. Zu, L., Zhao, Y., Liu, J., Su, F., Zhang, Y., Liu, P.: Detection and segmentation of mature green tomatoes based on mask R-CNN with automatic image acquisition approach. Sensors **21**(23), 7842 (2021). https://doi.org/10.3390/s21237842

25. Mimma, N.E.A., Ahmed, S., Rahman, T., Khan, R.: Fruits classification and detection application using deep learning. Sci. Program. **2022** (2022). https://doi.org/10.1155/2022/4194874

26. Mai, X., Zhang, H., Meng, M.Q.H.: Faster R-CNN with classifier fusion for small fruit detection. In: IEEE International Conference on Robotics and Automation (ICRA), pp. 7166–7172. IEEE (2018). https://doi.org/10.1109/ICRA.2018.8461130

27. Nagesh Appe, S.R., Arulselvi, G., Balaji, G.: Tomato Ripeness Detection and Classification using VGG based CNN Models. Int. J. Intell. Sys. Appli. Eng. **11**(1), 296–302 (2023). Retrieved from https://www.ijisae.org/index.php/IJISAE/article/view/2538

Adaptive Evolutionary Algorithm
for Maximizing Social Influence

Huda N. AL-mamory$^{(\boxtimes)}$ (iD)

Department of Information Networks, College of Information Technology, University of
Babylon, Hillah 51001, Iraq
`Huda.almamory@uobabylon.edu.iq`

Abstract. Influence Maximization (IM) is an issue that is represented by a predetermined group of users, sometimes referred to as the seed. The latter can have
an impact on their friends, who can then have an impact on other people, and
so on, until the network has the most users affected. According to it, the seeds
should be properly chosen to ensure widespread information dissemination. The
motivation for beginning this effort by building two models is the rate of user
interactions on one side and the density of relationships on another side using the
classic Independent Cascade (IC) as a diffusion model. IM has been modeled in
both models as a genetic algorithm optimization issue. In the first model, the population with a high rate of interaction is mostly used to represent the population.
In contrast, in the second, the population is represented by people who have a high
relationship density. The Higgs, Digg, and Twitter Dynamic networks were used
in an experimental framework to compare the two hypotheses with the standard
model. According to the findings, the suggested technique can boost the impact
spread from the baseline model by 6% to 200%.

Keywords: Social Networks · Influence Maximization · Independent Cascade ·
Genetic Algorithm · Information Diffusion

1 Introduction

In order to research information diffusion, methods from a wide range of fields are
employed. In this paper, we examine the subject from a sociological standpoint. During
the diffusion process, a node's ability to propagate information depends on how well-
liked it is by other nodes. More precisely, there is an edge u → v in the graph that
represents the social network when the activity of a user u implies an action of the user
v. This edge may represent, for instance, a supporter of a candidate in an election or
the likelihood that a product will be purchased by a user v if it is purchased by user u.
Generally speaking, the notation u → v denotes a likelihood of information spreading
from u to v. If successful, it is said that v is activated by u [1].

Diffusion rate explains how communities have become more interactive within the
context of the real graphs that have been selected. The recovery rate is the percentage
of social contact that is denied. The pace of associativity creation and consolidation is

known as the adoption rate. A social interaction between individuals constitutes real-world communication [2].

Users can communicate, collaborate, and exchange information on online social networks, which facilitates user communication. Online social networks are employed in many different industries because of their amazing capacity to reach huge audiences in a very short amount of time [3, 4]. These industries include marketing, healthcare, and political science. In this case, carefully selecting a subset of network nodes known as influencers who have a considerable impact on the flow of information is crucial [5].

Influential nodes have a great capacity for distributing information, and by initiating an information diffusion cascade, a piece of information coming from them can travel as far as possible within the network.

One of the most well-liked study issues in the field of network science is the identification of influential nodes for the aim of impact maximization [6, 7]. The majority of research on seed selection concentrates on node centrality measurements, but little attention is paid to node behavior or density of indirect relationships in addition to the direct one. We concentrate on the modeling of node acceptance to information or its relationships, which can aid in seed selection. IM has been shown to be an NP-hard problem [8, 9], so the evolutionary algorithms (EAs) is added to enhance the outcomes produced by the conventional methods, which have a complexity time.

To start developing the individuals as answers to the optimization problem, on which this study is focused, it may be a good idea to consider the density of relationships or the rate of node interactions. Several authors treated that issue as an optimization challenge. However, there are hardly any studies that propose solutions that consider indirect links or node behavior in addition to direct relationships. Most studies focus on the centrality measure as a means of selecting the seed nodes.

Regarding the estimate and maximizing of social impact, there are numerous ideas in the literature. In [10], authors employed the non-dominated sorting genetic algorithm II (NSGA-II), which is an enhanced version of the original method NSGA. In their work, a selection operator is introduced that combines the parent and offspring populations to produce a mating pool and chooses the best (in terms of fitness and spread) options. On the other hand, in the dynamic context, in order to create a dynamic seed set in social networks using IC models (i.e., a dynamic generalized genetic algorithm), the authors of [8] used soft computing techniques. They modeled a number of networks with edges and nodes that changed over time at specific timestamps.

In [11], a potential solution is specified by the EA as a fixed-length sequence of seed node identifiers. Adaptive initializers, mutation, and crossover operators produce prospective new population members. Additionally, the system employs graph-aware techniques like spread model approximations, domain-specific initialization and mutation of candidate solutions, and node filtering. The fitness of a prospective solution is determined using probabilistic spread model Monte-Carlo simulations. For IM in complex networks, Lijia Ma et al. [12] presented the evolutionary deep reinforcement learning method (referred to as EDRL-IM). First, EDRL-IM models the IM problem by continuously optimizing the weight parameters of the deep Q network (DQN). The EA is then combined with a deep reinforcement learning (DRL) technique to evolve the DQN (DRL). The DRL incorporates all of the available data and network-specific knowledge of DQNs to speed up their evolution while the EA simultaneously evolves a population of individuals, each of which represents a potential DQN and solves the IM problem through the use of a dynamic Markov node selection strategy.

Our work in [5] addressed the IM problem by adapting the NSGAII-based IM algorithm (NSGAII-IM), which is an improvement of NSGA-II. The nodes in the context of individual representation were selected using centrality measures in a pseudo-random manner (based on high centrality nodes as degree, closeness, and eigenvector). Increasing the influence's coverage size and minimizing the number of seed nodes as much as possible have been defined as the competing goals for the multi-objective function. The IC diffusion model has been recommended to be replaced with the proposed Weighted Integration Cascade (WIC).

Network discretization serves as the foundation for the improved differential evolution algorithm (IDDE) [13]. The algorithm enhances the differential evolution algorithm's variance rule, uses the discrete number and discrete granularity of the network that remains after the removal of the target node as an index to gauge the node's significance, and suggests a fitness function based on the network's robustness.

Although in a dynamic environment, the work in [13] takes user interaction behaviors into consideration; yet, our work is more closely tied to their work. In order to increase the effectiveness of problem-solving, the authors propose a two-stage IM solution algorithm (Outdegree Effective Link—OEL) based on node degree and effective links. They analyze the impact of node interaction on information dissemination in dynamic social networks, extend the classical IC model to a dynamic social network dissemination model based on effective links.

Graph-based long short-term memory (GLSTM) transfer learning has recently been proposed [14] as a novel approach to tackle the influence maximization problem as a typical regression task. The three popular node centrality approaches are first calculated as feature vectors for the network's nodes and each node's individual influence under the susceptible-infected-recovered (SIR) information diffusion model. This results in the construction of the labels for the network's nodes. The labels that correspond to each node of the produced feature vectors are then stored into a graph-based long short-term memory (GLSTM). The trained model is then used to calculate the predicted influence of each node in the target network.

2   Cascade Model Based EAs

Information and behavior can travel from person to person as content is shared, discussed, or voted on via online social networking sites. Informational and behavioral cascades that follow are extensive and potent. Understanding how technological innovation spreads, how individuals collaborate, and even how bad behaviors spread may be done using cascades [15].

A well-known optimization problem known as the "IM problem" involves finding the information spreaders who are most effective at spreading information across social networks [16]. There are several models have been introduced for studying cascades; game-theoretic, the IC and the linear threshold models that can be used to solve the IM problem [15].

Our analysis has focused on the IC model, which Kempe et al. first introduced in 2003 [17]. It was initially studied in the field of marketing to comprehend the effects of interpersonal word-of-mouth communication on marketing at the macro level [18]. It is the most fundamental cascade model. In actuality, IC are closely related to the models of epidemics and are part of the category of interacting particle systems models [19].

In the IC, each newly activated node n will successfully activate each currently dormant neighbor m with a fixed probability p; this property is system-wide and is thus shared by all edges n → m. The efforts at activation are carried out in any order when node n has many neighbors [17]. The threshold models depict a sophisticated type of agents' behavior in a social network, whereas IC models show how certain action spreads like a virus [19]. By first focusing on a select group of people to accept an invention, IM seeks to maximize the influence that spreads via a social network. Under the IC paradigm, IM is codified. [20], beginning with a collection of seed nodes. The weighted IC model was developed by enhancing the conventional independent cascade model. By tying qualities to the nodes, this innovative cascade model broadens the use of the IC paradigm. The goal of the IM problem in the WIC model is to increase the value of influenced nodes [21].

A number of methods, including heuristics with verifiable guarantees [22] or meta-heuristics [11] have been presented in the recent literature for the NP-hard [23] problem of determining the optimal (i.e., most impactful) set of seed nodes. Despite the fact that the runtime of both groups of approaches tends to rise with network size, they both produce good results. However, IM is a significant issue for many applications, including marketing and political campaigns, therefore its consequences are relevant.

Although it was discovered that EAs were capable of effectively exploring the enormous search space of all conceivable subsets of nodes [24], the user retained control over the quantity of seeds. Typically, a multiset of candidate solutions is evolved by EAs in order to find the optimal solution. A proposed solution for the issue at hand is, in practice, represented by a set of nodes that is a subset of the total nodes in the original network. In order to express the seeds of influence in the network, individuals are thus unordered sequences of integer node identifiers. Figure 1 reports a graphic illustration of the encoding issue [24].

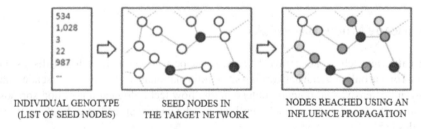

| INDIVIDUAL GENOTYPE | SEED NODES IN | NODES REACHED USING AN |
| (LIST OF SEED NODES) | THE TARGET NETWORK | INFLUENCE PROPAGATION |

Fig. 1. Encoding scheme described in [24]. Internally, a seed node is a collection of integers

3   Method

The selection of the seed nodes in set S is optimized via the influence-maximization issue. The expected size of the set A of active nodes as determined by a propagation model, is the influence of a given seed set S, indicated by the σ(S). This section presents two models which are proposed to formulate the problem based on the density or the interactions of nodes. Given graph G = (N, E), where N is the collection of nodes, E is the collection of directed edges.

3.1   The First Model

Users' past behavior can be modeled by their interactions. In other words, the frequency of users' interactions over time can be calculated. Suppose u and v, where u represents the set of users who tweet, vote or post (source nodes), and v (destination node) interacts with them.

$$I_v = \frac{|R_v|}{\sum_u |A|} \tag{1}$$

Let

$$I_m = \frac{\sum_{j \in N} I_j}{|N_m|} \tag{2}$$

where (1) represents the interaction rate of node v, I_v, „ by dividing v's response by the sum of all activities (A) for the set of users (u) with whom v interacts. As for the rate of interaction of a node m, I_m, $m \in u$, is the average of the interactions of all the neighbors (N) of node m as in (2). Consequently, identification a set of *influencers* (seeds set) for a network would be the users who have the highest rate of interactions as initial solution of the IM optimization problem. In other words, the K initial adopters are the nodes who have maximum rate of I_m:

$$S_{1...K} \leftarrow Max_{0 \leq I_u \leq 1} f(I_m) \tag{3}$$

The target node k in Fig. 2 is the first model's example, and it calculates the rate of interactions. It should be noted that the solid line in figure indicates that the node responded to the other node's post, while the dotted line indicates that there was no response. If the nodes v and h are assumed to be node k's neighbors, it is possible to determine the rate of interaction of its neighbors by using (1). The node v has three neighbors, m, a, and n, but only responds to m and a while ignoring n, resulting in an interaction rate, I_v, is 0.6. In a similar vein, h's interaction rate, I_h, is 0.5. Consequently, I_k = (0.6 + 0.5)/2 using (2). Worth to mention, in the example below, may the interactions between any pair of nodes is more than once, hence the matter is taking into account in (1).

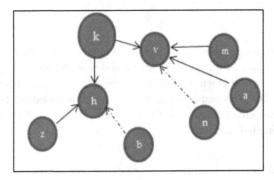

Fig. 2. Example of the first model.

3.2 The Second Model

The relationships density of each node can be calculated at the second level of its relationships. In other words, the direct relations and the links of the indirect relations of each node can be aggregated to determine the density level of the node.

$$D_u = \frac{|N_u \cup N_{uN}|}{|E|} \tag{4}$$

where D_u is the density of the relations of node u, $N_u and N_{uN}$ are the direct and indirect relations of u on the second level. The Fig. 3 illustrated an example to clarify the second model. The k node is the target, the density relationships of latter node on the second level is 0.7 using (4).

$$S_{1...K} \leftarrow Max_{0 \leq I_u \leq 1} f(D_u) \tag{5}$$

It's important to note that each solution (chromosome) consists of a number of nodes, each of which affects a number of other nodes. But what matters is that no one node appears more than once in the solution for both models. Therefore,

$$\bigcap_{i=1}^{n} g_i = \emptyset \tag{6}$$

where g_i represents the gen i in a chromosome with size n of an population, such that the intersections of all genes should be Phi.

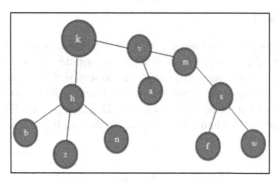

Fig. 3. Example of the second model.

4   Results and Discussion

Using real network datasets as the foundation for our underlying network structure, we have simulated information propagation. Twitter-Dynamic-Net, Digg, and Higgs datasets have been used to evaluate the proposed system. Only the tweet and retweet are extracted from Twitter Dynamic Net (https://www.aminer.cn/data-sna#Twitter-Dynami c-Net). Only the Digg network was utilized for Digg-vots (https://figshare.com/articles/ dataset/Digg_2009_social_news_votes_and_graph/2062467/1), and the Higgs network was used for retweet network. Edge list (https://snap.stanford.edu/data/higgs-twitter. html).

The attributes listed in Table 1 are those that our models used to find the best set of seeds for greatest influence propagation.

Table 1. List of attributes

Attributes	Values
Population size	50
Length of individual	10–50
Tournamet Selection size	2–5
No. of generations	50
Length of individual	10–50

In this work, the baseline which can be used as validation on which testing the proposed models. In the baseline, the individuals have been selected randomly without taking into account the rate of interactions or the density of relationships of a node.

4.1 The First Model

Figures 4 and 5 depict the best outcome after 50 generations for the Twitter Dynamic Net and Higgs networks for populations of 50 and individuals varying in length from 10 to 50. As seen in Fig. 4, it is very obvious that as the seed size increases, there are more active nodes. Figure 5 shows that a number with seed sizes of 10 and 50 is superior to one with a seed size of 30. For the networks of Twitter Dynamic Net and Digg, respectively, our model improved by 33% and 10% compared to the baseline.

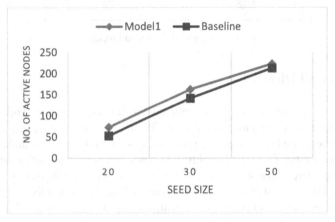

Fig. 4. Experimental results for the Twitter Network.

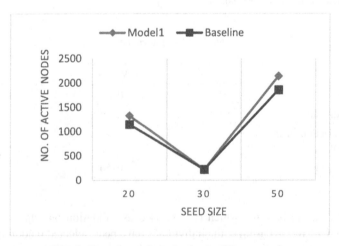

Fig. 5. Experimental results for the Digg network.

4.2 The Second Model

On the datasets from Twitter Dynamic Net, Digg, and Higgs, the second model has been used. We use a population size of 50 and run algorithm until 50 generations for all networks for different number of seed values ranging from 10 to 50. As opposed to other approach, our method revealed the greater influence maximization in networks, particularly for the Digg network. It is important to note that improvements began with earlier generations.

As seen in Fig. 6, our suggested algorithm increased influence propagation by an average of 70% across all seed sizes when compared to the Twitter Dynamic Net's baseline. Similar to Figs. 6, 7 and 8 show, respectively, the Higgs and Digg networks' experimental results. We deduced from the figures that our suggested approach performed better on Higgs and Digg as well. Our method demonstrated up to 200% better influence propagation in the Higgs network and 6% better influence propagation in the Digg network compared to the baseline for all seed sizes.

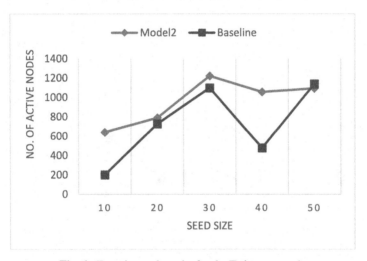

Fig. 6. Experimental results for the Twitter network.

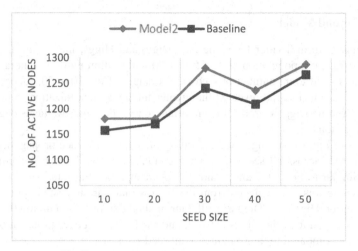

Fig. 7. Experimental results for the Higgs network

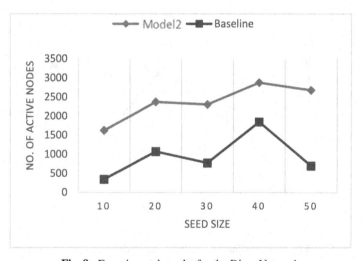

Fig. 8. Experimental results for the Digg. Network

5 Conclusion

In this study, we proposed a model based on interactions and node structure using the IC propagation model, which was improved using a genetic algorithm to pinpoint the optimal group of significant nodes in the network. We have assessed the performance of our proposed method in contrast to the baseline using experimental data. By determining the appropriate collection of seed nodes, the proposed strategy could maximize influence spread while boosting it by 6% to 200% over the baseline.

It can be concluded that user interactions may indicate their role in influencing others more effectively than just relying on their centrality measure. This is the focus of the first model.

From the second model it can be concluded that the density of nodes based on the first and second levels reveals the influence of a node. At the first level, degree centrality, only the direct relationships are considered, while these relationships can exist with important nodes that extend the sphere of influence.

It is possible to draw the conclusion that Model 2 outperforms Model1 and that relationship density significantly affects model superiority for all three networks, but particularly for Digg networks. Finally, The two models are characterized by their simplicity and lack of complexity compared to other methods.

The limitation of our work lies in the size of the individual representing the seed; as the size of the individual increases, the complexity of time increases.

As future work, we propose to control the rumor by the influencer nodes, this work can be implied by deep neural networks.

References

1. Konotopska, K., Iacca, G.: Graph-aware evolutionary algorithms for influence maximization. In: Proceedings of the Genetic and Evolutionary Computation Conference Companion, pp. 1467–1475. Association for Computing Machinery, New York, NY, USA (2021)
2. Kumar, P., Sinha, A.: Information diffusion modeling and analysis for socially interacting networks. Soc. Netw. Anal. Min. 11(1), 11 (2021)
3. Alasadi, M.K., Almamory, H.N.: Diffusion model based on shared friends-aware independent cascade. J. Phys: Conf. Ser. 1294(4), 042006 (2019)
4. Mohapatra, D., Panda, A., Gouda, D., Sahu, S.S.: A combined approach for k-seed selection using modified independent cascade model. In: Das, A.K., Nayak, J., Naik, B., Pati, S.K., Pelusi, D. (eds.) Computational Intelligence in Pattern Recognition, pp. 775–782. Springer Singapore, Singapore (2020)
5. Abbas, E.A., Nawaf, H.N.: Influence maximization based on a non-dominated sorting genetic algorithm. Karbala Int. J. Modern Sci. 7(2), 5 (2021)
6. Zhang, X., Zhu, J., Wang, Q., Zhao, H.: Identifying influential nodes in complex networks with community structure. Knowl.-Based Syst. 42, 74–84 (2013)
7. Seierstad, C., Opsahl, T.: For the few not the many? The effects of affirmative action on presence, prominence, and social capital of women directors in Norway. Scand. J. Manag. 27(1), 44–54 (2011)
8. Lotf, J.J., Azgomi, M.A., Dishabi, M.R.E.: An improved influence maximization method for social networks based on genetic algorithm. Phys. Statis. Mechan. Applicat. 586, 126480 (2022)
9. Wang, F., Zhu, Z., Liu, P., Wang, P.: Influence maximization in social network considering memory effect and social reinforcement effect. Future Internet 11(4), 95 (2019)
10. Deb, K., Pratap, A., Agarwal, S., Meyarivan, T.: A fast and elitist multiobjective genetic algorithm: NSGA-II. IEEE Trans. Evol. Comput. 6(2), 182–197 (2002)
11. Iacca, G., Konotopska, K., Bucur, D., Tonda, A.: An evolutionary framework for maximizing influence propagation in social networks. Software Impacts 9, 100107 (2021)
12. Ma, L., et al.: Influence maximization in complex networks by using evolutionary deep reinforcement learning. IEEE Trans. Emerg. Top. Computat. Intellig. 7(4), 995–1009 (2023)

13. Fu, B., Zhang, J., Li, W., Zhang, M., He, Y., Mao, Q.: A differential evolutionary influence maximization algorithm based on network discreteness. Symmetry **14**(7), 1397 (2022)
14. Kumar, S., Mallik, A., Panda, B.S.: Influence maximization in social networks using transfer learning via graph-based LSTM. Expert Syst. Appl. **212**, 118770 (2023)
15. Cheng, J.: Cascading Behavior in Social Networks. PhD. Thesis, Stanford University (2017)
16. Zhao, X., Liu, F.A., Xing, S., Wang, Q.: TSSCM: A synergism-based three-step cascade model for influence maximization on large-scale social networks. Plos one **14**(9), e0221271 (2019)
17. Kempe, D., Kleinberg, J., Tardos, É.: Maximizing the spread of influence through a social network. In: Proceedings of the ninth ACM SIGKDD international conference on Knowledge discovery and data mining, pp. 137–146. Association for Computing Machinery, Washington, D.C., USA (2003)
18. Bucur, D., Iacca, G.: Influence maximization in social networks with genetic algorithms. In: Squillero, G., Burelli, P. (eds.) Applications of Evolutionary Computation, pp. 379–392. Springer International Publishing, Cham (2016)
19. Chkhartishvili, A.G., Gubanov, D.A., Novikov, D.A.: Models of Information Influence, Control and Confrontation. Springer Cham, Warsaw, Poland (2019)
20. Yang, W., Brenner, L., Giua, A.: Influence maximization in independent cascade networks based on activation probability computation. IEEE Access **7**, 13745–13757 (2019)
21. Wang, Y., Wang, H., Li, J., Gao, H.: Efficient influence maximization in weighted independent cascade model. In: Navathe, S.B., Wu, W., Shekhar, S., Du, X., Wang, S.X., Xiong, H. (eds.) Database Systems for Advanced Applications, pp. 49–64. Springer International Publishing, Cham (2016)
22. Nguyen, H.T., Thai, M.T., Dinh, T.N.: A billion-scale approximation algorithm for maximizing benefit in viral marketing. IEEE/ACM Trans. Networking **25**(4), 2419–2429 (2017)
23. Singh, S.S., Srivastva, D., Verma, M., Singh, J.: Influence maximization frameworks, performance, challenges and directions on social network: A theoretical study. J. King Saud Uni.-Comp. Info. Sci. **34**(9), 7570–7603 (2022)
24. Bucur, D., Iacca, G., Marcelli, A., Squillero, G., Tonda, A.: Multi-objective evolutionary algorithms for influence maximization in social networks. In: Squillero, G., Sim, K. (eds.) Applications of Evolutionary Computation, pp. 221–233. Springer International Publishing, Cham (2017)

Automatic Identification of Ear Patterns Based on Convolutional Neural Network

Saba A. Tuama[1]($\boxtimes$) (iD), Jamila H. Saud[2] (iD), and Omar Fitian Rashid[3] (iD)

[1] Department of Engineering, University of Information Technology and Communications, Baghdad, Iraq
saba.ayad@uoitc.edu.iq
[2] Department of Computer Science, College of Science, Al-Mustansiriyah University, Baghdad, Iraq
[3] Department of Geology, College of Science, University of Baghdad, Baghdad, Iraq

Abstract. Biometrics represent the most practical method for swiftly and reliably verifying and identifying individuals based on their unique biological traits. This study addresses the increasing demand for dependable biometric identification systems by introducing an efficient approach to automatically recognize ear patterns using Convolutional Neural Networks (CNNs). Despite the widespread adoption of facial recognition technologies, the distinct features and consistency inherent in ear patterns provide a compelling alternative for biometric applications. Employing CNNs in our research automates the identification process, enhancing accuracy and adaptability across various ear shapes and orientations. The ear, being visible and easily captured in an image, possesses the unique characteristic that no two individuals share the same ear patterns. Consequently, our research proposes a system for individual identification based on ear traits, comprising three main stages: (1) pre-processing to extract the ear pattern (region of interest) from input images, (2) feature extraction, and (3) classification. Convolutional Neural Network (CNN) is employed for the feature extraction and classification tasks. The system remains invariant to scaling, brightness, and rotation. Experimental results demonstrate that the proposed system achieved an accuracy of 99.86% for all datasets.

Keywords: Biometric · Identification System · Ear Pattern · Convolutional Neural Network

1 Introduction

The high demand of secure information automated system and person identification accuracy lead to intensified the research in development of computer and machine vision, big data, artificial intelligence systems and biometric recognition technology. Most biometric human identification systems rely on identification technology and features that occupy a unique position in the realm of authentication and security systems [1]. Human ears have gradually garnered the attention of the scientific community due to their inherent stability, non-invasiveness, lack of expressiveness, and significant individual variations [2]. With the recent onset of the COVID-19 pandemic, ear recognition has gained particular significance as people frequently wear masks [3]. The distinction between human ear recognition and human ear detection lies in the former's ability not only to locate a human ear in an image but also to ascertain the identity of the individual to whom the ear belongs. Consequently, human identification can be accomplished through human ear recognition. Environmental changes and variations in posture can significantly impact the accuracy of ear recognition results, making deep learning-based methods more advantageous [4].

This study addresses the increasing demand for robust and efficient biometric identification systems by introducing an efficient approach to automatically identify ear patterns using Convolutional Neural Networks (CNNs). In contrast to conventional approaches relying on manually crafted features, our method harnesses the capabilities of deep learning to autonomously learn and extract distinctive features from ear images. This results in improved accuracy and heightened adaptability of the system to a wide range of ear shapes and orientations.

This paper is structured as follows: Sect. 2 presents the literature review. Section 3 the architecture of the proposed system. Section 4 discusses the experiments and results. Finally, conclusions are drawn in Sect. 5.

2 Literature Review

The human recognition based on Biometrics system, using deep learning methods have become a research interest. For example, Le et al. [3] have developed a small sample human ear database by considering the complex situations and changes of human ear images. This was leading to dynamic human ear recognition almost close to real-life status and increase the ability of human ear recognition. An ear recognition system is designed based deep learning method to test the small developed of DB, the result gave higher accuracy for small DB sample Alkababji & Mohammed [5] have used deep learning object detector "faster region CNN" for ear detection, where CNN extract feature. PCA with genetic algorithm have reduced and selected features l, the test of this model provided high accuracy with acceptable execution time. Six-layers deep convolutional neural network architecture for ear recognition have been proposed by Priyadharshini et al. [6], and the deep network model shows high recognition rate. Zhang et al. [7] suggested a human ear recognition model based on the convolutional neural network and tackle ear recognition problem by using few-shot learning based methods. Their model is flexible and can be quickly adapted to new identity and to perform fast recognition. Moniruzzaman & Islam [8] applied simple and deep raw conventional neural network to detect ear features as initial step of ear based biometric implications, where in the training dataset based on deep scratch architecture, has achieved promising results. Sanchez et al. [9] have introduced a novel approach that combines Geometric Morphometrics and Deep Learning to achieve automatic ear detection and feature extraction. This network possesses the capability to automatically generate morphometric landmarks on ear images. The practicality of utilizing ear landmarks as feature vectors introduces a new dimension to the field of biometrics applications.

3 Proposed Method

The aims of the proposed system are used for human identification by extracting the ear region from digital colour input images. It consists of three main stages: (i) image pre-processing, (ii) feature extraction, and (iii) classification. Figure 1 displays the layout of the proposed system. Each module of the presented system is described in the next sub-sections.

3.1 Pre-processing Stage

The purpose of this stage is to extract the ear region (region of interest) from digital color images by applying many sub-steps, which are:

HSV Color Model Conversion
The extraction of ear region from input images relies on color isolation. To do so, the input image is converted from RGB to HSV color space. So, three images are generated, which are: Hue image H () which contains the color, saturation image S () which determines how intense the color is, and the value image V () which determines the lightness of the image, as shown in Fig. 2. The following steps of the proposed system have used the Hue image H (). The generation of the three image is carried out using Eqs. (1–3) [10]:

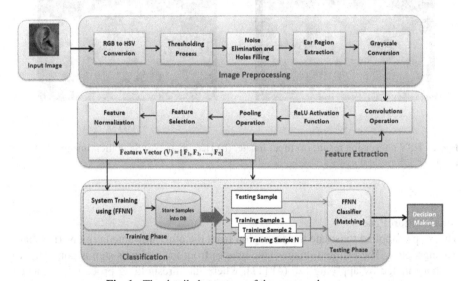

Fig. 1. The detailed structure of the proposed system

$$H = arccos \frac{\frac{1}{2}(2R - G - B)}{\sqrt{(R - G)^2 - (R - B)(G - B)}}$$ (1)

$$S = \frac{\max(R, G, B) - \min(R, G, B)}{max(R, G, B)}$$ (2)

$$V = max(R, G, B) \tag{3}$$

where R, G, and B represent red, green, and blue color values respectively, min represents the smaller value between R, G, and B, and max represents the larger value between R, G and B.

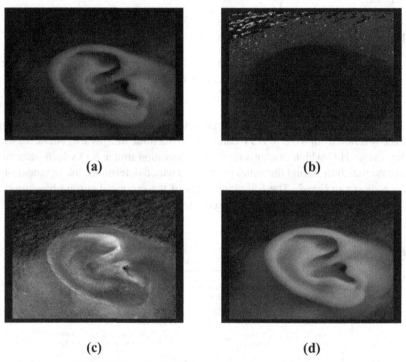

(a)   **(b)**

(c)   **(d)**

Fig. 2. HSV conversion result; (a) Input image, (b) Hue image, (c) Saturation image, (d) Value image

Binarization Process

This step is applied to isolate the ear region from the other unwanted regions (i.e., hair, earrings, etc.). The resulted image (Hue image) from the previous step is converted to a binary image by applying Eq. (4) [11], where the thresholding process is achieved by using more than one thresholding value. Figure 3 shows the resulting image after applying the binarization process.

$$\text{a)} \quad B_{(x,y)} = \begin{array}{l} 1 \quad ifThr1 \geq H(x, y) \geq Thr2 \\ 0 \qquad otherwise \end{array} \tag{4}$$

ere Thr1 and Thr2 stands for the lower and higher threshold values, H(x,y) denotes the pixel value in the Hue image, and B(x,y) is the resulting pixel value.

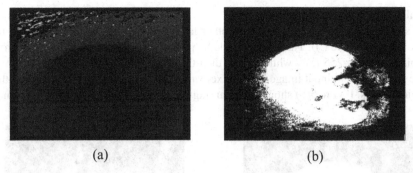

Fig. 3. Image thresholding results; (a) Hue image, (b) Result of applying thresholding process

Noise Elimination Process and Holes Filling

The seed filling technique [12, 13] is applied to remove small islands and holes that result from the thresholding process. This technique is applied more than once, in the first time, it is applied to remove small islands (i.e., white pixels) and in the second time, it is applied to remove holes (i.e., black pixels). Figure 4 shows the resulting image Ne (x, y) after applying the noise elimination process.

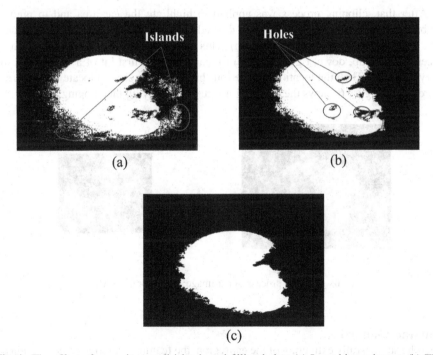

Fig. 4. The effect of removing small islands and filling holes: (a) Input binary image, (b) The image after removing small islands, (c) The image after filling holes

Localization of the Ear Region

This step aims to specify ear area from other face parts by checking each pixel in the input image (i.e., Ne (,)). This can be made by performing pixel-wise multiplication; if the pixel value is 255 (i.e., white pixel), then the value will be converted to the pixel value in the original input image. If the pixel value is 0 (i.e., black pixel), then its value remains as it is. Figure 5(b) shows the resulting image after the allocating ear region.

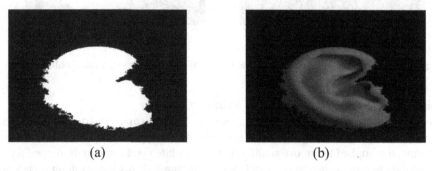

(a)   (b)

Fig. 5. Allocating ear region process; (a) Before, (b) After

After that, clipping process was applied to highlight the ear area and to ignore the background area to reduce the time desired to complete the system's next stages (i.e., feature extraction and classification). This is accomplished by scanning the four directions (i.e., up, down, right, and left) for capturing the first hit of a white pixel in every direction. Then the locations of the four-hit points are used to locate the ear area coordinates. Figure 6 shows the resulting image after applying the clipping process.

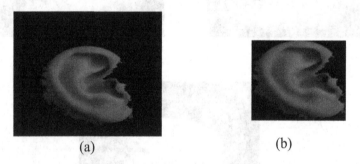

(a)   (b)

Fig. 6. Clipping process of ear image; (a) Before, (b) After

Grayscale Conversion

After the successfully extracting of the ear region, the feature extraction stage is applied to improve the efficiency and performance of the classification stage by extracting a set of discriminating features from the final processed image. Before applying the feature extraction stage, the resulting image from the pre-processing stage (i.e., Eimg (,)) is converted to a grayscale image by the process of gray image conversion [14] (Fig. 7).

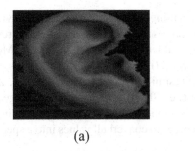 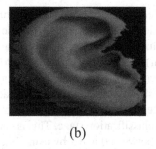

(a)   (b)

Fig. 7. Gray Image Preparation; (a) Before, (b) After

3.2 Feature Extraction

After the preparation of the gray image Gimg(,) CNN is utilized for extracting features by applying many operations, which are explained in the following paragraphs:

Convolutions Operation

This operation aims to extract features from the input image [15, 16] by performing an element-wise product between each element of the kernel mask with the corresponding pixel value in the input image (Gimg(,)) and summed out to compute the output value. The multiplying and summing process have been repeated until they are covering all pixels in the image. The output of this operation is called a feature map image Fimg(,). In this operation, the type of kernel mask that is used, is the edge law's mask [17, 18] to extract texture features.

Activation Function

The result of the convolution operation is passed to the rectified linear unit (ReLU) activation function for increasing the non-linearity in the images [17]. This operation is accomplished by using the following equation:

$$Relu = \begin{cases} 0 \; if & F_{Img}(x, y) \leq 0 \\ 1 & Otherwise \end{cases} \qquad (5)$$

whe Relu (x,y) represents the output image after the applied ReLU activation function, F_Img (x,y) represents the image generated by the convolution operation.

Pooling Operation

The pooling operation is also known as the sub-sampling layer which aims to reduce the dimensionality of the input image (reduce the number of parameters in the network) and consequently reduce the computational time. In the current system, the max pooling is utilized in pooling operations [20, 21]. The output of this operation would be a feature map containing the most distinguished features of the input feature map [22].

Feature Selection and Normalization

This operation is applied to remove the irrelevant features or less important in order to reduce the computational cost and to achieve a better accuracy. This operation is done by calculating some statistical features in the resulting features map from the previous operation (i.e., max-pooling). These statistical features are selected as texture features to be utilized for the classification stage. The statistical features vector SFv() are: (i) Mean (μ), (ii) Variance ($\sigma2$), (iii) Standard Deviation (Std), (iv) Energy, (v) Homogenous, (vi) Kurtosis, (vii) Skewness, (viii) Contrast, (ix) Dissimilarity, and (x) Entropy [23, 24]. Because the extracted statistical features vector (i.e., SFv()) from the previous stage are of different scales, the max-min normalization is applied to the SFv() before feeding it to the classification stage. This is carried out in order to convert all values into a specific range between 0 and 1 by using Eq. 6 [25].

$$X_{inew} = \frac{Xi - Xmin}{Xmax - Xmin} \tag{6}$$

where Xi represents the input feature value, Xmax represents the maximum value in the features vector, while X_min represents the minimum value in the features vector.

The first three operations of the feature extraction stage (Convolutions, ReLU activation function, and max-pooling) are repeated twice to obtain an acceptable result.

3.3 Classification Stage

After the application of the max-min normalization process, the SFv() is fed as inputs to the fully connected feedforward neural network [26, 28] for predicting the best labels describing the input image. The classification stage consists of two steps: (i) the training step, and (ii) the testing step (Fig. 8).

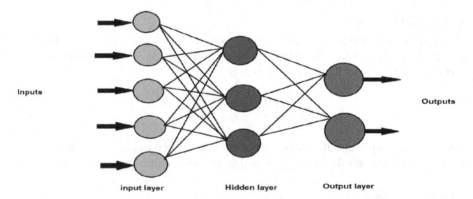

Fig. 8. A multi-layer feed forward neural network

4 Results and Discussion

The system was tested onto two datasets: (i) First Ear Dataset (ED1), (ii) The AMI Ear Database (AMI) [29]. ED1 has consisted of 30 images taken from different subjects aging ranging from 19 to 43 years old. It was collected by taking the snaps of ear from staff and students of the vision and Health Management Departments et al.-Mansour Technical Medical Institute. All the images were taken using Nikon camera under different lighting conditions. While, the AMI is publicly available consisting of 30 ear images with JPEG format. Each image was rotated six times in both clockwise and anticlockwise directions. Thus, a dataset consisting of 720 ear images is established. Figure 9 shows some of the ear samples of the considered datasets.

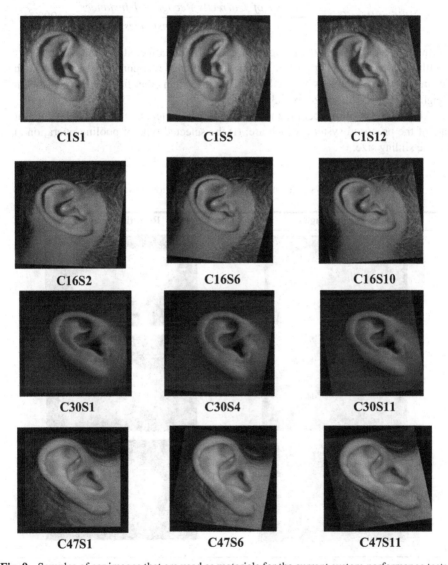

Fig. 9. Samples of ear images that are used as materials for the current system performance tests

The system gives in the segmentation stage an accuracy rate of 96.83% for extraction the ear region (ROI), while 0.9582% and 0.02284% for TPR and FPR respectively. These results indicated that the current proposed system in segmentation stage achieved high accuracy results when compare with the previous published methods. Table 1 shows some of the outcomes of the segmentation stage on some used samples.

In our system, the recognition rate is calculated using the following equation [30]:

$$Recognition\ Rate(\%) = \frac{Number\ of\ Correctly\ Recognized\ Instances}{Total\ Number\ of\ Instances} \times 100\% \quad (7)$$

In the present system, the recognition rate is highly affected by the type and size of the filters in convolution operation. The highest attained recognition rate is 99.86% that occurs when taken the Law's edge filter (EE) rather than other filter types (i.e., spot and weight) and apply filter size (3 × 3).

Also, in the pooling operation there are two parameters that altered the recognition rate of the proposed system, which are: (i) the selected type of pooling operation and (ii) the sliding size.

Table 1. Displays segmentation results of ear region on various datasets

Input image	Resulting image

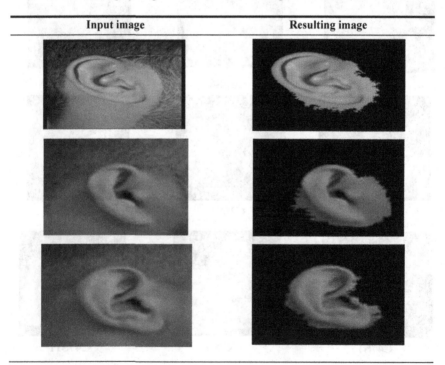

(*continued*)

Table 1. (*continued*)

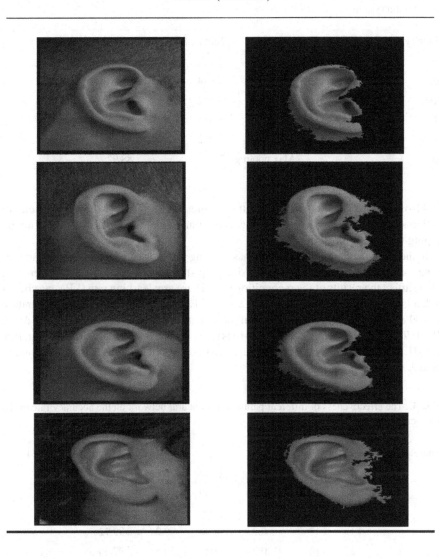

From the experiment test, the system shows that the highest recognition rate is 99.86% and has obtained when the max-pooling with sliding size is taken (2 × 2), as seen in Table 2.

Table 2. The effectiveness of different parameters in CNN operations on the accuracy rate of the identification system

Filter Type	Pooling Type	Sliding Size	Recognition Rate (%)
EE	Max pooling	2 × 2	99.86
EE	Max pooling	4 × 4	91.74
EE	Average pooling	2 × 2	92.83
EE	Average pooling	4 × 4	88.97
LL	Max pooling	2 × 2	90.41
SS	Max pooling	2 × 2	88.92

Also, the normalization process on the features vector is selected before the classification stage, in order to get a higher accuracy rate of an individual's identification by keeping features contrast in one level.

In the system, about 75% was taken as training samples and 25% as testing samples of the total images samples. In the classification stage, there are many parameters that will be affect the results of the recognition rate. These are: learning rate (LR), number of hidden layer (NHL), momentum value (Mo) and Epochs (NEp). The higher recognition rate (99.86%) is obtained when the configuration of neural network of ED1 is set to be: LR is 0.7, NHL is 15, NEp is 4000, and Mo is 0.2; while the neural network configuration of AMI is set to be: LR is 0.3, NHL is 17, NEp is 4000, and Mo is 0.2; as shown in Table 3.

Table 3. The effectiveness of different parameters in neural network on the accuracy rate of the identification system

Dataset Type	#	NHL	LR	Nep	Mo	Recognition rate (%)
ED1	1	15	0.7	4000	0.2	100
	2	14	0.4	4000	0.3	99.38
	3	18	0.2	4000	1	99.60
AMI	1	17	0.3	4000	0.2	99.72
	2	16	0.2	4000	0.5	97.40
	3	18	0.3	4000	0.6	98.61

Also, due to the extract of a small number of distinctive features, the method uses low computational complexity and so simple. Table 4 shows the comparison between the result of the identification accuracy obtained from the current proposed system and the results of the other methods of the recently published literatures.

Table 4. Comparison of the recognition rate of our proposed system with some other previously proposed methods

Reference	Biometrics Type	Biometric Traits	Recognition Rate (%)
[31]	Multimodal	Face, Ear, and Palm	97.4
[32]	Unimodal	Ear	97.5
[33]	Multimodal	Fingerprint and Ear	95.6
[3]	Unimodal	Ear	99.72
[4]	Unimodal	Ear	96.3
[34]	Unimodal	Ear	92.00
[35]	Unimodal	Ear	98.74
The Proposed System	Unimodal	Ear	99.86

5 Conclusions

Individual identification using ear biometrics has received important attention in recent years due to its permanence over time, easy acquisition process, non-intrusiveness, and highly discriminative nature. In this paper, a system is presented for human identification using ear biometrics via deep learning. The system consists of 3 main stages which are pre-processing, feature extraction and classification stage. In this system, the convolutional neural network is applied for feature extraction and classification. It is insensitive to rotation and scaling variation. The system was tested on two various datasets and obtained result has an accuracy of recognition rate equal to 99.86%% for all datasets.

Future work will be to integrate ear pattern with other biometric modalities, such as tongue or facial biometrics, which can offer various advantages, including enhanced robustness and accuracy in identification systems. Additionally, evaluating the model's performance on large and diverse datasets with varying acquisition conditions will help assess the generalization capabilities of the model and its robustness across different scenarios, potentially uncovering challenges and areas for improvement.

References

1. James, W., Jain, A., Zaltoni, M., Maio, D.: An introduction to biometric authentication systems. In: Biometric Systems, pp. 1–20. Springer, London (2005)
2. Pflug, A., Busch, C.: Ear biometrics: a survey of detection, feature extraction and recognition methods. IET biometrics **1**(2), 114–129 (2012)
3. Le, Y., Qian, J., Pan, D., Xu, T.: Research on small sample dynamic human ear recognition based on deep learning. Sensors, 22 February (2022)
4. Zhaobin, W., Gao, X., Yang, J., Yan, Q., Zhang, Y.: Local feature fusion and SRC-based decision fusion for ear recognition. Multimedia Systems **28**(3), 1117–1134 (2022)
5. Alkababji, A.M., Mohammed, O.H.: Real Time Ear Recognition using Deep Learning. Telkomnika Telecommunication, Computing, Electronics and Control **19**(2), (2021)
6. Priyadharshini, R.A., Arviazhagan, S., Arun, M.: A deep Learning Approach for Person Identification using Ear Biometrics. Springer Nature **28** (2020)

7. Zhang, J., Yu, W., Yang, X., Deng, F.: Few-shot learning for ear recognition. In: International Conference on Image, Video and Signal Processing (IVSP 2019), pp. 25–28. Shanghai, China (2019)
8. Moniruzzaman, M.D., Islam, S.M.S.: Automatic Ear Detection using Deep Learning. In: First International Conference on Machine Learning and Data Engineering, pp. 108–114 and 20–22 (2017)
9. Sanchez, M.Q., Alonzo, V.A., Paschetta, C.: Automatic ear detection and feature extraction using Geometric Morphometrics and Convolutional Neural Networks. IET Biomedics (2016)
10. Shaik, K.B., Ganesan, P., Kalist, V., Sathish, B.S., Jenitha, J.M.M.: Comparative study of skin color detection and segmentation in HSV and YCbCr color space. Procedia Computer Science **57**, 41–48 (2015)
11. Gonzalez, R.C., Woods, R.E.: Digital image processing (Book). Reading, Mass., Addison-Wesley Publishing Co., Inc.(Applied Mathematics and Computation, Third Edition (2008)
12. Palwinder, S., Amarbir S.: A study on image segmentation techniques. Int. J. Recent Trends in Eng. Res. (IJRTER) **2** (2016)
13. Gómez, O., González, J.A., Morales, E.F.: Image segmentation using automatic seeded region growing and instance-based learning" Iberoamerican Congress on Pattern Recognition, pp. 192–201. Springer, Berlin, Heidelberg (2007)
14. Gonzalez, R.C., Woods, R.E.: Digital Image Processing, 2nd edition. Prentice Hall (2002)
15. Yamashita, R., Nishio, M., Do, R.K.G., Togashi, K.: Convolutional neural networks: an overview and application in radiology. Insights Imaging **9**(4), 611–629 (2018)
16. Liu, Y., Pu, H., Sun, D.W.: Efficient extraction of deep image features using convolutional neural network (CNN) for applications in detecting and analyzing complex food matrices. Trends Food Sci. Technol. **113**, 193–204 (2021)
17. Dash, S., Jena, U.R.: Multi-resolution Laws' Masks Based Texture Classification. J. App. Res. Technol. **15**(6), 571–582 (2017)
18. Ghanbari, B., Atangana, A.: Some new edge detecting techniques based on fractional derivatives with non-local and non-singular kernels. Advances in Difference Equations 1–19 (2020)
19. Fan, C.: Survey of Convolutional Neural Network, pp. 1–22. Indiana University (2016)
20. Wu, H., Gu, X.: Max-pooling dropout for regularization of convolutional neural networks. In: International Conference on Neural Information Processing, pp. 46–54, Springer, Cham (2015)
21. Akhtar, N., Ragavendran, U.: Interpretation of intelligence in CNN-pooling processes: a methodological survey. Neural Comput. Appl. **32**(3), 879–898 (2020)
22. Ciresan, D.C., Meier, U., Masci, J., Gambardella, L.M., Schmidhuber, J.: Flexible, high performance convolutional neural networks for image classification. In: Twenty-second international joint conference on artificial intelligence, pp. 1237–1242 (2011)
23. Materka, A., Strzelecki, M.: Texture Analysis Methods –A Review. Technical University of Lodz, Institute of Electronics, COST B11 report, Brussels **10**(1), 1–33 (1998)
24. Rachidi, M., et al.: Application of Laws' Masks to Bone Texture Analysis: An Innovative Image Analysis Tool in Osteoporosis. In: 2008 5th IEEE International Symposium on Biomedical Imaging: From Nano to Macro. Pp. 1191–1194. IEEE (2008)
25. Han, J., Pei, J., Kamber, M.: Data mining: concepts and techniques, 3rd Edition. Elsevier (2011)
26. Sazli, M.H.: A brief review of feed-forward neural networks. Communications Faculty of Sciences University of Ankara Series A2-A3 Physical Sciences and Engineering **50**(1) (2006)
27. Baldi, P., Vershynin, R.: The capacity of feedforward neural networks. Neural Netw. **116**, 288–311 (2019)

28. Ketkar, N., Moolayil, J., Ketkar, N., Moolayil, J.: Feed-forward neural networks. Deep Learning with Python: Learn Best Practices of Deep Learning Models with PyTorch 93–131 (2021)
29. Alvarez, E.G.L.: And Mazorra. University of Las Palmas, L. CTIM. R & D Center for Image Technologies (2000)
30. Mala, S., Ambika, M.: Face Recognition: Demystification of Multifarious Aspect in Evaluation Metrics. Intechopen (2016). https://doi.org/10.5772/62825
31. Ali, M.M., Mariam, M.S.: Hybrid recognition system under feature selection and fusion. Int. J. Comp. Sci. Trends and Technol. (IJCST) **5**(4) (2017)
32. Mohammed, A.A.: Improved HMM by Deep Learning for Ear Classification. Int. J. Inno. Res. Comp. Sci. Technol. (IJIRCST) **6**(3) (2018)
33. Thivakaran, T.K., Padira, S.V., Kumar, A.S., Reddy, S.: Fusion Based Multimodel Biometric Authentication System using Ear and Fingerprint. Int. J. Intell. Eng. Sys. **12**(1), 62–73 (2019)
34. Booysens, A., Viriri, S.: Ear biometrics using deep learning: a survey. Applied Computational Intelligence and Soft Computing (2022)
35. Mehta, R., Kumar, K.R.: An Efficient Ear Recognition Technique Based on Deep Ensemble Learning Approach, pp. 1–17. Springer (2023)

Enhancing Data Security with a New Color Image Encryption Algorithm Based on 5D Chaotic System and Delta Feature for Dynamic Initialization

Hadeel Jabbar Shnaen$^{(\boxtimes)}$ and Sadiq A. Mehdi

Department of Computer Science, College of Education, Mustansiriyah University, Baghdad, Iraq
hadeel.albahadili@gmail.com

Abstract. This research introduces an innovative color image encryption algorithm that uses a 5D chaotic system and delta feature to modify the initial conditions during key generation. It employs diffusion and confusion processes across multiple stages. Furthermore, a single pixel value in the image is altered, and the delta features are recalculated to generate new encryption keys, Subsequently, the image is encrypted once again. The number of revised keys and the similarity between the encrypted images can be measured using the Hamming distance metric. The algorithm's effectiveness in encrypting color images is validated through simulation, testing, and comparisons with existing methods, demonstrating its effectiveness in terms of encryption quality, computational efficiency, and resistance against various attacks. The presented approach offers a promising solution for secure color image encryption, suitable for applications requiring reliable protection of sensitive visual information. The evaluation results of the algorithm, A high NPCR value (99.6099), suggest that a sizable part of the pixels has been altered during encryption, which is desirable for security. The PSNR value (7.738) shows minimal information loss during encryption, crucial to maintaining image quality. An ideal UACI value (31.131%–35.225%) suggests minimal visual artifacts and distortion in the encrypted image.

Keywords: Encryption Chaotic System · Sensitivity · Decryption · Initial Conditions · Correlation

1 Introduction

In today's technology era, in many applications such as digital medicine, government data exchange, electronic document interchange, and social media applications, protecting and preserving data, including color images, has become a greater challenge due to the increasing threats of unauthorized access and malicious activities. Hacking and cyber-attack techniques have evolved significantly, making data more vulnerable and targeted [1, 2]. Various encryption standards, such as the Advanced Encryption Standard

A. M. Al-Bakry et al. (Eds.): NTICT 2023, CCIS 2096, pp. 90–105, 2024.
https://doi.org/10.1007/978-3-031-62814-6_7

(AES) and Data Encryption Standard (DES), have been proposed. However, these standard methods are primarily designed for encrypting textual data and do not adequately meet the requirements for image encryption because of the redundancy, substantial data capacity, and robust correlation between pixels in images [3]. Then, researchers discovered an intriguing connection between encryption and chaos. Accordingly, the distinctive characteristics of chaotic systems offer an approach to enhance the security of image encryption techniques [4]. The research has two goals; firstly, to explain the framework and mathematical foundations of the NCS, highlighting its properties and potential for cryptographic purposes. Secondly, introduce a novel color image encryption algorithm that utilizes the behavior of the "New Chaotic System" NCS.

2 Related Works

Reviewing previous studies is an essential part of the research process as it contributes to identifying knowledge gaps and paves the way for deep understanding and comprehensive analysis. It helps in building a strong theoretical base for current research, identifying strengths and weaknesses, and directing focus toward areas that needs additional exploration.

Sadiq A. Mehdi (2018) used a sequence derived from the hyperchaotic system as an encryption key to enhance the Advanced Encryption Standard (AES) algorithm. Although the research addresses an important issue, it does not completely modify the cryptographic procedure in the original AES algorithm. In this study, deleting the Mix Columns operation was based on its time-consuming nature compared to other processes. This led to a notable decrease in the time required for encryption and decryption [5].

Dejian Fang and Shuliang Sun (2020) used a revolutionary image encryption technique based on a 5D hyperchaotic system to generate new chaotic sequences using initial secret keys. They are altered with the purpose of confusing and diffusing the image. A cycle shift is implemented to enhance the security of the cryptosystem; thus, it is very secure and may be used for secure communication [6].

In (2021), Zhongyue Liang, Qiuxia Qin, and Changjun Zhou proposed a novel approach that adheres to the diffusion-scrambling paradigm. Specifically, it incorporates the bit-level DNA mutation operation within the diffusion process to increase scrambling and diffusion effects. There have been enhancements made to the security and randomization of algorithms. This study evaluates the efficacy of a particular encryption technology for securing medical photographs and investigates its safety performance over time [7].

In 2022, Sarah, S. Ahmed, and Sadiq, A. Mehdi proposed a two-step method. During the initial stages, the pixel coordinates of an image undergo a chaotic process caused by a chaotic sequence formed by the permutation step. Next, the substitution phase involves applying the XOR operation to the pixel value obtained from the preceding step's image and a pseudo-random number sequence (referred to as key2) created by the proposed system [8].

Compared to the previous studies mentioned, the proposed system contributes to presenting a dynamic chaotic encryption algorithm for color images. Its dynamism is derived from the value of the delta feature that is calculated from the original image and is used to recreate the initial conditions of the chaotic system. Thus, each input image has its encryption keys, and with the presence of This feature, even if the attacker obtains the chaotic system and the initial conditions, will not be able to decode the image without obtaining the image itself and calculating the value of the delta feature to regenerate its keys.

3 Structure of New 5D Chaotic System

A unique chaotic system based on a 5D framework was used to raise the chaos inside a system with fourteen positive parameters and complicated, chaotic dynamics features. This system's fundamental attributes and dynamic properties are examined in equilibrium points, chaotic attractors, Lyapunov exponents, dissipative characteristics, symmetry, Kaplan-Yorke dimensions, waveform analysis, and sensitivity to initial conditions. According to the conclusions of the research, the new system has five Lyapunov exponents and two unstable equilibrium points. Estimated the values for Kaplan Yorke and maximum positive Lyapunov exponent (MLE) are 3.12204 and 4.45994, respectively. The novel system demonstrates unpredictably unstable, highly complicated, and inconsistent features.

The novel five-dimensional autonomous system is acquired by the following:

$$
\begin{aligned}
\frac{dx}{dt} &= ayu - byw + cy - \lambda x \\
\frac{dy}{dt} &= -bxz - exu + fx - gy \\
\frac{dz}{dt} &= h_1 xy + kuw - l(z - x) \\
\frac{du}{dt} &= -pw(y + z) - qu - bw \\
\frac{dw}{dt} &= bxy - rzu + t_1 u - lw
\end{aligned}
\tag{1}
$$

Where $x, y, z, u, w,$ and t  + referred to the states of system $a, b, c, \lambda, e, f, g, h_1, k, l, p, q, r$ and t_1 are positive system parameters.

The chaotic attractor of the 5D system "(1)" is observed when specific values are assigned to the system parameter.

$$a = 10, b = 2, c = 25, \lambda = 40, e = 0.5, f = 30, g = 0.1, h1 = 9, k = 3, l = 4$$

$p = 15, q = 19, r = 2.3$ and $t_1 = 34$. And we assume that $X(0) = 1, Y(0) = 0.5, Z(0) = 5, U(0) = 0.6,$ and $W(0) = 0.4$ are the initial circumstances.

3.1 Equilibrium Point

We can obtain that the system (1) has an equilibrium point:

$$
\begin{aligned}
0 &= ayu - byw + cy - \lambda x \\
0 &= -bxz - exu + fx - gy \\
0 &= h_1 xy + kuw - lz - x \\
0 &= -pw(y + z) - qu - bw \\
0 &= bxy - rzu + t_1 u - lw
\end{aligned}
\tag{2}
$$

When $a = 10, b = 2, c = 25, \lambda = 40, e = 0.5, f = 30, g = 0.1 h_1 = 9, k = 3, l = 4,$
$p = 15, q = 19, r = 2.3$ and $t_1 = 34$

The equilibrium point becomes: E0 {x = 0, y = 0, z = 0, u = 0, w = 0}, Then, the relative eigenvalues corresponding to equilibrium E0(0,0,0,0,0) are derived as follows: $\lambda_1 = -53.9322, \lambda_2 = 13.8322, \lambda_3 = -4, \lambda_4 = -11.5 + 3.42783$, and $\lambda_5 = -11.5-$ 3.42783i. The equilibrium E0 (0,0,0,0,0) is hence a saddle-focus point; at point E0. The hyperchaotic system is therefore unstable.

3.2 Lyapunov Exponents and Lyapunov Dimensions.

The calculation of the Lyapunov exponent is a quantitative measure approach within the framework of nonlinear dynamical theory [9], which aims to assess the sensitive dependency on beginning conditions. The quantity pertains to the mean rate at which two adjacent trajectories diverge or converge [10].

We take the initial conditions as $x(0) = 1, y(0) = 0.5, z(0) = 5, u(0) = 0.6, w(0) = 0.4$. The results are obtained as follows: L1 = 4.45994, L2 = 0.4620, L3 = -2.75930, L4 = -17.72055, L5 = -48.55465

The most significant Lyapunov exponent is positive, which indicates the system exhibits chaotic properties. THE EXPRESSION IS TRUE since L1 and L2 are positive Lyapunov exponents while the other three are negative. Thus, the system is very chaotic. The fractal dimension is also a property of chaos as computed by Kaplan-Yorke dimension using Lyapunov exponents, where j is the maximum value of Lyapunov exponent, and Li is in descending order of the sequence according to the sequence of Lyapunov exponents, and DKY may be stated as follows:

$$
D_{KY} = j + \frac{1}{|L_{j+1}|} \sum_{i=1}^{j} L_i
$$

$$
D_{KY} = 3 + \frac{1}{|L_{j+1}|} \sum_{i=1}^{3} L_i = 3 + \frac{L_1 + L_2 + L_3}{|L_4|}
$$

$$
= 3 + \frac{4.45994 + 0.46200 \pm 2.7593}{17.7206}
$$

$$
= 3.12204
$$

This indicates that the Lyapunov dimension of system (1) is fractional. In addition to having non-periodic orbits, the new system's neighboring paths diverge due to its fractal character. Consequently, this nonlinear system is chaotic [11].

4 Proposed Encryption Algorithm

There are strict requirements in many areas about maintaining data security and privacy, especially if they contain sensitive data. Image encryption helps comply with these standards and legislation, and is used to secure data transfers, over the Internet or internal networks. This protects images from being tampered with or hijacked during the transition. Because the process of encrypting color images faces a set of challenges in terms of increasing the size of files and compatibility with various image applications and software, our proposed system was therefore trying to achieve a balance between providing strong security and good performance, especially in the case of high-resolution images as shown in Fig. 1

4.1 Image Encryption Stage

The encryption phase of the proposed system is a significant phase consisting of the following separate processes.

4.2 Image Loading

The to-be-encrypted images are submitted to the suggested approach, where various color image kinds and sizes are uploaded.

4.3 Compute Feature from Image

In this step, an essential image characteristic, delta E, is computed to identify the color difference between adjacent image pixels. This feature is used to change the initial values of the chaotic system, as shown in Eq. (3). The delta feature operates on data having label-type characteristics. Where RGB values are converted to the CIELAB color space. This transformation consists of two steps: first, the RGB values are converted to the XYZ color space, then the XYZ values are converted continuously in CIELAB using a specific equation of this equation with x, y, and z values met on standard values associated with white light. This procedure occurs after the conversion of RGB values into the XYZ format.

4.4 Divide the Image into Blocks and Create Vectors

At this step of the encoding process, the image is divided into many blocks, and then three vectors (R, G, and B) are created from each block to encode the 2D image. These vectors correspond to the image's red, green, and blue color channels. These vectors are made by extracting the color values from each pixel in the image and grouping them into independent vectors. This step is necessary for following color channel encoding activities.

4.5   Generating Random Keys from 5D Hyper-Chaotic System

This approach uses a hyper 5D chaotic system to generate the image flip-ping and encryption keys. Initial values for the keys are modified by the image feature extracted in the preceding step, resulting in unique keys for each image. As shown in the following equations:

$$
\begin{aligned}
x(1) &= x(0) + (Image\,feature * 0.002)\\
y(1) &= y(0) + (Image\,feature * 0.002)\\
z(1) &= z(0) + (Image\,feature * 0.002)\\
u(1) &= u(0) + (Image\,feature * 0.002)\\
w(1) &= w(0) + (Image\,feature * 0.002)
\end{aligned}
\tag{3}
$$

The value (0.0002) was taken by default, to keep the value of the delta feature within the range [0.1], and to prove the sensitivity of the keys to the initial conditions. The process of generating chaotic sequences of the size of the image to be encrypted is done by creating two structures (arrays), one to store the chaotic key sequences based on the value of the delta feature, which is explained in Eq. (3). The other is to save the basic values after using (the remainder function) so that the values obtained are within the range [1,0]. Then they are converted into integers to be used as encryption keys.

4.6   Scrambling the Image Blocks

This stage of image encryption includes scrambling the pixel values semi-randomly to provide a layer of randomness to the image to enhance its security against tampering and reverse engineering. To do this, a 5D Hyper-Chaotic system is utilized to produce chaotic keys, which are then used to rearrange the values of the image's red, green, and blue color channels [12]. There are three steps in the scrambling process:

The First Phase: A vector of the same size as the color channels is produced with chaotic keys organized in a specific fashion. The unique values of this vector are used to create a new vector that is used to rearrange the pixel values depending on the image block's columns.

The Second Phase: Chaotic keys combine the color vector with the image, bringing an additional random element into the encryption process and enhancing its security. Each pixel value in the color channels is subjected to a sequence of bitwise operations using the key, with the resultant value being stored in the pixel location corresponding to the pixel value.

The Third Phase: A vector of the same size as the color channels are formed with chaotic keys ordered in a certain way. The unique values of this vector are then utilized to generate a new vector that is used to reorder the pixel values depending on the image block's rows. These steps make the proposed system more resistant to manipulation or alteration.

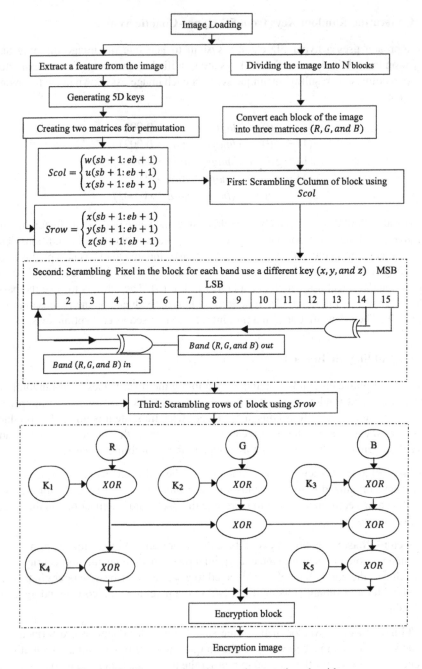

Fig. 1. The diagram of the Proposed encryption algorithm

4.7 Image Block Encryption

In the proposed image encryption system, Step 6 denotes the last stage of image encryption. An exclusive or (XOR) methodology is employed to modify the pixel values, preventing reverse engineering without the requisite cryptographic keys. The achievement of this task involves the submission of the red (R), green (G), and blue (B) vectors to a series of XOR operations. These operations utilize keys generated from chaotic values contingent upon the image's feature. These keys alter the pixel values, making determining their original values challenging.

5 Decryption Image Stage

A decryption algorithm extracts the original information from the encrypted data. The suggested image encryption technique showed how an input image may be encrypted using a hyperchaotic system and the XOR operation. In this case, the decryption method is the opposite of the encryption algorithm and entails reversing the processes made during encryption to recover the original image. Consequently, the suggested decryption technique is intended to recover the original image from the encrypted data using the same hyperchaotic system and XOR operation as the encryption algorithm.

6 Security Analysis

The security study of an image encryption technique entails evaluating its capacity to safeguard sensitive image data from unauthorized access or tampering. Evaluate the algorithm's encryption strength to guarantee that even if an attacker can access the encrypted image, they cannot decode its information without the correct decryption key. We must analyze the algorithm's computing efficiency. Although robust security is crucial, the method should be simple and easy since this might impede practical use cases [13] (Fig. 2).

6.1 Histogram Analysis

To mitigate the risk of extracting significant information from the plain image, avoiding any statistical correlation between the plain image and the encrypted image is imperative. Additionally, to enhance resistance against statistical attacks, the histogram of the encrypted image must exhibit a uniform distribution and be distinctly dissimilar from that of the plain image [4].

6.2 The Correlation Coefficient

It is a statistical measure employed to assess the association between two random variables or data sets and employed to measure the degree of resemblance between two images. In Table 1, a correlation coefficient close to 1 indicates a strong positive relationship. In the context of image encryption, this would mean that as one variable (e.g., encryption strength or security) increases, the other variable (e.g., the quality of the

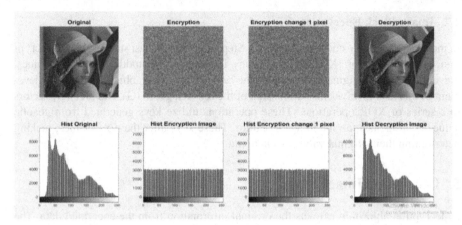

Fig. 2. Illustrates the histogram for each the Original/Encryption image and Decryption Image using the proposed algorithm

Table 1. Correlation values to the original images and the encrypted images using the proposed system

Image	Correlation coefficient for original image			Correlation coefficient for encryption image		
	H	V	D	H	V	D
Girl 256 × 256	0.9736	0.9654	0.9512	0.0024	0.001	0.00165
Lena 512 × 512	0.9669	0.975	0.9540	−0.000	−0.0007	0.0014
Peper512 × 512	0.9756	0.9778	0.9634	−0.0015	0.001	0.000
Bboon51 × 512	0.8626	0.75486	0.72434	−0.0017	0.0033	0.00005

encrypted image) also increases. In other words, a positive correlation suggests that more robust encryption results in better image security. When the calculation outcome is zero, there is no clear pattern or connection between the encryption strength and the quality of the encrypted image. A correlation coefficient close to −1 indicates a robust negative relationship [14].

6.3 Key Sensitivity Analysis

When a minute alteration is made to any individual parameter without any accompanying modifications, it will alter the encrypted image. The definition of this can be derived by analyzing the variation of pixel values between the encrypted image and its original counterpart on a pixel-by-pixel basis. The primary factor influencing the robustness of a cryptographic technique lies in its capacity to withstand an extensive range of potential key permutations, specifically on the order of 10^{15}, before the adversary's anticipated ability to ascertain the correct key with a high degree of certainty [15].

Table 2. Entropy values to the original/encrypted and decrypted images using the proposed system

Image	Entropy_Original image	Entropy_Encryption Image	Entropy_Decryption image
Girl 256 × 256	7.1835	7.9991	7.1835
Lena 512 × 512	7.4881	7.9998	7.4881
Peppers 512 × 512	7.7253	7.9998	7.7253
Baboon 512 × 512	7.7659	7.9997	7.7659

6.4   Information Entropy Analysis

Entropy measures the level of randomness or information content in an image [16]. Table 2 displays the entropy of the original image falls within (7.4881 to 7.7659). In this case, the original image appears to have moderate to high information content, which may contain diverse pixel values or patterns. In comparison, the entropy of the encrypted image (7.9881 to 9.9998) is significantly higher than that of the original image. This suggests that the encryption process has added a considerable amount of randomness or complexity to the image. A high entropy value in the encrypted image is often desirable as it indicates increased security and makes it more challenging for unauthorized parties to analyze or decrypt the image.

Table 3. Elapsed time to generate keys/Encryption and Decryption images using the proposed system

Image	Elapsed time to create keys	Elapsed time to Encryption	Elapsed time to Decryption
Girl 256 × 256	0.379064	0.468204	0.469224
Lena 512 × 512	1.642148	2.081405	1.946576
Peppers 512 × 512	0.965617	1.951043	1.954831
Baboon 512 × 512	1.012622	1.969275	1.982590

6.5   Time Analysis

Time analysis in the environment of encryption algorithms denotes measuring the computational time required to achieve encryption and decryption actions. Table 3 presents the encryption time for the image, which falls within (0.46820–1.96827 s). The times in this range appear to be reasonable and relatively efficient for encrypting images, and the decryption time for the image also falls within the same range as the encryption time. This suggests that the decryption process is symmetrical to the encryption process in terms of computational complexity and time. Robust encryption algorithms should require an impractical amount of time to decrypt without the correct key, making it computationally infeasible for attackers to break the encryption [17, 18].

Table 4. PSNR and MSE Using Proposed System

Image	Encryption image		Decryption image	
	MSE	PSNR	MSE	PSNR
Girl 256 × 256	9466.4318	7.3129	0	Inf
Lena 512 × 512	6341.1527	8.3964	0	Inf
Peppers 512 × 512	7365.1714	8.0995	0	Inf
Baboon 512 × 512	5536.7541	8.8116	0	Inf

6.6 Peak Signal-To-Noise Ratio (PSNR)

PSNR quantifies the quality of an image by comparing it to a perfect reference image. When PSNR is infinite, the decrypted image has no noise or distortion compared to the original. MSE (Mean Square Error): measures the average squared difference between the original image and the decrypted image on a pixel-by-pixel basis. When MSE is zero, there is no difference between the original and decrypted images, implying a perfect reconstruction, as shown in Table 4 [17].

Table 5. NPCR and UACI using the proposed system

Image	NPCR	UACI
Girl 256 × 256	99.6099	35.2259
Lena 512 × 512	99.5998	31.1313
Peppers 512 × 512	99.6045	33.1672
Baboon 512 × 512	99.6071	32.8193

6.7 Unified Average Changing Intensity (UACI)

Measures the average change in intensity between the original and ciphered images. It is used to assess the strength of the encryption technique. Table 5 includes the UACI values falling between 31 and 35% for encrypting images, suggesting varying degrees of quality reduction compared to the original images. The exact impact on image quality depends on factors such as the encryption algorithm, compression settings, and specific image content. Lower UACI values represent higher quality, while higher values indicate more significant quality loss, but it is resistant to differential attacks. NPCR measures the percentage of pixels that change when a single-bit change is made to the plaintext (original image) and then encrypted. A value of "99.6099" for NPCR suggests that when you change a single bit in the original image and encrypt it using the encryption system being evaluated, approximately 99.6099% of the pixels in the encrypted image are different from the original encrypted image [8].

6.8 Keyspace Analysis

The key space in cryptography refers to the number of possible keys used in a crypto-graphic algorithm. It plays a crucial role in ensuring the security of encrypted data. A larger key space makes it significantly more difficult for attackers to perform exhaustive search attacks (brute-force attacks) to guess the correct key [10].

$$\text{Key Space} = (10^{15})^{19} = 10^{285}$$

So, the key space for this cryptographic system is $10^{285} \sim 2^{947,}$ which is a considerable number, making it highly secure against brute-force attacks.

Table 6. Hamming distance, Similarity, and PSNR Between two color-coded images

Image	Hamming distance between Key and f Key	Hamming Weight	Similarity	PSNR
Girl 256 × 256	65,536	98,229	0.002675	7.738
Lena 512 × 512	262,144	392,954	0.005702	7.7514
Peppers 512 × 512	262,144	393,090	0.006574	7.7486
Baboon 512 × 512	262,144	392,529	0.007571	7.7567

7 Hamming Distance

The Hamming distance measures the difference between two strings of equal length by counting the positions at which the corresponding elements (usually bits) differ. It is used to evaluate the difference between two keys. Encryption's primary role is not in the encryption process but in assessing the integrity of transmitted or stored data [19]. Since the similarity function showed that there was no similarity between the two encrypted images and the keys, and the Hamming distance showed that all the keys were completely changed, this indicates that even a small change in the original image can lead to significant changes in the results due to the nature of the encryption and protection process.

These conclusions show the importance of key protection in cryptographic operations and the impact of small changes on security.

8   Hamming Weight

It is often not a direct component of encryption techniques such as symmetric or asymmetric approaches. Instead, it is more applicable to other cryptography and security-related elements [20].

Table 6 demonstrates that a hamming distance of either 262,144 or 65,536 indicates a notable level of dissimilarity or divergence between the two keys. Specifically, when comparing the first key to the fkey after altering a single pixel in a picture, this hamming distance reflects a major modification in the data these keys represent. The Hamming weight value of 392,954 indicates that among the 262,144 pixels in both images (512 × 512), a total of 392,954 pixels possesses color information or are non-zero. A similarity rating of approximately 0.0057024 between two encrypted photos suggests a relatively minimal level of similarity despite the alteration of only a single pixel. A PSNR value of 7.7514 obtained from the comparison of two encrypted photos indicates a considerably diminished image quality. This shows the presence of discernible dissimilarities between the two encrypted images, maybe accompanied by visual artifacts or distortions.

Table 7. Comparison between the proposed system and the others in the literature

Measurements	Proposed	Ref. [5]	Ref. [7]	Ref. [16]
Entropy	7.9998	7.9974	7.999	7.9972
Horizontal correlation	−0.00096	0.0003	−0.0023	0.0255
Vertical correlation	−0.00078	0.0075	0.0014	0.0022
Diagonal correlation	0.00141	0.0018	0.0025	0.0205
NPCR	99.6099	99.706	99.608	99.605
UACI	33.1672	33.461	33.482	33.455

9   The Proposed System and Another System Comparison

Testing and comparison with earlier studies have demonstrated that our suggested solution has a high level of security and the capacity to fend off attacks successfully. The system has a strong design and cutting-edge security mechanisms, which help to improve the protection of data and information. Because of this, it can be used in situations that call for high security and efficient defense against online threats as shown in Table 7.

Comparing a proposed encryption algorithm with well-established algorithms like AES (Advanced Encryption Standard) and another algorithm allows for a comprehensive security evaluation. AES has undergone extensive cryptanalysis and is considered highly secure. Comparing the proposed algorithm against AES helps identify any potential vulnerabilities or weaknesses. As shown in Table 8 the analytical and statistical results of our proposed system compared to the AES algorithm show that there is a convergence in the results and that this indicates the strength of the proposed algorithm.

Table 8. Comparison between the Proposed algorithm, AES algorithm, and Ref. [5] from (Baboon Image)

Measurements of Baboon Image (512 × 512)	Proposed	AES	Ref. [5]
Entropy	7.9997	7.9998	7.9998
Horizontal correlation	–0.00172	0.0012	0.0036
Vertical correlation	0.0033	0.0023	–0.0047
Diagonal correlation	0.00005	–0.0021	–0.0041
NPCR	99.6071	99.6075	99.7208
UACI	32.8193	32.847	33.020

The Conclusions

The efficacy and security of the method utilized for encrypting a color image based on a 5D chaotic system and incorporating the delta feature to modify the initial condition are substantiated by the subsequent outcomes. The entropy value of 7.9998 suggests a significant degree of randomness and complexity within the encrypted image, which is a favorable characteristic for encryption techniques. The correlation coefficient of 0.00141 indicates a significantly low correlation, implying that the encrypted image exhibits minimal redundancy. Consequently, this characteristic poses a considerable challenge for potential attackers as it hinders their ability to exploit patterns or establish linkages within the encrypted data. The Normalized Pixel Change Rate (NPCR) is calculated to be 99.6099. A higher NPCR number indicates that even slight modifications in the original image lead to substantial alterations in the encrypted image, hence improving the level of security. The value of the Unified Average Changing Intensity (UACI) is calculated to be 33.1672. A high UACI value signifies the algorithm's ability to effectively disperse the encryption key's impact across the image, hence ensuring robust diffusion. The key space, which has a size of 2^{947}, is characterized by its immense magnitude. This vast key space ensures an exceedingly high number of potential encryption keys, rendering brute-force attacks computationally unfeasible for potential attackers. The time rate of encrypting an image of size 256*256 to 1.951043, as compared to encrypting an image of size 512*512, indicates that the encryption process exhibits efficiency, as it maintains a respectable time rate even when dealing with bigger image dimensions. In summary, the algorithm exhibits resilience and appropriateness for image encryption, demonstrating a commendable degree of security, minimal correlation, and substantial diffusion. The practicality and resistance against attacks of the system are further enhanced by its huge key space and suitable time rate.

Recommendations

1. Evaluate the ability of the proposed image encryption algorithm to adapt to different types of images in terms of size and handle the case of different dimensions, as this procedure makes the algorithm applicable in various scenarios.

2. Focus optimization efforts on reducing memory requirements and computational load, making the encryption algorithm practical for resource-limited environments

References

1. Aljazaery, I.A., Alrikabi, H.S., Rabea, A.M.: Combination of hiding and encryption for data security. Int. J. Interact. Mobile Technol. **14**(9), 34–47 (2020). https://doi.org/10.3991/ijim. v14i09.14173
2. Amigó, J.M., Kocarev, L., Szczepanski, J.: Theory and practice of chaotic cryptography. Phys. Lett. Sect. A: General, Atomic and Solid-State Phys. **366**(3), 211–216 (2007). https://doi.org/ 10.1016/j.physleta.2007.02.021
3. Gatta, M.T., Al-Latief, S.T.A.: Medical image security using a modified chaos-based cryptography approach. J. Phys. Conf. Ser. **1003**(1) (2018). https://doi.org/10.1088/1742-6596/ 1003/1/012036
4. Shakir, H.R., Mehdi, S.A., Hattab, A.A.: A dynamic S-box generation based on a hybrid method of new chaotic system and DNA computing. TELKOMNIKA **20**(6), 1230 (2022). https://doi.org/10.12928/telkomnika. v20i6.23449
5. Mehdi, S.A., Jabbar, K.K., Abbood, F.H.: Image Encryption Based on the Novel 5d Hyper-Chaotic System Via Improved Aes Algorithm. Int. J. Civil Eng. Technol. **9**(10), 1841–1855 (2018). Article ID: IJCIET_09_10_183
6. Fang, D., Sun, S.: A new secure image encryption algorithm based on a 5D hyperchaotic map. PLoS ONE **15**(11), e0242110. https://doi.org/10.1371/journal.pone.0242110
7. Liang, Z., Qin, Q., Zhou, C.: Medical image encryption algorithm based on a new five-dimensional three-leaf chaotic system and genetic operation. PLoS ONE **16**(11), e0260014. https://doi.org/10.1371/journal.pone.0260014. eCollection 2021
8. Ahmed, S.S., Mehdi, S.A.: Image encryption algorithm based on a novel 5D chaotic system. In: 8th International Conference on Contemporary Information Technology and Mathematics, ICCITM (2022), pp. 249–255. https://doi.org/10.1109/ICCITM56309.2022.10031883
9. Bazzi, S., Ebert, J., Hogan, N., Sternad, D.: Stability and predictability in human control of complex objects. Chaos **28**(10) (2018). https://doi.org/10.1063/1.5042090
10. Jasim, O.A., Hussein, K.A.: A hyper-chaotic system and adaptive substitution box (S-Box) for image encryption. Corpus ID: 244841618 https://doi.org/10.1109/ACA52198.2021.962 6793
11. Wang, Z., Ma, J., Chen, Z., Zhang, Q.: A new chaotic system with positive topological entropy. Entropy **17**(8), 5561–5579 (2015). https://doi.org/10.3390/e17085561
12. Mehdi, S.A.: Image encryption algorithm based on a novel 4D chaotic system. Int. J. Inf. Secur. Priv. **15**(4), 118–131 (2021)
13. Lian, S.H., Sun, J., Wang, Z.: Security analysis of a chaos-based image encryption algorithm. Phys. A: Stat. Mech. Appl. **351**(2), 645–661 (2005). https://doi.org/10.1016/j.physa.2005. 01.001
14. Zeboon, H.T., Abdullah, H.N., Mansor, A.J.: Robust encryption system based on novel chaotic sequence. Res. J. Appl. Sci. Eng. Technol. **14**(1), 48–55 (2017). https://doi.org/10.19026/rja set.14.3988
15. Wu, X., Li, Y., Kurths, J.: A new color image encryption scheme using CML and a fractional-order chaotic system. PLoS One (2015). https://doi.org/10.1371/journal.pone.0119660
16. Munir, R.: Security analysis of selective image encryption algorithm based on chaos and CBC-like mode. In: 7th International Conference on Telecommunication Systems, Services, and Applications (TSSA) (2012). https://doi.org/10.1109/TSSA.2012.6366039

17. Qayyum, J.A.: Chaos-based confusion and diffusion of image pixels using dynamic substitution. https://doi.org/10.1109/ACCESS.2020.3012912
18. Petrenko, K., Mashatan, A., Shirazi, F.: Assessing the quantum-resistant cryptographic agility of routing and switching IT network infrastructure in a large-size financial organization. J. Inform. Secur. Appl. **46,** 151–163 (2019). https://doi.org/10.1016/j.jisa.2019.03.007
19. Ali, S.I., Kausar, F., Khan, F.A.: A cluster-based key agreement scheme using keyed hashing for Body Area Networks. Multimed. Tools Appl. **66**(2), 201–214 (2013). https://doi.org/10.1007/s11042-011-0791-4
20. Fan, X., Tong, J., Li, Y., Duan, X., Ren, Y.: Power analysis attack based on hamming weight model without brute force cracking. Secur. Commun. Netw. (2022). https://doi.org/10.1155/2022/7375097

Deepfake Detection Model Based on VGGFace with Head Pose Estimation Technique

Duha A. Sultan[1]([✉]) [iD] and Laheeb M. Ibrahim[2] [iD]

[1] Department of Biology, Education College for Girls, University of Mosul, Mosul, Iraq
`duhaasultan@uomosul.edu.iq`
[2] Department of Software, College of Computer Sciences and Mathematics,
University of Mosul, Mosul, Iraq

Abstract. Rapid developments in deepfake technology have produced hyper-realistic fake media such as images, audio, and video. Especially, the fabricated videos, which have gained a large widespread in social media sites. Because of the great harm inflicted by these videos, many researchers have increased their efforts to find a reliable method to identify and distinguish these fake videos from the actual ones. In this work, we suggested a new model to detect fake videos, based on two major techniques. First, the VGGFace model was used to extract the most important facial features combined with, second: the estimation of the head pose angle that represents the relative orientation of the human face in video frames. All these calculations are done on human faces detected and cropped from video frames, where 10, 20, and 30 frames were extracted from each video. FF++ dataset was used to train and test the model, which produced a max test accuracy of 0.885. The code was written using Python version 3.9.

Keywords: CNN · VGGFace · Deeplearning · Deepfake Detection · Head Pose

1 Introduction

Deep learning-generated forgery images or videos, known as deepfake, have become widespread with the coming of the precipitately improved deep generative models, such as autoencoders [1] that employ a face-swapping technique, generative adversarial nets (GANs) [2], which detects any flaws in the created deepfake and treat them through multiple rounds, making it very difficult for the detectors to decode them. GANs are also used as a popular method to create deepfakes, by training large amounts of data to "learn" how to develop new examples that mimic reality and provide highly accurate results. Over time, many types of GANs appeared, like FCGAN, DCGAN, StyleGAN, and so on, see [3] for more details about GAN variations. As a result, several applications have been built, such as DeepFaceLab, FaceApp, and FaceSwap making deepfake generation easier even for beginners. Depending on the user intent, some of these applications are just used for pure amusement purposes, but others may be used in a malicious way. That is why there is an urgent need for methods or techniques for detecting this fake content. Many experts believe that as technology advances, deepfakes could become

A. M. Al-Bakry et al. (Eds.): NTICT 2023, CCIS 2096, pp. 106–117, 2024.
https://doi.org/10.1007/978-3-031-62814-6_8

more sophisticated in the future, leading to more serious threats to the public relating to political tension, election interference, and additional criminal activity. For this, many researchers have attached great importance to this topic and many studies published dealing with the methods of detecting such fabricated videos.

The major contribution to this work:

1) Designing a new hybrid model to detect fake videos, based on two major techniques. First, the VGGFace model is used to extract facial features from video frames, along with determining head pose angles to enhance detection accuracy.
2) Assessing the efficacy of several machine learning classifiers by employing a diverse set of performance metrics, and choosing the one with the highest accuracy to improve model performance.

The research is divided into six parts. Section 1 is the introduction. Section 2 lists most of the previous works that are related to the topic. The Proposed model is described in Sect. 3, while the practical works and model results are explained in detail in Sects. 4 and 5 respectively. Finally, in Sect. 6, a brief paragraph summarizing the conclusion deduced after the completion of the work.

2   Related Works

In deepfake videos there is always something off, that is inconsistence with the complete video. The objective is to find these inconsistencies within video frames. Depending on the type of the extracted features, video detection methods can be classified into two varieties:

– Methods that benefit visual features inside frames.
– Methods that extract temporal features to benefit from the relationship among successive video frames.

2.1   Visual Features Within a Frame

After doing the required preprocessing, such as Viola_Jones to extract the face region within each frame, Afchar D. et.al.[4] submitted a model based on two CNNs to extract features, called Meso-4 and MesoInception-4. These networks were trained on two datasets: FaceForensic, with another dataset, containing real and fake videos with the same resolution collected from the internet. The model achieved a high detection rate for deepfake and Face2Face datasets. Another architecture introduced by Aya I. et al. [5] named YOLO-CNN-XGBoost which integrated the benefits of both XGBoost and CNN. YOLO face detector was used to determine faces in video frames, while the InceptionResNetV2 model was used to extract the discriminant spatial-visual features. These features are then fed to the XGBoost classifier to decide if the video is fake or real. Amerini et al. [6] presented a method using optical flow vectors obtained from two subsequent frames. Two CNNs were used to extract features, which are ResNet and VGG16. The model was trained on the FF++ dataset, with the resulting accuracy of 75.46% for VGG16 and 81.61% for ResNet50. To increase classification accuracy and enable the use of a light model, Tran V.N. et al. [7] suggested an architecture based on classifier networks with manual attention to target-specific areas to form distillation. Faces are detected and extracted from each frame utilizing facial landmarks and Multitask cascaded convolutional neural networks (MTCNN). A distillation set is a collection of patches. The classifier uses the distillation sets as input to decide which parts of the face will be trained. The whole face was trained using Inception_v3, whereas face patches were trained using MobileNet. Celeb_DFv2 and DFDC datasets were used to evaluate the model's performance, yielding high evaluation results. Matern F. et al.'s [8] research focused on the area around the eyes. The traits that the researchers used are: 1- The distance from the iris and the eye centers for the right and left eyes should be the same. 2- Both eyes should have equal radii and colors. 3- The eyes and teeth parts lack reflection and fine details. Therefore, before feature extraction, the Hough circle transform and Canny edge detection were employed to identify the eye region. For categorization, logistic regression and neural networks were utilized. The authors demonstrated that classifying using the features of the eyes and teeth together produced better results than using the features independently. 9.Wodajo D. et al. [9], submitted a model with a CNN_Vision Transformer (CViT). Where VGG_like CNN was used to extract features, while the ViT was used for classification. The model has achieved 91.5% accuracy, and a loss value of 0.32, after its training on a DFDC dataset.

2.2 Temporal Features Across Multiple Frames

To benefit the temporal relationships between successive frames and to increase accuracy, Guera D. et al. [10], proposed a deeplearnig model with two-stage of training. The model combines between CNN and LSTM to utilize the capability of CNN for features extraction, and the classification and memorization ability of LSTM. The dataset consists of 300 real videos from the HOHA dataset and 300 fake videos gathered from different websites.

Similar work was done by [11, 12]. Abdul Jamsheed V. et. al. [11] proposed a model that fuses between ResNeXt and LSTM neural networks. The model was trained on a dataset collected from FF++, DFDC, and Celeb_DF. While Yadav P. et al. [12] submitted a model that merges between InceptionResNet v2 and 2048-layer LSTM. DFDC dataset was used as the training set for the model for 20 and 40 epochs respectively. Li Y. et al. [13] were concerned in their research on the eye region, which transformed into discriminative features using a VGG16_CNN framework. The output is then fed to LSTM. To train and evaluate the model, the CEW dataset, which contains 1232 images of open eyes and 1193 images of closed eyes, was used. To fully utilize spatiotemporal information, Zhang D. et.al.[14] provide Temporal Dropout 3-Dimensional Convolution Neural Network (TD-3DCNN). Here, the video frame volumes are sampled by temporal dropout operation and fed into a 3DCNN. A collection of DFDC, FF++, and Celeb-DFv2 datasets were used. The model completed six detectors used in the research for comparison and outperformed them. A more generalized approach was achieved by Zhou and Lim [15], who submitted a combined visual/auditory deepfake detection model. Through the research, they observed that, when humans speak, there is a significant correlation between lip motion and pronounced syllables. The model can be implemented on various language videos, as it is language-agnostic, and achieved an accuracy of 81.96%.

In Agarwal S. et al. [16], concentrated on the shape of the mouth when pronouncing words having the sounds of M, B, or P where the lips should be totally close. Analysis was done using three approaches: 1-CNN: where classification is done by Xception CNN, 2-profile, and 3- manually by the analyst. Profile and CNN techniques achieved high performance.

3 Proposed Model

In this work, we suggested a new method for detecting fake videos, based on two major techniques. VGGFace model [17] was used to extract facial features along with the values of x_,y_, and z_ coordinates representing the head pose in each video frame. The general pipeline for manipulation of deepfake video detection can be divided into three major phases: 1-Preprocessing, 2-Feature extraction, and 3-Classification. Below (Fig. 1) expresses the steps of the overall process, (note that HP stands for Head Pose).

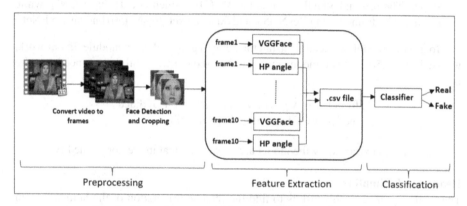

Fig. 1. General structure of deepfake video detection model

3.1 Preprocessing

From Fig. 1, it is clear that, before extracting features, all videos need to be manipulated first. The manipulation process includes: 1- converting each video into a certain number of frames, 2- detecting then cropping the face region within each frame. Although there are many methods for detecting faces, in this work we chose the MTCNN method because it is considered the most effective and accurate algorithm to detect the face region.

But MTCNN detects all faces and semi-faces found in the frame background. To preclude this, all detected faces are discarded except the face region with the largest area. That is, for both real and fake videos, only one face was taken from each frame and then saved in a test or train folder for later usage.

3.2 Feature Extraction

The second step in our method is to extract features from the cropped faces. Two techniques were combined and used:

1) The VGGFace model was used to extract 1024 features, which represent the most important features of a face.
2) estimating the pose of a head in the frames of all videos.

VGGFace Model
VGGFace refers to a series of models developed for face recognition. There are two versions of VGGFace:

1) VGGFace developed in 2015, trained on 2.6 million face images, a total of 2622 people.
2) VGGFace2 developed in 2017, trained on 3.31 million face images, a total of 9131 people. The original VGGFace uses the VGG16, which is a 16-layer CNN, while VGGFace2 is trained using ResNet50, and SqueezeNet_ResNet50 (known as SENet).

To use this model, we need to install the keras_vggface Python module. In our work, we used Resnet50, so to define the model to be used correctly, all that we need to write is:

```
from keras_vggface.vggface import VGGFace
model=VGGFace(model='resnet50')
```

The features extracted by the model are stored in a feature vector, named fv.

Head Pose Estimation
The main objective of this task is to find the relative orientation of the human face in video frames. So, the head pose estimate can provide information on which direction the human face is in the frame. To do this, two major steps are used: 1-the position of a set of facial landmarks is located in the cropped face region. Mediapipe python library was used for this purpose. Mediapipe's face landmarks detection algorithm catches around 468 key points from a face. FaceMesh python function from the mediapipe library was used to extract these key points, but we don't need to use all of them. We chose only 6 points that can represent a face. Those points are on the edge of the eyes, the nose, the chin, and the edges of the mouth, (Fig. 2) represent the key points landmarks indices are: 1 for the nose, 33 and 263 for eye edges, 61 and 291 for mouth edges, and 199 for the chin.

Using these key points, the face orientation angle, including $x\_$, $y\_$, and $z\_$ coordinates, was calculated using the Python function RQDecomp3 $\times$ 3 from the OpenCV library and stored in the same feature vector (fv). So, the values stored in this vector represent the features of the face in a video frame.

The processes in 3.2.1 and 3.2.2. Were applied on 10, 20, and 30 frames for each video. That is, (for 10 extracted frames), 10 feature vectors were obtained for each video. To get only one feature vector per video, mean and standard deviation(std) values were

Fig. 2. Detected face landmarks

calculated for the corresponding values on these 10 feature vectors associated with a specified video, (Fig. 3) illustrates the idea. As a result, each video is represented by a vector of 2054 values: 1024 mean and 1024 std from resnet50 vector plus 3 mean and 3 std values calculated for face pose angle. All these values were saved in a.csv file.

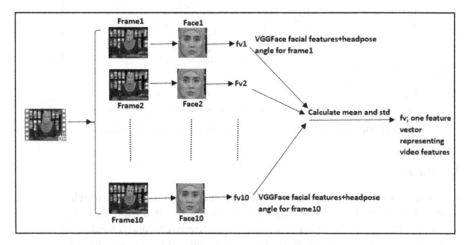

Fig. 3. Generating feature vector process

3.3 Classification

The final stage is the classification process, which determines whether a video is real or fake based on the features that were extracted and described in step 3.2. LazyClassifier, a Python function from the lazypredict library, was utilized for the classification problem. LazyClassifier applies several classifiers using little code and aids in identifying which models perform best without the need for more parameter tuning.

The three main stages of the model are shown in (Fig. 4).

4 Practical Work

In this section, we discussed the suggested model's training and testing results on the chosen datasets, along with the metrics used to assess the model. (Fig. 4) is a flowchart showing the stages of the work, starting from entering a video until the video is classified as fake or real.

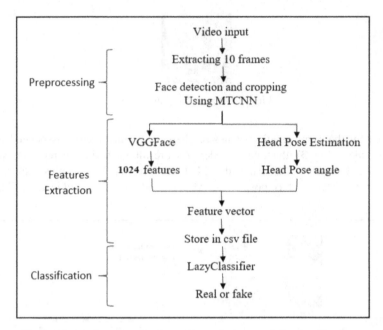

Fig. 4. Model phases: methods used in each phase and the classifier layers

4.1 Dataset

FaceForensics++ (FF++) [18] dataset was utilized to train the model, which comprises 1000 real and 1000 fake videos. The fake videos were created using four manipulation techniques: Deepfake (DF), Face2Face (F2F), FaceSwap (FS), and Neural Textures (NT). That is the dataset contains 1000 fake videos for each manipulation technique. Only the first three techniques for generating the fake video (DF, F2F, and FS) are used in this work.

4.2 Evaluation Metrix

The model was evaluated and tested using two metrics, they are accuracy and F1_score. The accuracy metric evaluates how often a classifier produces correct classifications for the entire dataset. It is calculated as the ratio of correctly classified cases to the total number of classified cases. Precision, on the other hand, focuses on the ratio of correctly classified positive cases to all cases that were classified as positive, whether correctly or incorrectly. It specifically measures the proportion of positive cases that were classified accurately.

Sensitivity, also known as recall or true positive rate, assesses the classifier's ability to correctly identify positive cases. It is determined by dividing the number of correctly classified positive cases by the total number of positive cases, including those mistakenly classified as negative.

Combining precision and sensitivity gives us the F1-score metric, which is used to evaluate a classifier's correctness. The F1-score balances both precision and sensitivity, providing a single value that considers the trade-off between accurately classifying positive cases and minimizing false negatives [19].

These metrics can be calculated after testing the model, Eq. 1 and Eq. 4 represent the equations of the approved evaluation metrics:

$$\text{Accuracy} = \frac{no.\ of\ correct\ predictions}{total\ no.\ of\ predictions} = \frac{TP + TN}{TP + TN + FP + FN} \tag{1}$$

$$\text{Precision} = \frac{TP}{TP + FP} \tag{2}$$

$$\text{Recall} = \frac{TP}{TP + FN} \tag{3}$$

where:

TP: no. of fake videos that are correctly predicted.
TN: no. of real videos that are correctly predicted.
FP: no. of fake videos that are incorrectly predicted.
FN: no. of real videos that are incorrectly predicted.

$$\text{F1\_score} = \frac{(2 * Precision * Recall)}{(2 + Precision + Recall)} \tag{4}$$

F1_score is the harmonic mean of the precision and recall scores and can be computed after calculating both Eq. 2 and Eq. 3.

5 Result

Most of the previous works rely on the use of one of the CNN networks pre-trained on 'imagenet', whether using CNN only to extract visual features or CNN with LSTM to extract both visual and temporal features). However, when we applied these methods to the FF++ dataset, the obtained test accuracy was very poor. It did not exceed 0.56%, while the proposed model achieved much better results as shown in Table 1. It is obvious from the table that the VGGFace model was first performed alone on the dataset, and then combined with the head pose estimation process. The two techniques were implemented on 10, 20, and 30 frames extracted from each video. Initially, VGGFace was used to extract features from 10 frames only and make a feature vector that was sent to the classifier, Fig. 4. After making a prediction, the obtained test accuracy was 81%, which can be considered a good result but not the best.

Secondly, generating the feature vector by using the same VGGFace features combined with the x_, y_, and z_values calculated through the head pose estimation process. There was some improvement in the test accuracy value by about 0.01. The same calculations were reapplied to 20, 30 frames and we noticed an increase in the value of test accuracy by an amount range between 0.03–0.04.

Table 1. Results of the proposed model in the form of accuracy and F1_score

		F2F		DF		FS	
	No.of frames	VGGFace	VGGFace + Head pose	VGGFace	VGGFace + Head pose	VGGFace	VGGFace + Head pose
accuracy	10 frames	0.8065	0.81407	0.8140	0.8165	0.8517	0.8517
	20 frames	0.8477	0.8654	0.8469	0.8622	0.8753	0.8753
	30 frames	0.8414	0.84399	0.8367	0.8392	0.8804	0.8854
F1_score	10 frames	0.8064	0.81406	0.8140	0.8165	0.8516	0.8516
	20 frames	0.8475	0.8654	0.8467	0.8622	0.8753	0.8753
	30 frames	0.8413	0.84385	0.8365	0.8391	0.8802	0.8854

All these processes were applied to the FF++ dataset with the three techniques to produce fake videos: F2F, DF, and FS, and the best result was obtained with VGGFace+ head pose estimation when applied to the FS fake dataset with 30 frames per video.

There is almost no change in accuracy if only 10 frames are used. However, with the increase in the number of frames extracted from the video, the obtained test accuracy increased, whether by using the VGGFace model alone or combined with head pose estimation. This gives us an explanation that: by increasing the number of frames, the accuracy of the temporal features extracted from each video increases. The values of test accuracy and F1_score are close to each other, because the dataset is balanced, meaning that the real and fake data are almost equal.

LasyClassifier was used for classification. LazyClassifier is a function in lazypredict machine learning Python library. It implicitly executes several classifiers, and through this, we can know which classifier gives the best result. Figure 5 shows the results after implementing the classifier on VGGFace+ head pose features extracted from 30 frames.

LazyClassifier displays the results sorted in descending order with the classifier of the highest accuracy value being the first one. So, the result indicates that the classifier NuSVC gave the best result with a test accuracy of more than 0.885.

In Table 2, the proposed model was compared with five different CNN models. The comparison was done using the value of test accuracy of the six models. The test accuracy of the first model was obtained from the self-practical experience of the ResNet CNN after training on the imagenet dataset. It was noted that the first two models were both ResNet but with different depths. The proposed model (using ResNet-50) gave better results despite being with the same depth as in the first one.

Model	Accuracy	Balanced Accuracy	ROC AUC	F1 Score	Time Taken
NuSVC	0.885496	0.885567	0.885567	0.885415	1.45297
SVC	0.882952	0.883042	0.883042	0.882814	1.24986
Perceptron	0.86514	0.865068	0.865068	0.865025	0.140613
PassiveAggressiveClassifier	0.854962	0.85489	0.85489	0.854838	0.171852
CalibratedClassifierCV	0.849873	0.849826	0.849826	0.849818	1.17759
SGDClassifier	0.849873	0.849762	0.849762	0.849575	0.124991
LogisticRegression	0.842239	0.842135	0.842135	0.841961	0.209284
LinearSVC	0.839695	0.839609	0.839609	0.839506	0.381709
GaussianNB	0.826972	0.826984	0.826984	0.82697	0.109288
LGBMClassifier	0.819338	0.819214	0.819214	0.818893	3.14189
NearestCentroid	0.798982	0.79891	0.79891	0.79881	0.0961978
RandomForestClassifier	0.793893	0.793782	0.793782	0.793484	2.13113
XGBClassifier	0.791349	0.791166	0.791166	0.790246	3.7191
BernoulliNB	0.788804	0.78877	0.78877	0.788763	0.159948
AdaBoostClassifier	0.778626	0.778489	0.778489	0.777962	9.47213
ExtraTreesClassifier	0.768448	0.768297	0.768297	0.767617	0.626025
RidgeClassifierCV	0.73028	0.730239	0.730239	0.730207	0.531188
BaggingClassifier	0.720102	0.71975	0.71975	0.714612	12.9871
RidgeClassifier	0.697201	0.697141	0.697141	0.697024	0.259451
LinearDiscriminantAnalysis	0.671756	0.671682	0.671682	0.671475	1.31826
DecisionTreeClassifier	0.628499	0.628341	0.628341	0.62705	1.51565
KNeighborsClassifier	0.618321	0.617684	0.617684	0.59272	0.171821
ExtraTreeClassifier	0.575064	0.574951	0.574951	0.57422	0.0937359

Fig. 5. LazyClassifier results on VGGFace+ Head pose features extracted from 30 frames

Table 2. Comparison of the proposed model with equivalent ones

Reference	Model	Accuracy
Self-experiment	ResNet50	0.56
Chandani et al. [20]	ResNet-152	0.767
Mittal et al. [21]	AlexNet	0.556
Sven van A. [22]	Exception	0.707
Afchar et al. [5]	MesoInception4	0.813
Proposed model	VGGFace+ head pose	0.8854

Additionally, the table contains the results obtained from the works of [20–22] and [5]. The superiority of the proposed model over the others was noted, referring that mixing more methods for extracting features gives better results. The worst result was gained from 'imagenet' pre-trained ResNet (first model), indicating that the weights adopted in it did not serve the task of distinguishing fake faces from real ones.

6 Conclusion

In social media, videos with fake content is increasing precipitously and there is a persistent need for designing models that detect such content. The feature extraction process is an influential factor in the success of such de-signed models. In this paper, we are concerned with both visual and temporal features of video frames and concluded that: 1- for visual features, better results can be obtained through merging two techniques for extracting those features than using each technique alone. 2- for temporal features, better results were obtained by increasing the number of frames extracted from each video. Combining these two ideas improved the model performance and test accuracy to an acceptable one. This led to the superiority of the proposed method when compared with a number of typical and well-known CNN models.

References

1. Kingma, D.P., Welling, M.: Auto-encoding variational bayes. In: International Conference on Learning Representations (ICLR), vol. 1 (2014)
2. Goodfellow, I., et al.: Generative adversarial nets. Adv. Neural Inf. Process. Syst. **27** (2014)
3. Sultan, D.A., Ibrahim, L.M.: A comprehensive survey on deepfake detection techniques. Int. J. Intell. Syst. Appl. Eng. **10**(3s), 89–202 (2022)
4. Afchar, D., Nozick, V., Yamagishi, J., Echizen, I.: Mesonet: a compact facial video forgery detection network. In: 2018 IEEE international workshop on information forensics and security (WIFS), vol. 11, pp. 1–7. IEEE (2018)
5. Ismail, A., Elpeltagy, M.S., Zaki, M., Eldahshan, K.: A new deep learning-based methodology for video deepfake detection using xgboost. Sensors **21**(16), 5413 (2021)
6. Amerini, I., Galteri, L., Caldelli, R., Del Bimbo, A.: Deepfake video detection through optical flow based CNN. In: Proceedings of the IEEE/CVF International Conference on Computer Vision Workshops (2019)
7. Tran, V.N., Lee, S.H., Le, H.S., Kwon, K.R.: High Performance deepfake video detection on CNN-based with attention target-specific regions and manual distillation extraction. Appl. Sci. **11**(16), 7678 (2021)
8. Matern, F., Riess, C., Stamminger, M.: Exploiting visual artifacts to expose deepfakes and face manipulations. In: 2019 IEEE Winter Applications of Computer Vision Workshops (WACVW), pp. 83–92. IEEE (2019)
9. Wodajo, D., Atnafu, S.: Deepfake video detection using convolutional vision transformer. arXiv preprint arXiv:2102.11126 (2021)
10. Güera, D., Delp, E.J.: Deepfake video detection using recurrent neural networks. In: 2018 15th IEEE International Conference on Advanced Video and Signal Based Surveillance (AVSS), pp. 1–6. IEEE (2018)
11. Abdul Jamsheed, V., Janet, B.: Deep fake video detection using recurrent neural networks. Int. J. Sci. Res. Comput. Sci. Eng. **9**(2), 22–26 (2021)

12. Yadav, P., Jaswal, I., Maravi, J., Choudhary, V., Khanna, G.: DeepFake Detection using Inception ResNetV2 and LSTM. In: International Conference on Emerging Technologies: AI, IoT, and CPS for Science Technology Applications (2021)
13. Li, Y., Chang, M.C., Lyu, S.: In ictu oculi: Exposing ai created fake videos by detecting eye blinking. In: 2018 IEEE International workshop on information forensics and security (WIFS), pp. 1–7. IEEE (2018)
14. Zhang, D., Li, C., Lin, F., Zeng, D., Ge, S.: Detecting deepfake videos with temporal dropout 3DCNN. In: IJCAI, pp. 1288–1294 (2021)
15. Zhou, Y., Lim, S.N.: Joint audio-visual deepfake detection. In: Proceedings of the IEEE/CVF International Conference on Computer Vision, pp. 14800–14809 (2021)
16. Agarwal, S., Farid, H., Fried, O., Agrawala, M.: Detecting deep-fake videos from phoneme-viseme mismatches. In: Proceedings of the IEEE/CVF Conference on Computer Vision and Pattern Recognition Workshops, pp. 660–661 (2020)
17. Parkhi, O., Vedaldi A., Zisserman A.: Deep face recognition. In: BMVC 2015-Proceedings of the British Machine Vision Conference 2015. British Machine Vision Association (BMVC) (2015)
18. Rossler, A., et al.: Faceforensics++: learning to detect manipulated facial images. In: Proceedings of the IEEE/CVF International Conference on Computer Vision, pp. 1–11 (2019)
19. Obayes, H.K., Al-Shareefi, F.: Secure heart disease classification system based on three pass protocol and machine learning. Iraqi J. Comput. Sci. Math. **4**(2), 72–82 (2023)
20. Chandani, K., Arora, M.: Automatic facial forgery detection using deep neural networks. In: Kumar, N., Tibor, S., Sindhwani, R., Lee, J., Srivastava, P. (eds.) Advances in Interdisciplinary Engineering. LNME, pp. 205–214. Springer, Singapore (2021). https://doi.org/10.1007/978-981-15-9956-9_21
21. Mittal, H., Saraswat, M., Bansal, J.C., Nagar, A.: Fake-face image classification using improved quantum-inspired evolutionary-based feature selection method. In: 2020 IEEE Symposium Series on Computational Intelligence (SSCI), pp. 989–995. IEEE (2020)
22. van Asseldonk S., Önal, I.: Deepfake video detection using deep convolutional and hand-crafted facial features with long short-term memory network. In: 33rd Benelux Conference on Artificial Intelligence and the 30th Belgian Dutch Conference on Machine Learning (2021)

Multi-models Based on Yolov8 for Identification of Vehicle Type and License Plate Recognition

Mustafa Noaman Kadhim, Ammar Hussein Mutlag,
and Dalal Abdulmohsin Hammood[✉]

Electrical Engineering Technical College, Department of Computer Technical
Engineering, Middle Technical University (MTU), Al Doura, 10022 Baghdad, Iraq
{bbc0080,ammar_alqiesy,dalal.hammood}@mtu.edu.iq

Abstract. Embedded systems with cameras and deep learning techniques have been shown to be flexible and good at finding different targets in the areas of intelligent monitoring and urban mobility. These use cases are present in diverse situations and regions. The collection of pertinent data from the deployment site is of utmost importance. This study introduces an innovative methodology for a comprehensive system that integrates vehicle category identification with license plate recognition using the YOLOv8 algorithm. The system comprises three main components: vehicle type detection and recognition, detection of the license plate, and detection of the license plate characters and numbers. The suggested approach intends to enhance the identification system's applicability in the unique context of Iraqi vehicles, particularly on roadways and in cities and their environments. The dataset used in this study was obtained from various areas inside Iraq. The detection system employed in our research successfully identified three distinct vehicle classes as well as detected and recognized license plates in both Arabic and English. The mean average precision achieved for the aforementioned tasks was 97.5%, 98.94%, 98.6%, and 98.4%, respectively. Through the use of visual data, such as images and videos, our system successfully identified license plates with reduced dimensions. It is posited that our technology has the potential to be used in densely populated areas in order to cater to the substantial requirements for improved visual acuity in smart urban environments.

Keywords: Vehicle Type Detection · License Plate Recognition · License Plate Detection · YOLOv8

1 Introduction

Applications for computer vision are used to automate operations that involve repeated actions, which often require the human capacity to consistently observe and make prompt judgements. Numerous applications have been created to

A. M. Al-Bakry et al. (Eds.): NTICT 2023, CCIS 2096, pp. 118–135, 2024.
https://doi.org/10.1007/978-3-031-62814-6_9

locate, recognize, and monitor a diverse range of items of interest. The rapid progress in smart city technologies has facilitated the deployment of various visual sensors in intelligent environments and smart infrastructures. These sensors include closed-circuit TV (CCTV), visual networks of sensors, intelligent monitoring systems, smart traffic systems, cameras for security, and vehicle black boxes [1–4].

A collection of cutting-edge deep learning methodologies designed to address complex computer vision tasks [5]. The system has the capability to detect and accurately identify a wide range of items from various categories on a large scale. The recognition of individuals and their cars is a topic of considerable importance in urban areas, where smart cameras are used for this purpose. Numerous methods for license plate identification [6–8] and recognition of make and model [9–11] were devised with the purpose of alleviating human operators from the laborious duty of manually recognizing, detecting, and identifying a diverse array of automobiles.

Handling the complexity associated with our endeavor to achieve accurate vehicle categorization and license plate recognition necessitates overcoming a multitude of challenges. The presence of different atmospheric conditions, especially changes in illumination, creates challenges that require robust algorithms capable of adjusting to various situations. The issue lies in fitting the wide range of vehicle types, each with unique sizes and designs for its registration plates. Additionally, the variety of typefaces used makes it more difficult to distinguish between letters and numbers on license plates. To tackle these issues, a comprehensive strategy is needed that takes into consideration not just the technical components of detecting and recognizing them but also factors in the diverse range of vehicle types, ambient circumstances, and license plate designs seen in the real world. We are committed to understanding and solving the complexities of these difficulties in order to improve the strength and precision of our system.

The classification of vehicle types is crucial for improving the efficiency of monitoring systems. Through the precise classification of vehicles into distinct types, such as cars, trucks, and buses, law enforcement and traffic management authorities may get vital information about traffic trends, detect possible security risks, and enhance the efficiency of road safety measures. Additionally, the process of identifying vehicles goes beyond just classifying them and also plays a crucial part in the subsequent task of recognizing license plates. The ability to visually see and classify vehicles serves as the fundamental basis for precise identification using license plate recognition systems.

In the context of a more intricate urban setting, we provide a comprehensive system called IVT-LPRs, as shown in Fig. 1, an acronym for Iraqi Vehicle Type Recognition System and License Plates. This system has the capability to discern and classify both vehicle types and license plates. The contributions presented in this work are as follows:

- In this study, we provide a novel two-phase architecture that uses YOLOv8 [12] as the foundation for identifying vehicle types and recognizing Iraqi license plates.

- A custom dataset including diverse Iraqi vehicle types and license plates obtained from different parts of Iraq was compiled and constructed. This dataset was used for the purpose of training and validating two custom detectors within the context of Iraqi Vehicle Type and License Plate Recognition Systems (IVT-LPRs).
- In this research, we demonstrate the efficacy of the IVT-LPRs in accurately detecting miniature license plates from images, thereby presenting an improvement over existing detectors.

The structure of the current paper is as follows: The study's introduction is thoroughly reviewed in the first part. A review of the pertinent research is given in the second part. In the third section, the approach is presented. The fourth section of the article provides a comprehensive analysis of the conducted experiments and presents the corresponding findings. Following this, the fifth section serves as the concluding part of the study.

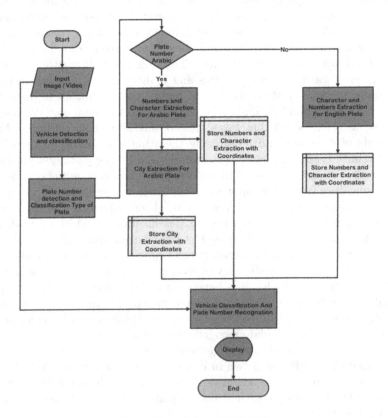

Fig. 1. Flowchart of IVT-LPRs

2   Related Work

This section presents a comprehensive analysis of previous studies undertaken on the identification and recognition of license plates in the context of Iraq. Numerous efforts have been undertaken to develop LPR systems that exhibit enhanced speed and precision. In recent years, there has been an increasing use of deep learning-based methodologies, including convolutional neural networks (CNN) [13,14], single shot detector (SSD) [15], and models based on the You Only Look Once (YOLO) framework [16–19], for the purpose of detecting and recognizing license plates.

The method provided by Naaman Omar [20] involves the first segmentation of the number plate area in a particular image of a vehicle. Subsequently, the city and plate number are retrieved from the segmented license plate. The use of deep feature extraction based on the Residual Neural Networks (ResNet) architecture is being studied. The fully linked layer of the ResNet model is employed to acquire the profound characteristics for the cropped Arabic numerals and city areas, correspondingly. The collected characteristics are inputted into the consecutive input section of the LSTM (long-short-term memory) classifier. The accuracy ratings obtained for Arabic numerals and city areas were 98.51% and 100%, respectively. Dhuha Habeeb et al. [21] a method was devised, relying on YOLOV2, that employs deep learning techniques to enable comprehensive recognition of license plate numbers in an end-to-end manner. A comprehensive analysis was conducted to compare the deep-learning-based model with conventional approaches, namely Support Vector Machines and Neural Networks. This analysis included the examination of Malaysian license Plates featuring black backdrop plates with Latin alphabet inscriptions as well as Iraqi LPs exhibiting three distinct Arabic writing styles. The approach that was developed demonstrated a higher level of performance in LP recognition, reaching an accuracy level of 88.86% and 85.56% for the datasets from Malaysia and Iraq, respectively.

The work conducted by Omar et al. [22] the present study used data gathered from the northern region of Iraq to investigate Iraqi LPs using a Convolutional Neural Network model. Firstly, the dataset was collected, and then image processing methods were used to enhance the contrast. The license plate was then divided into segments, revealing the nation and its provinces (cities). These features were then used to train the network to perform classification. With accuracy rates of 92.10%, 94.43%, and 91.01%, respectively, the results showed a high degree of success in the tasks of identification, segmentation, and recognition.

Dunya A. Abd Alhamza et al. [23] suggested identifying plates with a tripartite method. Character recognition, pre-processing methods targeted at recognizing license plates, and segmentation, which divides text into discrete units, are among the procedures. Using a camera to take a image and pre-process it is the first step. The division of numbers by individual splittings is called segmentation. The last phase involves the use of the K-Nearest Neighbours algorithm, a fundamental machine learning technique employed to associate numbers with training data in order to provide correct predictions. The system was constructed using the Python 3.5 programming language and the OpenCV package. The accuracy performance of the system was evaluated using a dataset consisting of 50 images, resulting in an accuracy rate of 90 percent.

In their study, Yousif et al. [24] in this study, a model is proposed with the objective of accurately identifying Iraqi license plates using the Single Shot MultiBox Detector. The system is distinguished by its rapid processing capabilities and has shown a high level of accuracy in detecting licence plates, achieving a detection accuracy rate of 98%. Additionally, image-processing methods were used to partition the automobile licence plate into several sections, including the left side portion, the top part, the bottom part, and the lower right part. The segmentation accuracy achieved a commendable 94%, indicating a high level of precision. This level of accuracy is particularly significant since it directly impacts the subsequent stages of the process. The last phase involves the differentiation of the color situated on the left side of the vehicle licence plate via the use of histogram analysis and the application of K-Nearest Neighbours methodologies. The accuracy rate achieved for color discrimination reached 94%.

Table 1 presents a comprehensive overview of the preceding studies. A comparison between the suggested method and current methods is shown in Table 2.

Table 1. Comparison of state of art work

Authors, year	Approach	Objective	Evaluation Metrics	Accuracy	Dataset	Limitation
Naaman, O et al. [20], 2022	ResNet + LSTM.	To get the profound characteristics of the clipped Arabic numerals and cities, respectively. To achieve higher performance in contrast to other ways.	Accuracy score	The percentages for Arabic numbers and city regions are 98.51% and 100%, respectively.	The dataset utilized in this study was gathered from the northern region of Iraq.	I. The dataset was insufficient, making it unable to correctly separate the Arabic numbers and characters. II. The dataset exclusively comprises three provinces, namely Erbil, Sulaymaniyah, and Dohuk
Habeeb, D et al. [21], 2021	YOLOv2 + SVM + NN.	The objective is to develop a recognition system for license plates containing Arabic and Latin alphabets by employing a deep-learning methodology.	Average Precision (AP50).	The average recognition rate of 85.56% and 88.86% on the datasets from Iraq and Malaysia, respectively.	A dataset comprising 404 images was collected from Iraqi vehicles and 681 images from Malaysian vehicles.	I. The dataset did not encompass all the provinces within the country of Iraq. II. The dataset lacks the truck plate
Omar, N et al. [22], 2020	CNN.	The goal is to build a reliable ALP detection and identification system for northern Iraqi automobiles.	Recall, Precision and F-measure scores.	92.10%, 94.43% and 91.01%, in Recall, Precision and F-measure scores, respectively.	The northern region of Iraq encompasses three prominent cities, namely Erbil, Duhok, and Sulaymaniyah.	I. Dataset is Limited. II. An increase in compute time. III. The final recognition was directly impacted by the semantic segmentation

(continued)

Table 1. (*continued*)

Authors, year	Approach	Objective	Evaluation Metrics	Accuracy	Dataset	Limitation
Abbass, G et al. [24], 2021	SSD + KNN.	The primary aim of this study is to focus on the detection and segmentation of Iraqi licence plates.	Accuracy	The obtained outcomes demonstrate a high level of success, with a 98% accuracy rate in plate detection and a 96% accuracy rate in the segmentation operation.	———	I. The category of a car depends on the left-side segment of its license plate. II. Complexity
Abd Alhamza, D et al. [23], 2020	KNN.	The primary goal of this study is to develop a method for effectively detecting and recognizing license plates on vehicles, with a specific focus on precisely identifying Arabic numerals and characters that are shown on the plate.	Accuracy	The system demonstrated an accuracy rate of 90%.	Own dataset	I. The performance of the system is suboptimal when dealing with large datasets. II. The algorithm exhibits a high level of sensitivity towards noisy data, missing values, and outliers. III. The provinces are not recognized.

Table 2. Comparison of the suggested method with current methods.

Authors, year	Detection Type		Recognition of Plate Number		Input Data	
	Vehicle Type	Plate Number	Arabic	English	Image	Video
Naaman, O et al. [20], 2022	X	✓	✓	X	✓	X
Habeeb, D et al. [21], 2021	X	✓	✓	✓	✓	X
Omar, N et al. [22], 2020	X	✓	✓	X	✓	X
Abd Alhamza, D et al. [23], 2020	X	✓	X	X	✓	X
Abbass, G et al. [24], 2021	X	✓	✓	X	✓	X
Our Proposal	✓	✓	✓	✓	✓	✓

3   Proposed Methodology

A typical license plate recognition system (LPR) aims to provide textual representations of numbers and characters seen on license plates. The objective of our study was to develop a License Plate Recognition (LPR) system capable of accurately detecting and classifying license plates from Iraq, while also being able to distinguish between different kinds of vehicles often seen in Iraq. We provide an extensive system for Iraqi vehicle type and license plate recognition, referred to as IVT-LPRs, in this work. YOLOv8 serves as the system's primary object detection model.

A summary of the YOLOv8 algorithm-powered Iraq Vehicle Type-License Plate Recognition (IVT-LPR) system is shown in Fig. 2. Accurate vehicle classification as well as the effective detection and interpretation of license plate information from both images and videos are the goals of the IVT-LPRs. A detailed description of the IVT-LPRs, including the YOLOv8-based object detector and the steps involved in data collection, is included in the next subsections.

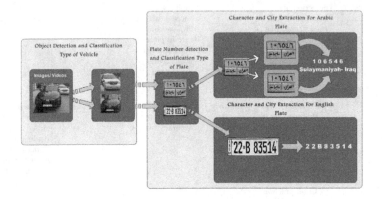

Fig. 2. The general description of the IVT-LPRs

3.1   Vehicle Type and License Plate Recognition Based on YOLOv8

The IVT-LPRs process an input video or image. We collected real Iraqi vehicle types and licence plate data in order to construct our own dataset. Subsequently, the custom dataset was used for training purposes, using YOLOv8 to construct four custom detectors. The vehicle type detector is the first one employed, which identifies and classifies three distinct vehicle types (car, truck, and bus) within the given input image. Following the categorization of the vehicle, a subsequent detector is used to identify and classify the license plate, discerning between Arabic and English plates. Upon successfully identifying the license plate and ascertaining its language type (Arabic or English), the subsequent step involves the implementation of the third detector specifically designed for Arabic plates. This detector has two distinct components: the first component is responsible for recognizing numerals and characters, while the second component is responsible for detecting provinces. However, in the case of English plates, the fourth detector is used to identify and analyses numerical digits and alphabetic characters. The rationale for the production of many detectors is to enhance accuracy rather than depend on a single detector to do all duties.

After recognizing the plate and determining its language type, namely Arabic or English, it is thereafter extracted and sent to the corresponding model that corresponds to its designated kind. The model exhibits a non-sequential (random) pattern of detecting characters and numbers, as seen in Fig. 3.

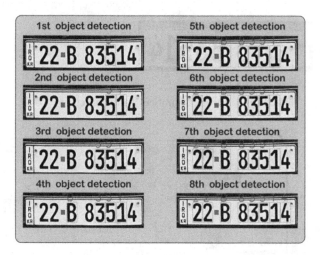

Fig. 3. Labeling of LP numbers and characters

In order to read the numbers and character correctly, the image's dimensions [25, 26] (namely, its height and breadth) that were previously truncated were measured. Additionally, a reference point was established at the midpoint of the image's left side. The center-left point is defined as (0, height/2) [27]. Subsequently, the computation was performed to determine the distance between the midpoint of the left side and the coordinates of the boundary box for each number or character [28]. As shown in the following equation.

Point coordinates:

$$(Xp, Yp) = (0, \frac{height}{2}) \tag{1}$$

Bounding box coordinates of numbers and character:

$$(Xmin, Ymin) and (Xmax, Ymax) \tag{2}$$

The center of the bounding box, C, can be calculated as:

$$C = (\frac{(Xmin + Xmax)}{2}, \frac{(Ymin + Ymax)}{2}) = (Cx, Cy) \tag{3}$$

The Euclidean distance [29] between the point and the bounding box center, D, can be calculated as:

$$D = \sqrt{(Xp - Cx)^2 + (Yp - Cy)^2} \tag{4}$$

After performing calculations to determine the distances of each number and letter, The objects are arranged in ascending order of distance in order to precisely display them from left to right. As demonstrated in Fig. 4. The VT-LP detection and recognition approach for Iraqi vehicles is shown in Fig. 5.

Fig. 4. Extract numbers and characters

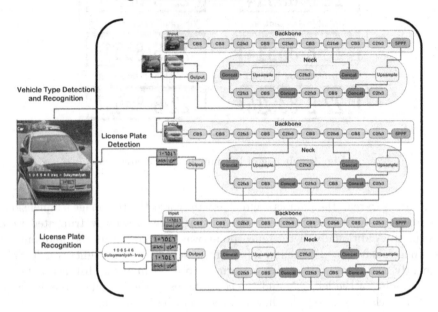

Fig. 5. Proposed method.

3.2 Dataset Collection

Vehicle Types. The dataset was collected manually and from different places, with the aim of including a wide range of vehicle types and license plate image that accurately depict the contextual and environmental conditions in Iraq. The collected dataset is manually labelled by using Roboflow [30]. The vehicles were categorized into three distinct groups, namely automobiles, trucks, and buses. The dataset we used in our study includes three distinct vehicle kinds, as seen in Fig. 6. Table 3 presents a data set that has been gathered pertaining to the types of vehicles in Iraq as well as the corresponding licence plates in both Arabic and English.

Fig. 6. Collected data for vehicle types.

Table 3. Collected data for vehicle types and license plate.

Class	Training	Validation	Test	Total
Car	3,863	754	431	5,048
Bus	4,048	804	523	5,375
Truck	3,822	776	489	5,087
license plate Arabic	2,679	506	217	3,402
license plate English	2,409	489	201	3,099

LP Numbers and Characters. Datasets for Arabic and English paintings were collected manually under different conditions and angles in order to train the model well and achieve a high level of accuracy.

Numerous problems pertaining to the license plate on the Iraqi car were covered in this study. The majority of these difficulties had to do with the plate's incorporation of various patterns and hues as well as the use of Arabic and English in its layout. It's also crucial to note that the Arabic and Latin alphabets are used in conjunction with capital and lowercase characters on the Iraqi license plate. This may be seen in Fig. 7.

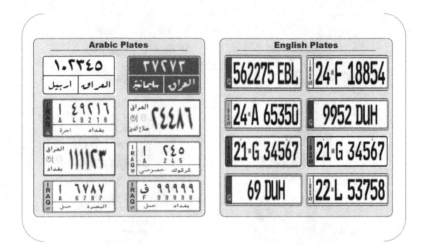

Fig. 7. Different Iraq LP styles

4   Results and Discussion

In this part, a comprehensive analysis of the simulation findings is presented. The experiments are conducted using the Google Colaboratory (Colab) platform [31]. Colab offers users the opportunity to use a Virtual Machine with exceptional performance capabilities. Several NVIDIA GPU types, such as K80s, T4s, P4s, and P100s, as well as Random Access Memory (RAM), may be assigned to this virtual machine.

In order to determine the feasibility and effectiveness of our Iraq vehicle type and license plate recognition system (IVT-LPRs), we performed an assessment specifically targeting its capacity to properly categorize a wide range of vehicle types. In addition, we thoroughly examined the system's performance in the demanding job of accurately detecting and recognizing license plates, which include letters from both the Arabic and English alphabets. The purpose of this extensive assessment was to measure the overall efficiency and resilience of the system in managing the many intricacies related to various vehicle types and license plate forms in several languages and sizes.

4.1   Implementation

We implemented our proposed IVT-LPRs using the YOLOv8 [12] object detector as the base architecture. In this research, YOLOv8 was trained using an open-source PyTorch [32, 33] framework to identify a customized collection of classes, such as different kinds of vehicles, license plates, and Arabic and English numbers and characters. In each class, 70% of the dataset was utilized for training, 20% for validation, and 10% for testing the license plate recognition and vehicle type detector.

4.2 Performance of Vehicle Type

The vehicle type detector in the Iraq vehicle type and license plate recognition system (IVT-LPRs) can recognize and categorize vehicles into three distinct groups: cars, buses, and trucks. And to evaluate the effectiveness of the vehicle-type detector, we employed established metrics widely recognized for evaluating object detection systems. The metrics include Mean Average Precision (mAP) [34], precision [35], and recall [36], which are defined by Eqs. 5, 6, and 7, respectively.

$$Mean Average Precision(mAP) = \frac{1}{N} \sum_{i=1}^{n} AP_i \qquad (5)$$

$$Precision(P) = \frac{TP}{TP + FP} \qquad (6)$$

$$Recall(R) = \frac{TP}{TP + FN} \qquad (7)$$

Fig. 8(a) presents a collection of performance measures, including training loss, accuracy, recall, and mAP, shown as curves. The aforementioned criteria are used for the purpose of assessing the efficacy of the YOLOv8 model in the task of vehicle type detection. As seen in Fig. 8(a), the losses exhibit a progressive decline, while the metrics of accuracy, recall, and mean average precision demonstrate improvement as the number of epochs increases.

Finally, Fig. 8(c) presents a collection of test sample images that have been used to evaluate the effectiveness of the proposed YOLOv8 model for vehicle type detection. The figure includes the predicted classes for each image, along with their respective confidence scores. The accuracy of vehicle type recognition achieved a mAP of 97.5%.

4.3 Performance of License Plate Detection

Iraq offers a diverse range of automobile licence plates, including both English and Arabic. The plate recognition system first identifies the presence of a license plate and then classifies it as either Arabic or English. This classification is crucial for further processing, as it allows for the extraction of the plate's content and its transmission to the appropriate module responsible for reading the characters and numbers. The results of the vehicle license plates are shown in Figs. 8(d). The performance of our detector was consistently stable when tested on various images, exhibiting a mAP of 98.94%.

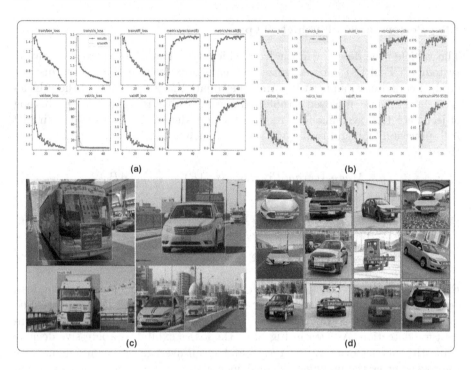

Fig. 8. (a) Curves for several performance metrics for the YOLOv8 model. (b) curves for various metrics of performance for car license plates. (c) Successfully identified /detected the type of vehicle. (d) Successfully detected Arabic and English license plates.

4.4 Performance of License Plate Recognition

Upon identification of the license plate and classification of its type, it is then processed and sent to the designated model responsible for detecting the arrangement of the characters and numbers. As previously said, there are numerous types and diverse colors of license plates in Iraq. The detector used for the identification of Arabic license plates has two separate models. The first model is designed to identify a total of 27 categories, including numerical digits ranging from 0 to 9, as well as Arabic characters. Regarding the second model, it has the capability to identify 21 distinct categories. These categories include Iraqi provinces as well as various plate kinds such as private, taxi, and truck.

The first model achieved a mAP of 98.5% in identifying letters and numbers, while the second model achieved a mAP of 98.8%. Figures 9(a) and 9(b) show the effectiveness of the first and second models using the same measures. Figure 9(c) shows the effectiveness of the first model in identifying and distinguishing letters and numbers and the second model in accurately identifying interruptions and distinguishing the specific type of license plate.

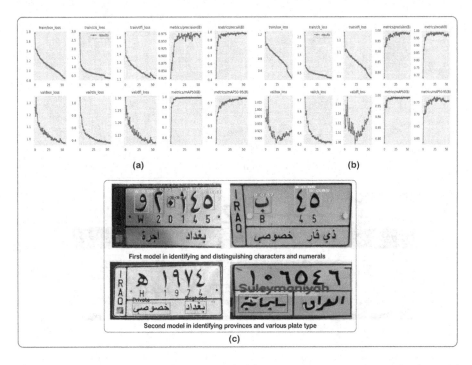

Fig. 9. (a) Performance of the first model. (b) Performance of the second model. (c) Results of the first and second models.

The detector designed to identify English license plates of a certain model is capable of recognizing 36 distinct categories, including numerals 0 to 9 and the English alphabet. The detection accuracy for numbers and characters on English license plates was found to be a mAP of 98.4%. Figure 10(a) depicts the performance of the model, while Fig. 10(b) illustrates the model's proficiency in identifying numerals and letters included on English license plates.

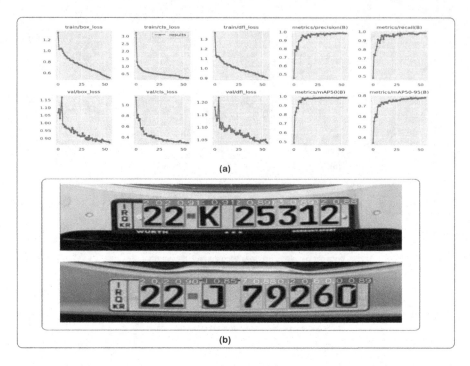

Fig. 10. (a) Performance of the English license plate model. (b) Successfully detected/recognized characters and numbers in English license plates.

The suggested approach exhibits a minor decrease in performance under conditions of low lighting. There are many potential factors that may contribute to this phenomenon. One possible explanation is that the reduced presence of colors at night poses a challenge for the models to extract distinctive characteristics. This difficulty arises from the similarity of the surrounding hues, which are almost indistinguishable from one another. This phenomenon becomes evident when the model fails to recognize automobiles located in areas lacking streetlight illumination. Furthermore, the intense luminosity emitted by the headlights contributes significantly by augmenting the overall brightness of the captured image. Consequently, this excessive brightness hinders both the camera and the model from accurately perceiving the true form and contours of the vehicle.

5   Conclusion

This study presents a novel IVT-LPR system, including two distinct phases. The first stage involves the identification of both the vehicle type and the license plate, while the subsequent stage focuses on the detection and differentiation of alphanumeric characters present on the license plate. This work offered a novel dataset, including images that were gathered from diverse locations and captured under varying situations and viewpoints. The empirical findings demonstrate that the strategy described in this study has shown effectiveness in both the tasks of detection and recognition. The mAP attained by vehicle type detection was 98.6%, while the license plate detection achieved a mAP of 98.94%. The detection accuracy for Arabic and English panels in terms of numerals and letters obtained mAP of 98.6% each. Regarding the limits of the proposed methodology, it is worth noting that the technique demonstrates a marginal decline in indicated performance when subjected to poor illumination circumstances. However, we have successfully shown the advantages of our proposed IVT-LPR system in efficiently tackling Iraq's license plate recognition challenges, namely vehicle type identification and licence plate recognition. This system has the potential to be used in diverse applications for smart cities with intricate requirements.

References

1. Gharaibeh, A., Salahuddin, M.A., Hussini, S.J., Khreishah, A., Khalil, I., Guizani, M., Al-Fuqaha, A.: Smart cities: A survey on data management, security, and enabling technologies. IEEE Communications Surveys & Tutorials **19**(4), 2456–2501 (2017)
2. Winkler, T., Rinner, B.: Security and privacy protection in visual sensor networks: A survey. ACM Computing Surveys (CSUR) **47**(1), 1–42 (2014)
3. Won, M.: Intelligent traffic monitoring systems for vehicle classification: A survey. IEEE Access **8**, 73340–73358 (2020)
4. Baran, R., Rusc, T., Fornalski, P.: A smart camera for the surveillance of vehicles in intelligent transportation systems. Multimedia Tools and Applications **75**, 10471–10493 (2016)
5. Liu, L., Ouyang, W., Wang, X., Fieguth, P., Chen, J., Liu, X., Pietikäinen, M.: Deep learning for generic object detection: A survey. Int. J. Comput. Vision **128**, 261–318 (2020)
6. Du, S., Ibrahim, M., Shehata, M., Badawy, W.: Automatic license plate recognition (alpr): A state-of-the-art review. IEEE Trans. Circuits Syst. Video Technol. **23**(2), 311–325 (2012)
7. Anagnostopoulos, C.-N.E., Anagnostopoulos, I.E., Psoroulas, I.D., Loumos, V., Kayafas, E.: License plate recognition from still images and video sequences: A survey. IEEE Trans. Intell. Transp. Syst. **9**(3), 377–391 (2008)
8. Shashirangana, J., Padmasiri, H., Meedeniya, D., Perera, C.: Automated license plate recognition: a survey on methods and techniques. IEEE Access **9**, 11203–11225 (2020)
9. Siddiqui, A.J., Mammeri, A., Boukerche, A.: Real-time vehicle make and model recognition based on a bag of surf features. IEEE Trans. Intell. Transp. Syst. **17**(11), 3205–3219 (2016)
10. Manzoor, M.A., Morgan, Y., Bais, A.: Real-time vehicle make and model recognition system. Machine Learning and Knowledge Extraction **1**(2), 611–629 (2019)
11. Hsieh, J.-W., Chen, L.-C., Chen, D.-Y.: Symmetrical surf and its applications to vehicle detection and vehicle make and model recognition. IEEE Trans. Intell. Transp. Syst. **15**(1), 6–20 (2014)
12. G. Jocher, A. Chaurasia, and J. Qiu, "Yolo by ultralytics," URL: https://githubcom/ultralytics/ultralytics, 2023
13. K. He, G. Gkioxari, P. Dollár, and R. Girshick, "Mask r-cnn," in *Proceedings of the IEEE international conference on computer vision*, pp. 2961–2969, 2017

14. R. Girshick, "Fast r-cnn," in *Proceedings of the IEEE international conference on computer vision*, pp. 1440–1448, 2015
15. W. Liu, D. Anguelov, D. Erhan, C. Szegedy, S. Reed, C.-Y. Fu, and A. C. Berg, "Ssd: Single shot multibox detector," in *Computer Vision–ECCV 2016: 14th European Conference, Amsterdam, The Netherlands, October 11–14, 2016, Proceedings, Part I 14*, pp. 21–37, Springer, 2016
16. J. Redmon and A. Farhadi, "Yolov3: An incremental improvement," arXiv preprint arXiv:1804.02767, 2018
17. A. Bochkovskiy, C.-Y. Wang, and H.-Y. M. Liao, "Yolov4: Optimal speed and accuracy of object detection," arXiv preprint arXiv:2004.10934, 2020
18. J. Redmon, S. Divvala, R. Girshick, and A. Farhadi, "You only look once: Unified, real-time object detection," in *Proceedings of the IEEE conference on computer vision and pattern recognition*, pp. 779–788, 2016
19. J. Redmon and A. Farhadi, "Yolo9000: better, faster, stronger," in *Proceedings of the IEEE conference on computer vision and pattern recognition*, pp. 7263–7271, 2017
20. Naaman, O.: Resnet and lstm based accurate approach for license plate detection and recognition. Traitement du Signal **39**(5), 1577 (2022)
21. D. Habeeb, F. Noman, A. A. Alkahtani, Y. A. Alsariera, G. Alkawsi, Y. Fazea, A. M. Al-Jubari, *et al.*, "Deep-learning-based approach for iraqi and malaysian vehicle license plate recognition," *Computational intelligence and neuroscience*, vol. 2021, 2021
22. Omar, N., Sengur, A., Al-Ali, S.G.S.: Cascaded deep learning-based efficient approach for license plate detection and recognition. Expert Syst. Appl. **149**, 113280 (2020)
23. D. A. Abd Alhamza and A. D. Alaythawy, "Iraqi license plate recognition based on machine learning," *Iraqi Journal of Information and Communication Technology*, vol. 3, no. 4, pp. 1–10, 2020
24. Abbass, G.Y., Marhoon, A.F.: Iraqi license plate detection and segmentation based on deep learning. Iraqi Journal for Electrical and Electronic Engineering **17**(2), 102–107 (2021)
25. S. T. Ahmed, D. A. Hammood, R. F. Chisab, A. Al-Naji, and J. Chahl, "Medical image encryption: A comprehensive review," *Computers*, vol. 12, no. 8, 2023
26. R. S. Jebur, C. S. Der, and D. A. Hammood, "A review and taxonomy of image denoising techniques," in *2020 6th International Conference on Interactive Digital Media (ICIDM)*, pp. 1–6, IEEE, 2020
27. Willson, R.G., Shafer, S.A.: What is the center of the image? JOSA A **11**(11), 2946–2955 (1994)
28. A. Mousavian, D. Anguelov, J. Flynn, and J. Kosecka, "3d bounding box estimation using deep learning and geometry," in *Proceedings of the IEEE conference on Computer Vision and Pattern Recognition*, pp. 7074–7082, 2017
29. Wang, L., Zhang, Y., Feng, J.: On the euclidean distance of images. IEEE Trans. Pattern Anal. Mach. Intell. **27**(8), 1334–1339 (2005)
30. Roboflow, "Roboflow: Your machine learning data pipeline," Year of Access. Accessed on Date of Access
31. E. Bisong and E. Bisong, "Google colaboratory," *Building machine learning and deep learning models on google cloud platform: a comprehensive guide for beginners*, pp. 59–64, 2019
32. A. Paszke, S. Gross, F. Massa, A. Lerer, J. Bradbury, G. Chanan, T. Killeen, Z. Lin, N. Gimelshein, L. Antiga, *et al.*, "Pytorch: An imperative style, high-performance

deep learning library," *Advances in neural information processing systems*, vol. 32, 2019

33. S. Imambi, K. B. Prakash, and G. Kanagachidambaresan, "Pytorch," *Programming with TensorFlow: Solution for Edge Computing Applications*, pp. 87–104, 2021

34. P. Henderson and V. Ferrari, "End-to-end training of object class detectors for mean average precision," in *Computer Vision–ACCV 2016: 13th Asian Conference on Computer Vision, Taipei, Taiwan, November 20-24, 2016, Revised Selected Papers, Part V 13*, pp. 198–213, Springer, 2017

35. Xie, L., Ahmad, T., Jin, L., Liu, Y., Zhang, S.: A new cnn-based method for multi-directional car license plate detection. IEEE Trans. Intell. Transp. Syst. **19**(2), 507–517 (2018)

36. Li, H., Wang, P., You, M., Shen, C.: Reading car license plates using deep neural networks. Image Vis. Comput. **72**, 14–23 (2018)

Face Identification System in Transform Domains Over Secure Communication Channel

Taif Alobaidi[1][(✉)] [iD] and Wasfy B. Mikhael[2]

[1] Department of Mobile Communications and Computing Engineering, College of Engineering, University of Information Technology and Communications (UOITC), Baghdad, Iraq
taif.alobaidi@uoitc.edu.iq
[2] Department of Electrical and Computer Engineering, University of Central Florida, Orlando, FL 32816, USA

Abstract. In a Face Recognition (FR) system, the facial pose image is utilized as the information based upon which a person can be identified. Enrollment and recognition are the two modes in FR system where poses are processed after pre-processing step. In FR, it is accustomed to employ the same pre-processing step(s) in both enrollment and recognition modes. In addition, it is crucial to have an accurate, fast, and less resources demanding system. In this paper, a new pre-processing approach is presented in which the training and testing face poses have different dimensions. The proposed technique aims at improving the overall recognition system performance by increasing both accuracy and final decision revealing time and decreasing model storage space requirements. The presented work is exhaustively tested on a FR system. Two Dimensional Discrete Haar Transform (2D DHT), and Cosine (2D DCT) are employed in an adaptive algorithm to calculate the optimal weights of each coefficient in each domain by diminishing residual portion of the input pose signal. After utilizing pre-processing step (that includes image crop and resize), the adaptive algorithm is utilized to calculate the weights in both DWT and DCT domains. Only a predefined number of dominant DWT coefficients, as observed by the corresponding higher weight of that coefficient, are retained. The classifier employed in this work is based on discriminative sparse representation technique for FR with employment of ℓ_2 regularization. As the results show, the new technique met three-parameter model criteria (storage, computational complexity, and accuracy rate) and achieved an average of 95.56% for ORL dataset with five training poses. The candidate application of the proposed technique is in communications systems utilized by FR systems.

Keywords: Digital Signal Processing · Face Recognition · Discrete Cosine Transform · Discrete Haar Transform · Mixed Transforms

1 Introduction

Pattern recognition (PR) is a field of study that focuses on extracting meaningful information from raw data to identify and categorize patterns [1]. Preprocessing plays a crucial role in pattern recognition systems, as it involves transforming and enhancing

A. M. Al-Bakry et al. (Eds.): NTICT 2023, CCIS 2096, pp. 136–149, 2024.
https://doi.org/10.1007/978-3-031-62814-6_10

the data. This process aims to remove noise, normalize data, and extract relevant features, ultimately improving the accuracy and efficiency of pattern recognition systems. One of PR applications is Facial Recognition (FR) in which people are identified by images of their faces. Applications of FR systems include: airport and border control, banking, healthcare, and enhanced Cyber security.

Table 1 shows a summary of the related works stated in this article. In [2], a study to introduce Hexagonal pixel-based image processing (HIP) versions of three fundamental texture extraction techniques in square pixel-based image processing (SIP) which are Gray-Level-Co-occurrence-Matrices (GLCM), Local Binary Pattern (LBP), and the recent local-holistic graph-based descriptor (LHGPD). The images undergo a transformation from the SIP domain to the HIP domain. The corresponding HIP domain equivalents Gray-Level Co-occurrence Matrix (HexGLCM), Local Binary Pattern (HexLBP), and local-holistic graph-based descriptor (HexLHGPD) of their counterparts in the SIP domain are established. Facial recognition performances of both SIP and HIP domain versions of GLCM, LBP, and LHGPD are evaluated and compared using primary data sets. The experimental results demonstrate that the HIP domain GLCM, LBP, and LHGPD have equal performance levels with their SIP domain counterparts, with certain instances where they outperform them in terms of face recognition accuracy. The data sets utilized were CASPEAL-R1 [3], EXTENDED YALE B [4], FACES95 [5], ORL [6], and Lab2 [7]. The features were extracted and the three classifiers were employed, namely, Optimizable Discriminant Analysis (ODA), Optimizable Ensemble (OE), and Optimizable Neural Network (ONN). The available data were split into 80% training, and 20% testing and five trails were averaged to obtain the results. The maximum reported results were: 96.25%, 99.9%, 99.31%, 98.61%, and 97.5% for CASPEAL-R1 [3], EXTENDED YALE B [4], FACES95 [5], ORL [6], and Lab2 [7] respectively. Recently, several sparse representation approaches have been reported for face recognition [8–14]. Dictionary-based, and naive training sample-based are the two classes of these methods. In the dictionary-based category, a dictionary is generated using training samples to represent each class of input data. On the other hand, the naive training sample-based approach directly uses training samples as representations for each class [15]. Sparse representation methods in FR typically introduce constraints in the objective function to regularize the solution. The commonly employed constraints include the ℓ_1 norm [16], ℓ_2 norm [17, 18], and ℓ_2 norm regularization [19].

In [20], a new sparse representation model with ℓ_2-norm regularization was proposed for FR. The model was evaluated using five databases: ORL [6], Extended YaleB [4], Georgia Tech Face Database [21], Subset of the FERET [22, 23], and CMU Multi-PIE [24]. Computational efficiency and high recognition accuracies were achieved as shown in results in Table 1. However, the conclusion of the study mentioned that the balance parameter in the objective function was selected from a set of candidate values.

Table 1. Summary of Reviewed Articles

Reference Number	Approach	Database(s)	Training Poses: Testing Poses	Number of Runs	Maximum Identification Accuracy
[2]	Hexagonal pixel-based image processing	CASPEAL-R1[3]	8:2	5	96.25%
		EXTENDED YALE B [4]			99.9%
		FACES95 [5]			99.31%
		ORL [6]			98.61%
		Lab2 [7]			97.5%
[20]	ℓ_2 regularization sparse representation	ORL [6]	6:4	1	95.00%
		Extended YaleB [4]	18:44		82.16%
		Georgia Tech Face Database [21]	9:6		78.00%
		Subset of the FERET [22, 23]	5:6		80.50%
		CMU Multi-PIE [24]	10:17		99.86%
[25]	Hybrid Orthogonal Polynomials	ORL [6]	8:2	20	98.23%
		FEI1 [26]			97.5%
[28]	Cosine Domain Transformation of approach in [20]	ORL [6]	5:5	252	96.1%
		YALE	5:6	462	67.5%
		FERET	5:6	462	64.96%
		FEI [26]	10:4	1001	54.5%
		Cropped AR	6:7	1716	82.72%
		Georgia Tech Face Database [21]	7:8	6435	78.34%
[40]	Haar Domain Transformation of approach in [20]	ORL [6]	5:5	252	96.21%
		YALE	5:6	462	63.66%
		FERET	5:6	462	64.84%
		Georgia Tech Face Database [21]	7:8	6435	77.77%
		Cropped AR	6:7	1716	82.66%

In [25], a new scheme for FR was presented, wherein features are extracted using hybrid orthogonal polynomials. The complexity of the extraction step was reduced using the embedded image kernel scheme. Subsequently, these features were classified using a Support Vector Machine (SVM). Additionally, a fast-overlapping block processing approach for feature extraction is employed to reduce computation time. Several evaluation of the proposed method were conducted on two different face image databases, ORL

[6] and FEI [26][1]. To assess its accuracy, the approach there was presented with other recently reported FR approaches. The highest obtained results for the noise-free cases (the average of 20 runs, the samples were not specified but the *Training* : *Testing* ratio was 8:2) were 98.23%, and 97.5% for ORL, FEI respectively. In [27], they proposed an automated approach to calculate the balance parameter in the objective function in [20]. Furthermore, in [28] they improved the recognition accuracy, speed, and storage requirements of that system by transform the input image from spatial to other transform domains.

In this work, a new preprocessing approach is utilized in a FR system in which the dimensions of the input data sample to the system during the Enrollment mode is different, i.e., larger, than the size of input sample to the system during recognition mode. Due to the minimization of the required input dimensions during recognition, the proposed technique is a very good solution to the recognition system where secure communications (wire and wireless) is essential to successful recognition. The extracted features are computed through an adaptive algorithm (that is based on perfectly analyzing the pose into two transform domains). As shown in the results, the size reduction improves the overall FR system performance.

The organization of the next sections is: Sect. 2 (a description about FR modules (preprocessing, feature extraction, and classifier)), Sect. 3(the FR system with the proposed approach is explained in details), Sect. 4 (contains all the obtained results), and the conclusion is presented in the last part.

2   Background on FR System Modules

Raw data obtained from various sources often contain imperfections, inconsistencies, and irrelevant information that can hinder pattern recognition algorithms [29]. Preprocessing addresses these challenges by cleaning and transforming the data, making it more suitable for analysis. The primary objectives of preprocessing include Noise Removal, Data Normalization, Feature Extraction, and Data Transformation [30]. Noise refers to random variations or errors that distort the original data. It can arise due to measurement errors, sensor limitations, or environmental factors. Firstly, Noise removal techniques, such as filtering and smoothing, help eliminate or reduce noise, enhancing the reliability of the data. Secondly, raw data often exhibit variations in scale, units, and ranges (this work exploits this point in the advantage of the system). Data normalization techniques, such as re-scaling or standardization, ensure that the data is consistent and comparable. This step is crucial as it prevents certain features from dominating the analysis due to their larger magnitudes. Thirdly, in pattern recognition, relevant features are essential for accurate classification and identification. Feature extraction techniques aim to identify and extract the most informative attributes from the data, reducing dimensionality and computational complexity. These techniques include statistical measures, transform methods, and domain-specific feature extraction algorithms. Finally, Data transformation techniques modify the data to meet certain assumptions or requirements of pattern recognition algorithms. For example, logarithmic or power transformations can be applied to

---

[1] Please note that other literature (as in [27]) cited the FEI database as a 200 people (14 poses each) database and not 100 as in [25].

data that exhibits skewed distributions. Transformations can also be used to convert data into a different representation, such as converting images into frequency domains using Fourier or wavelet transforms. The common preprocessing Techniques include Filtering, Smoothing, Standardization, Principle Component Analysis (PCA), and Image Enhancement. Filtering is a widely used technique in preprocessing to remove noise and unwanted components from the data. It involves applying filters, such as low-pass, high-pass, or band-pass filters, to attenuate or eliminate specific frequency components. Filtering is commonly used in signal processing applications, such as audio or image recognition. Smoothing techniques aim to reduce noise and eliminate small variations in the data. Moving average, Gaussian smoothing, and median filtering are commonly used methods to achieve this. Smoothing can be particularly useful in time-series analysis or image processing tasks. Standardization, also known as z-score normalization, data transformation to have $\mu = 0$, and $\sigma^2 = 1$. . This technique is widely employed to bring the data onto a common scale, ensuring that each feature contributes equally to the analysis. Standardization is commonly used in machine learning algorithms that rely on distance-based calculations, e.g., k-nearest neighbors (KNN)[31] or Support Vector Machines (SVM) [32]. PCA [33] (please note that PCA algorithm can be employed as a preprocessing and a feature extraction step simultaneously in some PR systems) is a dimensionality reduction technique that transforms the data into new group of variables that are uncorrelated called principle components. These components capture the most significant variations in the data and allow for a compact representation of high-dimensional data. PCA is commonly used to reduce the dimensionality of feature vectors while preserving the essential information. Image Enhancement: In image recognition tasks, preprocessing techniques like contrast enhancement, histogram equalization, and edge detection are employed to improve the quality of images, enhance important details, and highlight relevant features. These techniques play a vital role in applications such as object recognition, facial recognition, and medical imaging.

2.1 Discrete Haar Transform (DHT)

The 2D-DHT [34] procedure produces 4 frequency bands, LL, LH, HL, and HH (please see[2]), which are combined in a data formation of a matrix represented as C_2. When employed on images, which can be regarded as 2D signals, a one-level DHT decomposition utilizes a Scaling known as $\varphi(x, y)$ in addition to 3 wavelets denoted as $\psi(x, y)$. The following equations show the calculations of the wavelets:

$$\varphi(x, y) = \varphi(x)\varphi(y) \tag{1}$$

$$\psi(x, y)^H = \psi(x)\varphi(y) \tag{2}$$

$$\psi(x, y)^V = \varphi(x)\psi(y) \tag{3}$$

$$\psi(x, y)^D = \psi(x)\psi(y) \tag{4}$$

---

[2] L is Low Pass, and H is High Pass.

where $\varphi(x, y)$ is the scaling (or the LL Band), column variations (the LH Band) are captured by measuring variations along columns. Row variations (the HL Band) are detected by the sensitivity of $\psi(x, y)^V$ to variations along rows. Lastly, diagonal variations, or the HH Band, are simulated by $\psi(x, y)^D$ to emulate variations along the diagonal. The 2D-DWT of a facial pose $g(x, y)$ of size M * M is:

$$W_\varphi(j_0, m, m) = \frac{1}{\sqrt{MM}} \sum_{x=0}^{M-1} \sum_{y=0}^{M-1} g(x, y) \varphi_{j_0, m, m}(x, y) \tag{5}$$

$$W_\psi^i(j, m, m) = \frac{1}{\sqrt{MM}} \sum_{x=0}^{M-1} \sum_{y=0}^{M-1} g(x, y) \psi_{j, m, m}^i(x, y), i = \{H, V, D\} \tag{6}$$

j_0 is a random initial scale value and the $W_\varphi(j_0, m, m)$ elements is the Approximation $g(x, y)$ at j_0. The $W_\psi^i(j, m, m)$ elements add parallel, perpendicular, and diagonal specifics for scales $j \geq j_0$. Normally j_0 equals to zero and select the value of $M = 2^J$ and j ranges from zero to $(J-1)$ while m ranges from zero to $2^j - 1$.

2.2 Discrete Cosine Transform (DCT)

The DCT representation of signals such as speech and facial images contains a concentrated amount of energy within a limited range. Consequently, a smaller number of coefficients can effectively represent the unprocessed signal compared to the temporal domain. This transform is fundamental in lossy compression techniques for widely used formats like Joint Photographic Experts Group and MPEG-2 Audio Layer III. Forward equation, as described in [35] can be utilized to compute the 2D DCT:

$$C(m, n) = \frac{2}{\sqrt{M*N}} \sum_{u=0}^{M-1} \sum_{v=0}^{N-1} g(u, v) *c_m* \cos\left(\frac{m(2u+1)\pi}{2M}\right) *c_n* \cos\left(\frac{m(2v+1)\pi}{2N}\right) \tag{7}$$

where $c(u, v)$ is the temporal domain signal while $C(m, n)$ is the element in row m, column n for u ranges from 0 to M $-$ 1 (with one step increase), and v follows the same fashion of u with the other limit as N $-$ 1. The other pair of the transform is:

$$cs(u, v) = \frac{2}{\sqrt{M*N}} \sum_{u=0}^{M-1} \sum_{v=0}^{N-1} C(m, n) *c_m* \cos\left(\frac{m(2u+1)\pi}{2M}\right) *c_n* \cos\left(\frac{m(2v+1)\pi}{2N}\right) \tag{8}$$

where $c_m, and c_n$ are defined as:

$$c_m = \begin{cases} \frac{1}{\sqrt{2}} \ for \ m = 0 \\ 1 \ \ otherwise \end{cases} \tag{9}$$

2.3   Face Pose Representation Using Adaptive Error Minimization Algorithm

The adaptive algorithm to extract features from a face pose, after a preprocessing step, is shown Fig. 1 and it is implemented in the following manner:

A weight matrix, $W_2 = [\alpha]$, is initialized with 0.5 and multiplied, in particular a Schur Product (that is an entry-wise product [36]) by the 2D DCT Coefficients. The residual after retaining only dominant coefficients is transformed using 2D Inverse DCT. Next, A 2D-DWT is applied in the spatial domain. Another weight matrix, $W_3 = [\beta]$, is initialized with 0.3 and entry-wise multiplied with the coefficients. The residual after retaining only dominant coefficients is transformed using 2D Inverse DWT. The residual of the pose energy, $\Phi(\alpha, \beta)$, the energy (in each domain) is subtracted from total spatial energy and the result is considered as a cost function that have to reach minimum. In particular, $\Phi(\alpha, \beta)$ is calculated in the following manner:

$$\Phi(\alpha, \beta) = [\text{Spatial Elements}]^2 - [\text{Retained DCT}]^2 - [\text{Retained DWT}]^2 \qquad (10)$$

where $[]^2$ is the element-wise square.

The residual is minimized using a Steepest Descent Algorithm [37]. Once the iteration ends, weight matrices $W_2 = [\alpha]$, $W_3 = [\beta]$ is finalized. The following updating formulas (repeated each iteration) are as follows [38]:

$$\alpha_{i,j}(n+1) = \alpha_{i,j}(n+1) - \mu_{\alpha_{i,j}} \nabla_{\alpha_{i,j}} \Phi \qquad (11)$$

$$\beta_{i,j}(n+1) = \beta_{i,j}(n+1) - \mu_{\beta_{i,j}} \nabla_{\beta_{i,j}} \Phi \qquad (12)$$

where i, and j take all possible values in its domain and depending on $\alpha_{i,j}$ (elements in $[\alpha]$), $\beta_{i,j}$ (elements in $[\beta]$). The index of iteration is n, and the converging factor is μ. Face pose dimensions are M * M. The adaptive loop halts when $\Phi < 0.05\%$ of the input spatial energy. The converging factors, $\mu_{\alpha_{i,j}}$ and $\mu_{\beta_{i,j}}$, are calculated in the following manner:

$$\mu_\alpha = \frac{\Phi(n)}{\sum_{i=0}^{N-1} \sum_{j=0}^{N-1} [\nabla_{\alpha_{i,j}} \Phi]^2} \qquad (13)$$

$$\mu_\beta = \frac{\Phi(n)}{\sum_{i=0}^{N-1} \sum_{j=0}^{N-1} [\nabla_{\beta_{i,j}} \Phi]^2} \qquad (14)$$

2.4 FR Utilizing Sparse Representation

The feature matrix X (training) has N elements (where N is L * p), where L is the number of people in a database and p is count of poses allocated for training.

2.4.1 ℓ_1 Norm Model

All techniques in this group aim to solve the optimization problem [10]

$$\min B_1 \quad s.t. \quad y = XB \tag{15}$$

where B is coefficients vector, y is the test sample, i.e., y = XB (noise-free case) or for noisy case:

$$\min B_1 \quad s.t. \quad y - XB_2 < \epsilon \tag{16}$$

where coefficients vector $B = [b_1, \ldots \ldots b_N]^T$, and error margin of equals to ϵ. The ℓ_1-norm regularization based representation algorithms include Orthogonal Matching Pursuit, Homotopy and Augmented Lagrangian, the Fast Iterative Shrinkage and Thresholding algorithm [39], and ℓ_1 Regularized Least Squares (L1LS).

2.4.2 ℓ_2 Norm Model

One of the ℓ_2-norm representations is the Collaborative Representation Classification (CRC). Hence, B is calculated as follows:

$$B = (X^T X + \gamma I)^{-1} X^T y \tag{17}$$

where I is the unit matrix, and γ is a balance coefficient [18].

3 Proposed System

The details about the proposed technique is shown in Fig. 2. In the enrollment mode, the training pose image are resized to 32 * 32. Then, the adaptive algorithm explained in the previous section is utilized to find the optimal weights of coefficients in 2D DCT and 2D DWT. A predefined count of the DHT elements are retained. This process is repeated for all the poses allocated for training. In the recognition mode, the end user terminal sends an image to the remote database. All processing steps are implemented on that remote database platform. The final decision about the identity is sent back to the original end user. The following subsections explain in details all parts of the FR system.

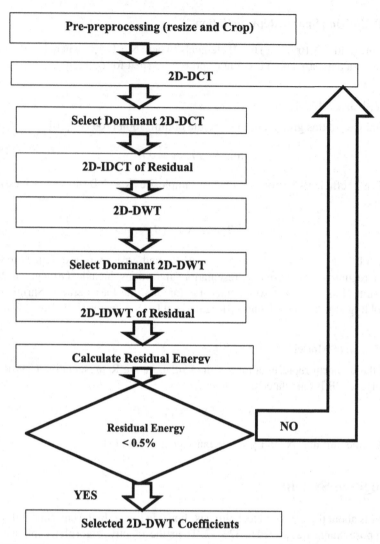

Fig. 1. Flow Chart of the Adaptive Algorithm Employed to Obtain the Weights of Transform Domains Coefficients

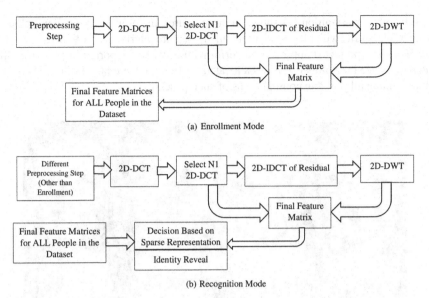

(a) Enrollment Mode

(b) Recognition Mode

Fig. 2. Recognition System Employed to Evaluate the Proposed Pre-Processing Technique

4 Experimental Results

The proposed technique is employed as a part of the FR system that is assessed using a publicly available data set called ORL [6]. The data set includes 400 samples from which 10 poses are allocated for each person of the 40 people. Unprocessed samples from this data set are displayed in Fig. 3. The original image dimensions are 112 * 92. To evaluate the technique in FR system, it is compared with [40]. All available combinations of train/test sets are attempted in all FR systems. Average identification Rate (AIR) in the figures is computed as follows:

$$AIR = \frac{Total\ Correct\ Identifications}{Total\ Attempts} \tag{18}$$

As an example, there are 252 sets in one trial when half of the available poses are utilized for training. The maximum of these 252 is considered as Maximum Identification Rate. The maximum and average recognition accuracy rates obtained are shown in Figs. 4 and 5 respectively. The input dimensions of the training poses are 32 * 32 which results in 1024 pixels. The testing samples are either 8 * 8 (referred to as P8 in Figs. 4, and 5), 16 * 16(referred to as P16 in Figs. 4, and 5), or 32 * 32 (referred to as P32 in Figs. 4, and 5). Only half of the testing dimensions, i.e., 32, 128, or 512 values are retained. Therefore, even when the testing images are 32 * 32, a 50% reduction is achieved. Furthermore, the first two proposed dimensions are reduce the data size by 96.875%, and 87.5%, respectively. The coefficients selections is based on their relative weights in the Wavelet domain as indicated by the output of the adaptive algorithm. The obtained results can be easily compared with [2], since they use 8 poses for training and 2 for testing. On the other hand, the system presented here utilized only 5 poses to achieve the full correct recognition. As shown in those figures, the proposed technique achieved comparable results with [40] while reducing the feature matrix size by 96.875%, 87.5%, and 50%. The proposed approach outperforms the systems reported in [25] (for ORL database) since the number of poses allocated for training were far less the ones in [25] while running all possible combinations of such poses.

Fig. 3. ORL Raw Samples

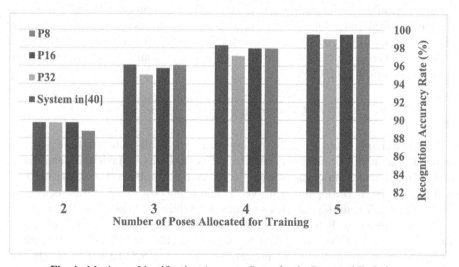

Fig. 4. Maximum Identification Accuracy Rates for the Proposed Technique

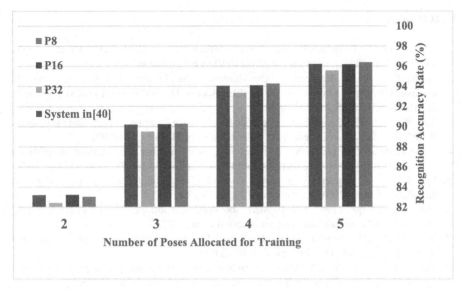

Fig. 5. Average Identification Accuracy Rates for the Proposed Technique

5 Conclusion

In this work, a new pre-processing technique for Face Recognition (FR) is presented. The proposed technique is based on the use of dissimilar input image sizes in the two modes of the system. In the Enrollment mode, the dimensions of the input face pose images are larger than the ones employed in the resize step in the recognition mode. After the resize step, an iterative approach is employed to calculate the weights of coefficients in two non-spatial domains, namely, Cosine, and Haar. The feature extraction step is finalized by selecting only half of the reduced training face image size. The classifier is based on the ℓ_2 regularization sparse representation. The results (for ORL database with mean recognition accuracy rate of 96.38%, and a maximum rate of 99.5%, and a maximum of 96.875% reduction in feature matrix size) reveal that the new approach met three-parameter FR favorable criteria (storage, computational complexity, and accuracy rate) and achieved a better performance compared with the latest reported technique. An improvement to the overall recognition system performance was shown in the results by increasing both accuracy and final decision revealing time and decreasing model storage space requirements and communications channel bandwidth between a remote database and end users. The proposed technique can be utilized in communications systems utilized in a recognition in which the data sent by the end user is less than what it is used to be. Future work can include employing different transforms rather than Cosine, and Haar domains.

References

1. Li, S., Deng, W.: Deep facial expression recognition: a survey. IEEE Trans. Affect. Comput. **13**, 1195–1215 (2022). https://doi.org/10.1109/TAFFC.2020.2981446
2. Cevik, T., Cevik, N., Rasheed, J., Abu-Mahfouz, A.M., Osman, O.: Facial recognition in hexagonal domain—a frontier approach. IEEE Access. **11**, 46577–46591 (2023). https://doi.org/10.1109/ACCESS.2023.3274840
3. Gao, W., et al.: The CAS-PEAL large-scale Chinese face database and baseline evaluations. IEEE Trans. Syst. Man Cybern.-Part A: Syst. Humans **38**, 149–161 (2007)
4. Vishwakarma, V.P., Dalal, S.: A novel non-linear modifier for adaptive illumination normalization for robust face recognition. Multimedia Tools Appl. **79**, 11503–11529 (2020)
5. Spacek, D.L.: Computer vision science research projects. https://www.essex.ac.uk/mv/all faces/faces95.html (2008). Last accessed 21 Sep 2023
6. Database, O. (April (1992–1994)). At&t laboratories Cambridge database of faces. https://cam-orl.co.uk/facedatabase.html. Last accessed 21 Sep 2023
7. Xu, Y., Zhong, A., Yang, J., Zhang, D.: Bimodal biometrics based on a representation and recognition approach. Opt. Eng. **50**, 037202 (2011)
8. Zhang, D., Guo, Z., Gong, Y.: Multispectral biometrics systems. In: Multispectral biometrics, pp. 23–35. Springer, Cham (2016). https://doi.org/10.1007/978-3-319-22485-5_2
9. Duc, B., Fischer, S., Bigun, J.: Face authentication with sparse grid Gabor information. In: 1997 IEEE International Conference on Acoustics, Speech, and Signal Processing, pp. 3053–3056. IEEE (1997)
10. Wright, J., Yang, A.Y., Ganesh, A., Sastry, S.S., Ma, Y.: Robust face recognition via sparse representation. IEEE Trans. Pattern Anal. Mach. Intell. **31**, 210–227 (2008)
11. Keinert, F., Lazzaro, D., Morigi, S.: A robust group-sparse representation variational method with applications to face recognition. IEEE Trans. Image Process. **28**, 2785–2798 (2019)
12. Qiu, H., Pham, D.-S., Venkatesh, S., Liu, W., Lai, J.: A fast extension for sparse representation on robust face recognition. In: 2010 20th International Conference on Pattern Recognition, pp. 1023–1027. IEEE (2010)
13. Chen, C.-F., Wei, C.-P., Wang, Y.-C.F.: Low-rank matrix recovery with structural incoherence for robust face recognition. In: 2012 IEEE conference on computer vision and pattern recognition, pp. 2618–2625. IEEE (2012)
14. Sun, Y., Wang, X., Tang, X.: Sparsifying neural network connections for face recognition. In: Proceedings of the IEEE conference on computer vision and pattern recognition, pp. 4856–4864 (2016)
15. Yang, J., Chu, D., Zhang, L., Xu, Y., Yang, J.: Sparse representation classifier steered discriminative projection with applications to face recognition. IEEE Trans. Neural Netw. Learn. Syst. **24**, 1023–1035 (2013)
16. Wright, J., Ma, Y., Mairal, J., Sapiro, G., Huang, T.S., Yan, S.: Sparse representation for computer vision and pattern recognition. Proc. IEEE **98**, 1031–1044 (2010)
17. Xu, Y., Zhu, Q., Fan, Z., Zhang, D., Mi, J., Lai, Z.: Using the idea of the sparse representation to perform coarse-to-fine face recognition. Inf. Sci. **238**, 138–148 (2013)
18. Zhang, L., Yang, M., Feng, X.: Sparse representation or collaborative representation: Which helps face recognition? In: 2011 International conference on computer vision, pp. 471–478. IEEE (2011)
19. Ren, C.-X., Dai, D.-Q., Yan, H.: Robust classification using $\ell 2$, 1-norm based regression model. Pattern Recogn. **45**, 2708–2718 (2012)
20. Xu, Y., Zhong, Z., Yang, J., You, J., Zhang, D.: A new discriminative sparse representation method for robust face recognition via l_2 regularization. IEEE Trans. Neural Netw. Learning Syst. **28**, 2233–2242 (2017)

21. Goel, N., Bebis, G., Nefian, A.: Face recognition experiments with random projection. In: Biometric technology for human identification II, pp. 426–437. SPIE (2005)
22. Phillips, P.J., Wechsler, H., Huang, J., Rauss, P.J.: The FERET database and evaluation procedure for face-recognition algorithms. Image Vis. Comput. **16**, 295–306 (1998)
23. Phillips, P.J., Moon, H., Rizvi, S.A., Rauss, P.J.: The FERET evaluation methodology for face-recognition algorithms. IEEE Trans. Pattern Anal. Mach. Intell. **22**, 1090–1104 (2000)
24. Database. Cmu multi-pie. https://www.cs.cmu.edu/afs/cs/project/PIE/MultiPie/Multi-Pie/Home.html. Last Accessed 21 Sep 2023
25. Abdulhussain, S.H., Mahmmod, B.M., AlGhadhban, A., Flusser, J.: Face recognition algorithm based on fast computation of orthogonal moments. Mathematics **10**, 2721 (2022)
26. Thomaz, C.E., Giraldi, G.A.: A new ranking method for principal components analysis and its application to face image analysis. Image and Vision Comput. **28**, 902–913 (2010). http://fei.edu.br/cet/facedatabase.html
27. Alobaidi, T., Mikhael, W.B.: A modified discriminant sparse representation method for face recognition. In: 2018 IEEE 8th Annual Computing and Communication Workshop and Conference (CCWC), pp. 727–730. IEEE (2018)
28. Alobaidi, T., Mikhael, W.B.: A transform domain implementation of sparse representation method for robust face recognition. Circ. Syst. Signal Process. **38**, 4302–4313 (2019)
29. Hassaballah, M., Aly, S.: Face recognition: challenges, achievements and future directions. IET Comput. Vis. **9**, 614–626 (2015)
30. García, S., Ramírez-Gallego, S., Luengo, J., Benítez, J.M., Herrera, F.: Big data preprocessing: methods and prospects. Big Data Anal. **1**, 1–22 (2016)
31. Kramer, O.: Dimensionality Reduction with Unsupervised Nearest Neighbors. Springer (2013)
32. Jakkula, V.: Tutorial on support vector machine (svm). School of EECS, Washington State University **37**, 3 (2006)
33. Maćkiewicz, A., Ratajczak, W.: Principal components analysis (PCA). Comput. Geosci. **19**, 303–342 (1993)
34. Mallat, S.: A Wavelet Tour of Signal Processing. Elsevier (1999)
35. Burger, W., Burge, M.J.: Digital Image Processing: An Algorithmic Introduction. Springer Nature (2022)
36. Davis, C.: The norm of the Schur product operation. Numer. Math. **4**, 343–344 (1962)
37. Widrow, B., McCool, J.: A comparison of adaptive algorithms based on the methods of steepest descent and random search. IEEE Trans. Antennas Propag. **24**, 615–637 (1976)
38. Ramaswamy, A., Mikhael, W.B.: Multi-transform multi-dimensional signal representation. In: Proceedings of 36th Midwest Symposium on Circuits and Systems. pp. 1255–1258. IEEE (1993)
39. Gregor, K., LeCun, Y.: Learning fast approximations of sparse coding. In: Proceedings of the 27th International Conference on International Conference on Machine Learning, pp. 399–406 (2010)
40. Alobaidi, T., Mikhael, W.B.: A Wavelet domain implementation of sparse representation method for face recognition. In: 2018 IEEE 61st International Midwest Symposium on Circuits and Systems (MWSCAS), pp. 214–217. IEEE (2018)

Predicting Covid-19 Protein Interactions Through Sequence Alignment

Ali K. Abdul Raheem[1,2](✉) 🆔 and Ban N. Dhannoon[3] 🆔

[1] College of Information Technology, University of Babylon, Hillah, Babil, Iraq
alikareem.sw.phd@student.uobabylon.edu.iq
[2] University of Warith Al-Anbiyaa, Kerbala, Iraq
[3] Department of Computer Science, College of Science, Al-Nahrain University, Baghdad, Iraq

Abstract. The global outbreak of COVID-19 necessitates a profound understanding of the molecular aspects of the causative agent, SARS-CoV-2. This study employs the Basic Local Alignment Search Tool (BLAST) to discern crucial insights into COVID-19 proteins. Leveraging BLAST, our investigation identifies and analyzes COVID-19 proteins with precision. Through sequence alignment, we unveil significant details about the functions of viral proteins, contributing to our understanding of their roles in disease progression. Our findings also shed light on potential interactions between viral and host proteins, enriching our comprehension of the virus's impact on cellular processes. Additionally, we explore the evolutionary relationships between COVID-19 proteins and other coronaviruses, providing context for their genetic evolution. These key findings provide a foundation for further research, guiding the development of targeted therapies and diagnostics in the ongoing battle against COVID-19.

Keywords: Bioinformatics · Sequence Alignment · Protein-protein interactions · BLAST · COVID-19 · Prediction

1 Introduction

The global outbreak of COVID-19, caused by the severe acute respiratory syndrome coronavirus 2 (SARS-CoV-2), has underscored the urgent need for an in-depth understanding of the virus's molecular structure and function [1]. A pivotal aspect in unraveling the biological intricacies of SARS-CoV-2 involves the identification and analysis of its constituent proteins, crucial players in the virus's lifecycle, pathogenesis, and interactions with the host [2]. Recent strides in computational biology have introduced innovative techniques for analyzing biological sequences, with sequence alignment emerging as a fundamental tool, particularly in the context of COVID-19, for identifying and characterizing viral proteins. Among various alignment algorithms, the Basic Local Alignment Search Tool (BLAST) stands out as a robust and widely used method [4]. In this paper, we delve into the application of the BLAST algorithm to identify and analyze COVID-19 proteins, showcasing how this approach contributes to a comprehensive understanding of the virus's biology. By aligning COVID-19 protein sequences with annotated databases, we uncover insights into their functions, potential interactions with host proteins, and evolutionary relationships with other coronaviruses. This research represents a significant contribution to the ongoing efforts to combat the COVID-19 pandemic, offering a foundation for further studies that guide the development of targeted therapies and diagnostics, leveraging the power of computational biology and the specificity of sequence alignment to mitigate the impact of COVID-19 and other viral diseases.

2 Related Work

In the pursuit of unraveling the complexities of COVID-19 and understanding the functional aspects of its constituent proteins, numerous studies have employed various computational and experimental approaches. This section presents an overview of the related work that has contributed to the field of COVID-19 protein identification and analysis.

Computational techniques have played a crucial role in dissecting the intricate molecular details of COVID-19 proteins. Sequence alignment, as demonstrated in this paper, has been widely used to identify conserved domains, functional motifs, and similarities with other viral proteins. Apart from BLAST, other algorithms such as HMMER and PSI-BLAST have also been instrumental in recognizing remote homologies and predicting protein structures [5]. These techniques have enabled researchers to deduce potential protein functions and infer evolutionary relationships, thereby providing insights into the virus's biology and pathogenicity.

Beyond sequence alignment, structural analysis has provided a deeper understanding of COVID-19 proteins. X-ray crystallography and cryo-electron microscopy have been employed to determine the three-dimensional structures of key viral proteins, including the spike protein and main protease [6, 7]. These structures have served as the basis for rational drug design and virtual screening, aiming to identify compounds that could disrupt protein-protein interactions critical for viral replication [8].

Predicting protein functions and interactions has been a focus of research in understanding the mechanisms of COVID-19 pathogenesis. Computational methods, such as network analysis and machine learning, have been applied to predict host-virus interactions and identify potential drug targets [9]. Integration of various omics data, including transcriptomic and proteomics, has facilitated the construction of comprehensive interaction networks that shed light on the complex interplay between viral proteins and host factors.

The race to develop vaccines and therapeutics for COVID-19 has driven extensive research into the functional properties of viral proteins [10]. Insights gained from structural and computational analyses have guided the design of vaccines targeting the spike protein, such as mRNA-based vaccines [11]. Small molecule inhibitors have also been identified through virtual screening and structure-based approaches, aimed at disrupting viral enzymes essential for replication.

3 Theoretical Background

3.1 Protein-Protein Interactions

Protein-protein interactions (PPIs) are critical for the functioning of biological systems and play a central role in various cellular processes [12]. PPIs are involved in diverse biological processes and have significant implications, including cellular signaling, complex assembly and regulation, and disease mechanisms [13]. Disruptions or dysregulations in PPIs can contribute to the development and progression of diseases [14].

Typically, proteins hardly act as isolated species while performing their functions in vivo [15]. It has been revealed that over 80% of proteins do not operate alone but in complexes [16]. The substantial analysis of authenticated proteins reveals that the proteins involved in the same cellular processes are repeatedly found to be interacting with each other [17].

Methods for detecting interactions between proteins can be grouped into three main categories: in vitro, in vivo, and silico techniques. In vitro methods involve conducting procedures in a controlled environment outside of a living organism. Conversely, in vivo techniques involve procedures carried out within the entire living organism. In the context of PPI detection, in silico methods encompass sequence-based and structure-based approaches [18].

Approaches for Predicting Interactions Based on Sequence Information Forecasts of PPIs have been executed by amalgamating established interaction data with details about sequence homology. The procedure for sequence-based prediction revolves around conveying annotations from functionally characterized protein sequences to the target sequence, hinging on their resemblances. Annotation grounded in similarity rests on the homologous disposition of the inquiry protein in annotated protein databases, facilitated by pairwise local sequence algorithms [19]. In the organism being investigated, numerous proteins might exhibit substantial parallels with proteins engaged in intricate associations within other organisms.

The prediction process is initiated by juxtaposing a test gene or protein with those annotated proteins present in different species. If the test gene or protein demonstrates noteworthy similarity with the sequence of a gene or protein possessing known functionality in another species, it is inferred that the test gene or protein likely shares either an equivalent function or akin attributes [20].

Sequence alignment involves arranging two or more sequences in such a way that their homologous positions are matched and aligned. The goal is to identify regions of similarity or conservation and detect differences or variations between sequences. By aligning sequences, researchers can gain insights into the evolutionary relationships, functional motifs, and conserved regions within biological sequences.

Sequence alignment algorithms can employ various scoring schemes and substitution matrices to assign scores to matches, mismatches, and gaps [21]. Additionally, gap penalties can be adjusted to control the extent of gaps allowed in the alignment.

Sequence alignment has numerous applications in bioinformatics and computational biology. Some notable applications include:

Phylogenetic Analysis: Sequence alignment is crucial for inferring evolutionary relationships between species or genes. By aligning homologous sequences from different organisms, researchers can construct phylogenetic trees that depict the evolutionary history and relatedness of species [22]. This information is valuable for understanding evolutionary processes, studying genetic diversity, and identifying common ancestors.

Protein Structure Prediction: Protein sequence alignment is used in comparative modeling and fold recognition to predict the three-dimensional structure of proteins [23]. By aligning the target protein sequence with known protein structures, researchers can infer the structural characteristics and potential functions of the target protein.

Functional Annotation: Sequence alignment helps in annotating functional elements in DNA, RNA, and protein sequences. By comparing a query sequence to a database of known sequences, researchers can identify conserved motifs, protein domains, or regulatory regions that contribute to specific functions [24]. This information is crucial for understanding gene function, identifying potential drug targets, and designing experiments.

Variant Calling: In genomic studies, sequence alignment is used to identify genetic variations, such as single nucleotide polymorphisms (SNPs) or insertions/deletions (indels), by aligning sequencing reads to a reference genome [25]. Accurate alignment allows for the identification and characterization of genetic variations associated with diseases, population genetics, and personalized medicine.

Mutations Prediction: detect whether the affected person (patient) has mutations or not by using sequence alignment by alignment for two DNA sequences (mutant and non-mutant sequences) [26].

4 Materials and Methods

PPIs play a crucial role in various biological processes and are essential for the functioning of cells. Understanding these interactions is important for elucidating the mechanisms of cellular processes, identifying potential drug targets, and designing therapeutic interventions. One approach to identifying PPIs is through sequence alignment.

Sequence alignment is a computational technique used to identify similarities and differences between two or more protein sequences. In the context of studying protein-protein interactions, sequence alignment can help identify and characterize proteins that are related to a specific disease, such as COVID-19.

The following steps are involved in this process to find COVID-19 related proteins using sequence alignment:

1. Download the SARS-CoV-2 reference genome sequence in FASTA format from the National Center for Biotechnology Information (NCBI) website.
2. Use the Biopython library to read the reference genome sequence into memory.
3. Translate genome sequence into amino acid sequence using the translate function.
4. Use the BLAST (Basic Local Alignment Search Tool) algorithm to search for similar protein sequences in the NCBI non-redundant protein database.
5. Identify the top hits from the BLAST search results, which are the protein sequences with the highest similarity scores to the query sequence.
6. Analyze the protein sequences using various bioinformatics tools and databases to determine their potential functions and interactions.

Overall, this process involves identifying potential proteins encoded by the SARS-CoV-2 genome, comparing them to known protein sequences in public databases, and analyzing their potential functions and interactions. This information can provide insights into the biology of the virus and potential targets for therapeutic interventions.

It is worth noting that finding proteins related to COVID-19 using sequence alignments does not directly count as finding a PPI, which refers to the physical interaction between two or more proteins, where they bind to each other to carry out specific biological functions.

While sequence alignments provide valuable insights into protein similarities and functional regions, they do not directly confirm physical interactions between proteins. Also finding proteins related to COVID-19 through sequence alignments is an important step in understanding the functional characteristics of the virus and its proteins. To identify PPIs, the process need experimental techniques or other protein interaction assays.

4.1 COVID-19: Genetic Information

The first step is to make sequence alignment by reading the DNA sequence for COVID-19. COVID-19, also known as SARS-CoV-2, is a novel coronavirus that caused a global pandemic in 2019–2020. The genome of the virus causing Covid-19 consists of 29903 nucleotides. It encodes several structural and non-structural proteins that are essential for viral replication and infection. The genetic information of COVID-19 was obtained from the NCBI GenBank public database.

As shown above, the COVID-19 genome has 29903 genetic letters (sequence of A T C and G). The information in DNA is stored as a code made up of four chemical bases: adenine A, guanine G, cytosine C, and thymine T. The order, or sequence, of these bases, determines the information available for building and maintaining an organism, similar to how letters of the alphabet appear in a certain order to form words and sentences.

Figure 1 shows COVID-19's first 500 genetic letters (out of 29903 letters). And Fig. 2 shows the distribution of the nucleotides (A, T, C, G) over the COVID-19's DNA.

```
Sequence length: 29903 nucleotides
'ATTAAAGGTTTATACCTTCCCAGGTAACAAACCAACCAACTTTCGATCTCTTGTAGATCTGTTCTCTAAACGAACTTTAA
AATCTGTGTGGCTGTCACTCGGCTGCATGCTTAGTGCACTCACGCAGTATAATTAATAACTAATTACTGTCGTTGACAGGA
CACGAGTAACTCGTCTATCTTCTGCAGGCTGCTTACGGTTTCGTCCGTGTTGCAGCCGATCATCAGCACATCTAGGTTTCG
TCCGGGTGTGACCGAAAGGTAAGATGGAGAGCCTTGTCCCTGGTTTCAACGAGAAAACACACGTCCAACTCAGTTTGCCTG
TTTTACAGGTTCGCGACGTGCTCGTACGTGGCTTTGGAGACTCCGTGGAGGAGGTCTTATCAGAGGCACGTCAACATCTTA
AAGATGGCACTTGTGGCTTAGTAGAAGTTGAAAAAGGCGTTTTGCCTCAACTTGAACAGCCCTATGTGTTCATCAAACGTT
CGGATGCTCGAACTG'
```

Fig. 1. COVID-19's DNA

4.2 Transcription and Translation

For us, to find proteins related to COVID-19, DNA must be converted into a protein, and this is done through transcription and translation. Transcription and translation, are

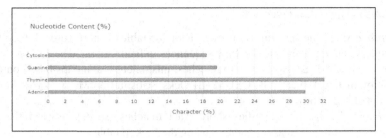

Fig. 2. Covid19's DNA

fundamental steps in converting the genetic information from the COVID-19 genome into functional proteins as shown in Fig. 3.

1. Transcription is the first step in gene expression. It involves the synthesis of messenger RNA (mRNA) molecules from a DNA template.
2. Translation is the process of protein synthesis, where the mRNA transcript is used as a template to synthesize proteins.

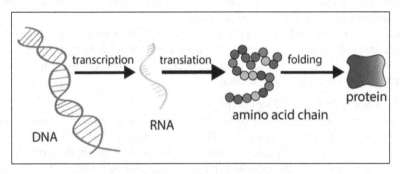

Fig. 3. Transcription and Translation functions

4.3 BLAST Algorithm

BLAST (Basic Local Alignment Search Tool) is a widely used algorithm for sequence alignment and similarity searching. BLAST rapidly identifies regions of similarity between a query sequence and a database of known sequences, enabling researchers to infer evolutionary relationships, identify homologous genes, and annotate functional elements in genomic sequences.

The BLAST algorithm employs several key techniques, including the construction of a lookup table of short words (k-mers or k-means) from the query sequence, the calculation of local alignment scores using a substitution matrix, and the use of statistical measures to assess the significance of sequence similarity. BLAST also utilizes optimizations such as indexing and database organization to improve search speed and efficiency.

BLAST Algorithm Overview:

a. *Preprocessing*: The algorithm constructs a lookup table of short words (k-mers) from the query sequence and the database sequences to facilitate rapid searching.
b. *Scoring*: BLAST assigns scores to matches, mismatches, and gaps based on a substitution matrix, typically BLOSUM (BLOcks SUbstitution Matrix) or PAM (Point Accepted Mutation).
c. *Seed Search*: BLAST identifies short exact matches (seeds) between the query sequence and the database sequences using the lookup table.
d. *Extension*: The algorithm extends the seeds by applying a dynamic programming approach to calculate local alignment scores and identify high-scoring segment pairs (HSPs).
e. *Scoring Statistics*: BLAST uses statistical measures, such as the E-value, to assess the significance of the sequence similarity.
f. *Output*: The algorithm reports the identified HSPs and generates an alignment between the query sequence and the matching database sequence.

BLAST Scoring:

a. *Match/Mismatch Score*: BLAST assigns scores to matches and mismatches using a substitution matrix, such as BLOSUM or PAM. The score represents the similarity or dissimilarity between two aligned residues.
b. *Gap Penalties*: BLAST employs gap penalties for introducing gaps (insertions or deletions) in the alignment. It assigns separate gap opening and extension penalties to control the gap lengths. These penalties penalize the introduction of gaps; as longer gaps are less likely than shorter ones.

Extension and Dynamic Programming:
BLAST uses a variant of the dynamic programming algorithm to extend the initial seeds (exact matches) and identify HSPs. It calculates local alignment scores by iteratively extending the alignment in both directions (left and right) from the seed. The algorithm optimizes the extension process to maximize speed while maintaining sensitivity.

Scoring Statistics:
BLAST provides statistical measures to assess the significance of sequence similarity. One commonly used statistic is the E-value, which estimates the expected number of false positive alignments by chance. A lower E-value indicates higher significance. The E-value is calculated based on the alignment score, database size, and composition of the database.

Equations:

a. *Alignment Score (S)*: The alignment score is the sum of the match/mismatch scores and the gap penalties for the aligned residues in the query and database sequences.
b. *E-value*: The E-value estimates the expected number of alignments with a score equal to or better than the observed score, purely by chance. It is calculated based on the alignment score (S), the effective database size (m), and a scaling parameter (K) using the formula:

$$E - value = K * m * n * exp(-lambda * S)$$

Here, n is the query sequence length, lambda is the scaling parameter, and K is a constant that depends on the scoring system and database size.

BLAST compares a query sequence with a database of known sequences to identify homologous regions and calculate sequence similarity scores. In the context of COVID-19, the BLAST algorithm can be applied to analyze the genetic information of the virus and compare it with existing sequence databases. By using BLAST, we can identify similar or closely related viral strains, and determine genetic variations. The algorithm calculates a sequence similarity score based on the alignment, allowing it to prioritize matches that have higher similarity scores.

The following example demonstrates the steps of the BLAST algorithm. In practice, BLAST operates on larger sequences and databases, employing more sophisticated scoring schemes, statistical measures, and optimizations to enhance performance.

Assume we have a query DNA sequence "ACGTGTC" and a database with two DNA sequences: "ACGTC" and "ATGCT".

Preprocessing: Construct a lookup table of k-mers (e.g., k = 3) from the query sequence and the database sequences. For example: Query sequence: ACGTGTC Database sequences: ACGTC, ATGCT Query k-mers: ACG, CGT, GTG, TGT, GTC Database k-mers: ACG, CGT, GTC, ATG, TGC, GCT.

Seed Search: Identify exact matches (seeds) between the query k-mers and the database k-mers. In this case, both "ACG" and "CGT" are exact matches found in the query and database sequences.

Extension: Use dynamic programming to extend the seeds and identify high-scoring segment pairs (HSPs). We'll focus on the alignment of "ACG" from the query with "ACG" from the database.

Alignment: Query: ACG--- Database: ACG---

Scoring: Match/Mismatch score= +1 (for matching bases), Gap penalty= -1 (for introducing gaps)

Alignment score: 3 (3 matches)

E-value: Calculated based on the alignment score, database size, and other factors.

Output: The BLAST algorithm reports the identified HSPs and generates an alignment.

In this case, the alignment score is 3, indicating a perfect match between the query and the database sequence. Algorithm (1) shown a flowchart for sequence alignment.

Algorithm 1: Sequence Alignment

Input: Two sequences Seq1 and Seq2

Output: Bitscore.

Begin

1. Step1: read Seq1 and Seq2

2. Step2:

    2.1: Let M = size of Seq1 and N = size of Seq2

    2.2: let Cell [,] Matrix = new Cell [N, M];

    2.3 let Gap = 1; Similarity = -2;

    2.3: for i=0 to M

       Matrix[0, i] = new Cell(0, i, i*Gap)

       End for

    2.4: for i=0 to N

       Matrix[i, 0] = new Cell(i, 0, i*Gap)

       End for

    2.5: for j=1 to N

       for i=1 to M

          Matrix[j, i] = Max(Diagonal, Left, Up)

       End for

       End for

3. Step 3:

    3.1: Cell CurrentCell = Matrix[Sq2.Length - 1, Sq1.Length - 1];

    3.2: while (CurrentCell.CellPointer != null)

         3.2.1: if (CurrentCell.Type == Diagonal)

            Seq1.Add(Sq1[CurrentCell.CellColumn]);

            Seq2.Add(Sq2[CurrentCell.CellRow]);

         3.2.2: if (CurrentCell.Type == Left)

            Seq1.Add(Sq1[CurrentCell.CellColumn]);

            Seq2.Add('-');

         3.2.3: if (CurrentCell.Type == Above)

            Seq1.Add('-');

            Seq2.Add(Sq2[CurrentCell.CellRow]);

    3.3: CurrentCell = CurrentCell.CellPointer;

    3.4 End while

End

---

5 Results

The paper focused on utilizing the Basic Local Alignment Search Tool (BLAST) algorithm to uncover insights into the proteins associated with COVID-19. Through the alignment of COVID-19 protein sequences with annotated protein databases, significant findings related to the functions of viral proteins, potential interactions with host proteins, and evolutionary connections with other coronaviruses were revealed as shown in Table 1.

The BLAST algorithm revealed notable matches between COVID-19 protein sequences and annotated proteins, providing information about the identity and potential roles of these proteins. The bitscore values indicate the significance of alignment matches, and the alignment column showcases a portion of the alignment between the COVID-19 protein sequences and the annotated proteins. Figure 4 displays alignment details for the top-scoring match, pdb|7MSW|A, showcasing the aligned regions and sequence similarities. Figure 5 illustrates alignment details for the second-highest scoring match, pdb|7FAC|A, highlighting aligned regions and sequence similarities.

Figures 4 and 5 provide more comprehensive insights into the two highest-scoring alignment results. These figures present detailed alignment information, emphasizing the regions of alignment and the sequence similarities between COVID-19 proteins and annotated proteins. The alignment patterns depicted in these figures offer visual confirmation of the sequence alignment results and contribute to a deeper understanding of the potential functional relationships between COVID-19 proteins and known proteins.

These figures serve as valuable visual aids that enhance the interpretation of the BLAST alignment results, aiding researchers in discerning the significance and implications of the identified protein matches in the context of COVID-19 pathogenesis and treatment.

The selection of the BLAST (Basic Local Alignment Search Tool) technique for this study is justified by its wide applicability to diverse biological data, computational efficiency in handling large datasets of COVID-19 proteins, and a balanced approach between sensitivity and specificity. BLAST's user-friendly interface accommodates researchers with varying levels of bioinformatics expertise, while its integration with extensive databases facilitates the accurate identification and annotation of COVID-19 proteins by comparing them to known sequences. The established reputation, community support, and previous successes in similar studies contribute to BLAST's reliability and make it a trusted tool for the specific objectives of this research, emphasizing its suitability for rapid, accurate, and comprehensive sequence alignment in the context of COVID-19 protein analysis.

Limitations of Sequence Alignment and BLAST Algorithm.

1. *Sensitivity to Sequence Variability:* sequence alignment methods, including BLAST, exhibit sensitivity to the variability in protein sequences. Highly divergent or rapidly evolving sequences may pose challenges, potentially compromising the accuracy of protein-protein interaction predictions.
2. *Incomplete Sequence Information:* the success of sequence alignment relies on the availability and completeness of sequence data. Incomplete or fragmented sequences can impede the reliability of identifying interactions, introducing a limitation based on the quality of the available data.
3. *Homologous Proteins and Functional Divergence:* homologous proteins may present challenges, as sequence similarity does not always correlate with functional similarity. Functional divergence in evolutionary processes may complicate the accurate prediction of protein-protein interactions solely based on sequence alignment.

Table 1. BLAST Results for COVID-19 Protein Identification

id	bitscore	alignment
pdb|7MSW|A	1328.540	((A, Y, T, R, Y, V, D, N, N, F, C, G, P, D, G,...
pdb|7FAC|A	746.503	((K, L, D, G, F, M, G, R, I, R, S, V, Y, P, V,...
pdb|6WUU|A	674.855	((L, R, E, V, R, T, I, K, V, F, T, T, V, D, N,...
pdb|7CMD|A	674.470	((E, V, R, T, I, K, V, F, T, T, V, D, N, I, N,...
pdb|7CJD|A	671.389	((E, V, R, T, I, K, V, F, T, T, V, D, N, I, N,...
pdb|6XAA|A	670.618	((R, E, V, R, T, I, K, V, F, T, T, V, D, N, I,...
pdb|6XA9|A	669.463	((R, E, V, R, T, I, K, V, F, T, T, V, D, N, I,...
pdb|7D47|A	669.078	((E, V, R, T, I, K, V, F, T, T, V, D, N, I, N,...
pdb|6W9C|A	668.692	((E, V, R, T, I, K, V, F, T, T, V, D, N, I, N,...
pdb|7NT4|A	668.307	((E, V, R, T, I, K, V, F, T, T, V, D, N, I, N,...
pdb|6WZU|A	668.307	((E, V, R, T, I, K, V, F, T, T, V, D, N, I, N,...
pdb|7NFV|AAA	667.922	((E, V, R, T, I, K, V, F, T, T, V, D, N, I, N,...
pdb|7LBR|A	667.922	((E, V, R, T, I, K, V, F, T, T, V, D, N, I, N,...
pdb|7JRN|A	667.922	((E, V, R, T, I, K, V, F, T, T, V, D, N, I, N,...
pdb|8CX9|A	667.152	((E, V, R, T, I, K, V, F, T, T, V, D, N, I, N,...
pdb|7D6H|A	666.766	((R, E, V, R, T, I, K, V, F, T, T, V, D, N, I,...
pdb|6YVA|A	665.226	((E, V, R, T, I, K, V, F, T, T, V, D, N, I, N,...
pdb|7CJM|B	665.226	((E, V, R, T, I, K, V, F, T, T, V, D, N, I, N,...
pdb|6WRH|A	665.226	((E, V, R, T, I, K, V, F, T, T, V, D, N, I, N,...
pdb|7QCG|A	664.070	((E, V, R, T, I, K, V, F, T, T, V, D, N, I, N,...
pdb|7UV5|A	663.300	((E, V, R, T, I, K, V, F, T, T, V, D, N, I, N,...
pdb|8E4J|A	659.062	((S, I, T, S, A, V, L, Q, S, G, F, R, K, M, A,...
pdb|7D7K|A	659.062	((T, I, K, V, F, T, T, V, D, N, I, N, L, H, T,...
pdb|7N6N|A	658.292	((S, A, V, L, Q, S, G, F, R, K, M, A, F, P, S,...
pdb|7KFI|A	652.514	((S, G, F, R, K, M, A, F, P, S, G, K, V, E, G,...
pdb|7VTH|A	652.129	((S, G, F, R, K, M, A, F, P, S, G, K, V, E, G,...
pdb|7W9G|A	652.129	((S, G, F, R, K, M, A, F, P, S, G, K, V, E, G,...
pdb|6XA4|A	652.129	((S, G, F, R, K, M, A, F, P, S, G, K, V, E, G,...
pdb|7VU6|A	652.129	((S, G, F, R, K, M, A, F, P, S, G, K, V, E, G,...
pdb|7CB7|A	652.129	((S, G, F, R, K, M, A, F, P, S, G, K, V, E, G,...
pdb|7CWC|A	652.129	((S, G, F, R, K, M, A, F, P, S, G, K, V, E, G,...
pdb|7BRO|A	652.129	((S, G, F, R, K, M, A, F, P, S, G, K, V, E, G,...

(continued)

Table 1. (*continued*)

id	bitscore	alignment
pdb|5R7Y|A	652.129	((S, G, F, R, K, M, A, F, P, S, G, K, V, E, G,...
pdb|7CBT|A	651.358	((S, G, F, R, K, M, A, F, P, S, G, K, V, E, G,...
pdb|7MPB|A	651.358	((S, G, F, R, K, M, A, F, P, S, G, K, V, E, G,...
pdb|7T2U|A	651.358	((L, Q, S, G, F, R, K, M, A, F, P, S, G, K, V,...
pdb|8EYJ|A	650.973	((Q, S, G, F, R, K, M, A, F, P, S, G, K, V, E,...
pdb|7ZB7|A	650.973	((S, G, F, R, K, M, A, F, P, S, G, K, V, E, G,...
pdb|7U29|A	650.973	((S, G, F, R, K, M, A, F, P, S, G, K, V, E, G,...
pdb|8D4L|A	650.973	((S, G, F, R, K, M, A, F, P, S, G, K, V, E, G,...
pdb|8D4N|A	650.973	((S, G, F, R, K, M, A, F, P, S, G, K, V, E, G,...
pdb|8E1Y|A	650.588	((S, G, F, R, K, M, A, F, P, S, G, K, V, E, G,...
pdb|8DZ6|A	650.588	((S, G, F, R, K, M, A, F, P, S, G, K, V, E, G,...
pdb|8DZA|A	650.588	((S, G, F, R, K, M, A, F, P, S, G, K, V, E, G,...
pdb|8DDM|A	650.203	((S, G, F, R, K, M, A, F, P, S, G, K, V, E, G,...
pdb|8DZ1|A	650.203	((S, G, F, R, K, M, A, F, P, S, G, K, V, E, G,...
pdb|6XMK|A	650.203	((L, Q, S, G, F, R, K, M, A, F, P, S, G, K, V,...
pdb|8E26|A	650.203	((S, G, F, R, K, M, A, F, P, S, G, K, V, E, G,...
pdb|7RVM|A	650.203	((S, G, F, R, K, M, A, F, P, S, G, K, V, E, G,...
pdb|7ZB8|A	649.818	((S, G, F, R, K, M, A, F, P, S, G, K, V, E, G,...

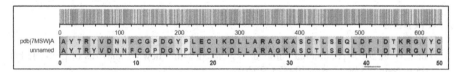

Fig. 4. Alignment Details for Highest Scoring Match (pdb\|7MSW\|A)

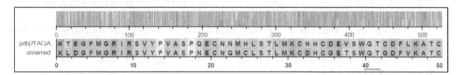

Fig. 5. Alignment Details for Second Highest Scoring Match (pdb\|7FAC\|A)

4. *Alternative Approaches and Integration:* while sequence alignment is a valuable tool, it should be perceived as part of a broader strategy for protein-protein interaction prediction. Integrating other computational methods or experimental data becomes

imperative to complement the limitations of sequence alignment and enhance the overall reliability of predictions.

6 Conclusions

In light of the global outbreak of COVID-19 caused by the severe acute respiratory syndrome coronavirus 2 (SARS-CoV-2), understanding the molecular structure and function of the virus has become paramount. This paper focused on the identification and analysis of COVID-19 constituent proteins, a critical aspect in unraveling the intricate biology of SARS-CoV-2. Protein-protein interactions (PPIs) were explored as they play a vital role in cellular processes. The approach of using sequence alignment, specifically the Basic Local Alignment Search Tool (BLAST), proved effective in shedding light on COVID-19 protein characteristics.

Through the application of BLAST, the alignment of COVID-19 protein sequences with annotated databases revealed significant insights into the functions of viral proteins, their potential interactions with host proteins, and their evolutionary connections with other coronaviruses. The alignment results were presented in a comprehensive table, showcasing the highest-scoring matches. Moreover, Figs. 4 and 5 provided visual representations of the alignment details for the two top-scoring matches, enhancing the clarity and depth of our findings.

The methodologies and findings outlined in this paper provide a foundation for further research in the field of COVID-19 protein characterization. By leveraging sequence alignment tools like BLAST, researchers can gain a deeper understanding of the intricate protein interactions associated with the virus. These insights hold potential implications for the development of targeted therapies and diagnostic tools, addressing the urgent need to combat the global pandemic.

References

1. Zhou, P., et al.: A pneumonia outbreak associated with a new coronavirus of probable bat origin. Nature **579**(7798), 270–273 (2020)
2. Gordon, D.E., et al.: A SARS-CoV-2 protein interaction map reveals targets for drug repurposing. Nature **583**(7816), 459–468 (2020)
3. Smith, T.F., Waterman, M.S.: Identification of common molecular subsequences. J. Mol. Biol. **147**(1), 195–197 (1981)
4. Altschul, S.F., et al.: Basic local alignment search tool. J. Mol. Biol. **215**(3), 403–410 (1990)
5. Eddy, S.R.: Accelerated Profile HMM Searches. PLoS Comput. Biol. **7**(10), e1002195 (2011)
6. Wrapp, D., et al.: Cryo-EM structure of the 2019-nCoV spike in the prefusion conformation. Science **367**(6483), 1260–1263 (2020)
7. Jin, Z., et al.: Structure of Mpro from SARS-CoV-2 and discovery of its inhibitors. Nature **582**(7811), 289–293 (2020)
8. Elfiky, A.A.: Anti-HCV, nucleotide inhibitors, repurposing against COVID-19. Life Sci. **248**, 117477 (2020)
9. Gysi, D.M., et al.: Network Medicine Framework for Identifying Drug Repurposing Opportunities for COVID-19. (2020) ArXiv, arXiv:2004.07229

10. Abdul Raheem, K., Ali, D., Ban, N.: Comprehensive Review on Drug-target Interaction Prediction - Latest Developments and Overview, Current Drug Discovery Technologies 21 (2023). https://doi.org/10.2174/1570163820666230901160043

11. Polack, F.P., et al.: Safety and Efficacy of the BNT162b2 mRNA Covid-19 Vaccine. N. Engl. J. Med. 383(27), 2603–2615 (2020)

12. Orchard, S., et al.: The MIntAct project—IntAct as a common curation platform for 11 molecular interaction databases. Nucleic Acids Res. 47(D1), D81–D85 (2019). https://doi.org/10.1093/nar/gky1128

13. Luck, K., et al.: A reference map of the human binary protein interactome. Nature 580(7803), 402–408 (2020). https://doi.org/10.1038/s41586-020-2188-x

14. Rolland, T., et al.: A proteome-scale map of the human interactome network. Cell 159(5), 1212–1226 (2014). https://doi.org/10.1016/j.cell.2014.10.050

15. Yanagida , M.: Functional proteomics; current achievements. Journal of Chromatography B. 771(1–2), :89–106. [PubMed] [Google Scholar] (2002)

16. Berggård, T., Linse, S., James, P.: Methods for the detection and analysis of protein-protein interactions. Proteomics. 7(16), 2833–2842. [PubMed] [Google Scholar] (2007)

17. von Mering, C., Krause, R., Snel, B., et al.: Comparative assessment of large-scale data sets of protein-protein interactions. Nature; 417(6887), 399–403. [PubMed] [Google Scholar] (2002)

18. Zhang, A.: Protein Interaction Networks-Computational Analysis. New York, NY, USA: Cambridge University Press [Google Scholar] (2009)

19. Lee, S.-A, Chan, C.-H, Tsai, C.H., et al.: Ortholog-based protein-protein interaction prediction and its application to inter-species interactions. BMC Bioinformatics; 9(supplement 12, article S11) [PMC free article] [PubMed] [Google Scholar] (2008)

20. Tatusov, R.L, Koonin, E.V, Lipman, D.J.: A genomic perspective on protein families. Science. 278(5338):631–637 (1997). [PubMed] [Google Scholar]

21. Henikoff, S., Henikoff, J.G.: Amino acid substitution matrices from protein blocks. Proc. Natl. Acad. Sci. 89(22), 10915–10919 (1992). https://doi.org/10.1073/pnas.89.22.10915

22. Hug, L.A., et al.: A new view of the tree of life. Nat. Microbiol. 1, 16048 (2016). https://doi.org/10.1038/nmicrobiol.2016.48

23. Yang, J., Roy, A., Zhang, Y.: BioLiP: A semi-manually curated database for biologically relevant ligand-protein interactions. Nucleic Acids Res. 41(D1), D1096–D1103 (2013). https://doi.org/10.1093/nar/gks966

24. The UniProt Consortium: UniProt: A worldwide hub of protein knowledge. Nucleic Acids Res. 47(D1), D506–D515 (2019). https://doi.org/10.1093/nar/gky1049

25. Poplin, R., Ruano-Rubio, V., DePristo, M. A., Fennell, T. J., Carneiro, M. O., Banks, E.: Scaling accurate genetic variant discovery to tens of thousands of samples. bioRxiv, 201178 (2018). https://doi.org/10.1101/201178

26. Abdul, A., Dhannoon, B.N.: Predication and classification of cancer using sequence alignment and back propagation algorithms in brca1 and brca2 genes. Int. J. Pharma. Res. 11, 1 (2019). https://doi.org/10.31838/ijpr/2019.11.01.062

Vehicle Logo Recognition Using Proposed Illumination Compensation and Six Local Moments

Nada Najeel Kamal[1]([⊠]) [iD], Loay Edwar George[2,3] [iD], and Zainab A. Yakoob[1] [iD]

[1] Computer Sciences Department, University of Technology- Iraq, Baghdad, Iraq
{Nada.N.Kamal,Zainab.A.Yakoob}@uotechnology.edu.iq
[2] University of Baghdad, Baghdad, Iraq
loayedwar57@uoitc.edu.iq, loay.george@sc.uobaghdad.edu.iq
[3] University of Information Technology and Communications, Baghdad, Iraq

Abstract. A logo is a significant sign of a vehicle, and it can be considered one of its identification characteristics which cannot be easily changed. Logo identification has many applications, especially in intelligent transportation systems. This paper presents a new method to identify the manufacturer of a vehicle despite challenging images of logos that differ in size, rotation angle and brightness conditions. Three major steps are made to accomplish the recognition task. Preprocessing includes converting images to a grayscale level, then the proposed local illumination compensation is applied. For the feature extraction stage, six local spatial moments are extracted from the sub images of the logo. Finally, a classification process is done using distance measurements. Different tests are made to measure the accuracy rates of the proposed recognition system and to determine the effect of every single feature on the recognition rate, as well as to understand the effect of the proposed local illumination compensation. A dataset containing 544 images for 34 different vehicle logo classes was used, in which each class contains 16 samples. The results show that the success rate reached 86.94853% when all test samples were used in a comprehensive test.

Keywords: Logo Recognition · Feature Extraction · Local illumination · Spatial Moments · Classification · VLR

1 Introduction

As it is well known that automated transportation systems play an important role in modern society. Also, it is well known that computer vision and intelligent applications are considered the backbone of such systems. Vehicle license plate recognition (LPR), Vehicle logo recognition (VLR), vehicle colour recognition, and vehicle model recognition are basic techniques for gathering vehicle information [1, 2]. A License plate recognition system (LPR) is nowadays a very common method that is used to identify the vehicle, but a license plate can be easily removed [3], also factors like staining, color removal, plate position, and plate tampering may affect the efficiency of LPR system. In

A. M. Al-Bakry et al. (Eds.): NTICT 2023, CCIS 2096, pp. 164–178, 2024.
https://doi.org/10.1007/978-3-031-62814-6_12

recent years, for those reasons vehicle logo recognition systems are getting more attention, because the vehicle logo, which represents a key symbol of the car manufacturer that is difficult to counterfeit [4–6].

VLR framework basically includes two levels: detection level to detect the logo position in the vehicle image, and classification level to identify the logo company. In general there are few literature that tackles the VLR issue, and the main challenge in creating VLR systems is that logos range in size, shape, colour, brightness and skewness angle, and there is a huge intra-class variety accompanied with external complicated environment [7–9].

Different methods were used to identify logos Vehicle Manufacturer Recognition [VMR] [2, 10] or in other words using VLR. For example, the traditional method that trains a detector by providing features like feature points, edges, and invariant moments [11] using classification methods like HOG [12] and SIFT and SVM as the classifier [13] Also, deep learning-based methods: with the rapid development, many deep learning methods have been used successfully in logo classification [14, 15], many of the used method are discussed in Sect. 2.

Despite challenging differences in size, rotation angle and brightness conditions the main goal of this work is to create a reliable system that recognizes vehicle logo images using six local moments: local mass, Moment X, Moment Y, Moment X2, Moment Y2, and Moment XY. Another essential contribution is to create a local illumination compensation that overcomes the brightness problems in logo images and leads to higher recognition success rates gained from different conducted extensive experiments. As logo recognition helps in vehicle management and traffic monitoring [16], the importance of the proposed work is that it can be integrated with LPR systems to identify vehicles on roads and create a more secure intelligent transportation system.

The rest of this of this paper is organized as follows: Sect. 2 includes a review for a collection of recent literatures. Section 3 clarifies the used dataset, while Sect. 4 presents the layout of the system. Section 5 includes preprocessing stage. Feature extraction, classification, and experimental results can be seen in Sects. 6, 7 and 8 respectively.

2 Literature Review

This section reviews several literatures that tackle the issue of vehicle logo recognition in the last decade.

Zhao and Wang produce logo recognition using different methods. They used Hu invariant and support vector machine for classification and applied the system to 9 kinds of vehicle. They got 92% rate for recognition, but it works better only with low illumination conditions. Also, they found that Cross-validation is weaker than Gray Wolf Optimize to get the recognized logo [1].

Cyganek and Woźniak [17] use the multi-linear tensor classifier to create a VLR, in which not enough documentation about the gained rate of 92% with about 12 classes. Sam and Tian [18] classify 200 logo image using Tchebichef Moments after locating it using machine learning based on AdaBoost and normalized its shape, their highest success rates reached to 92%.

Yu Y. et al. [4] applied overlapping on images and then extracted features from each block using overlapping enhanced patterns of oriented edge magnitudes to improve the quality of each block. HFUT-VL data set used which has 80 vehicle classes 200 images for each with size 64*64 took in various weather conditions. Their paper got the best rate of 97.3% using various overlap rates. Ansari I., et al. [19] used faster region-based convolutional neural network and preprocessing in logo recognition and they tried to recognize logo without preprocessing and found that the accuracy rate is better if the image quality is high. The paper concentrates on street CCTV camera which provides high-quality images. The system tested in day and night images but they found that the low quality of night images affects the recognition rate.

Xiao K. and Pengqiang D. [20] in their paper used YOLO and Faster R-CNN frameworks to recover vehicle logo, and they used for the detection of Non maximum Suppression method to enhance results. They concluded that their method works better in one stage method that immediately finds the logo image and the performance of YOLO is stronger than Faster R-CNN. They use data set with small size which limit their training in the complex scene.

Soon, F.C. et al. [21] produced a paper to train the Convolutional Neural Network (CNN) via searching and finding the best architecture for the CNN. Depending on the stochastic technique of particle swarm optimization gaining the best hyper-parameters to train and test the data to get the best network convergence and classification performance. They used 14,950 logo images both for training and testing. Segmentation for each image was performed and a multiclass Softmax classifier for classification. In this paper, no obvious processing time is explained because of using real-time during execution. They get 99% accuracy rate on 13 vehicle manufactures.

Kadhm and Yun [22] proposed VLR by using OpenCV library for the detection and extraction of images with size 800*600. These images are converted to grayscale image level after that converted it to binary image depending on a specific threshold taken by special formula. After that apply Gaussian Blur to resize the image's size into half of its size. For the detection logo Mustafa and Lim used Cany or Sobel algorithms. The accuracy rate was high but there is no obvious accuracy rate number mentioned in the paper.

Qiang and Wenhao [23], improved YOLOF to detect a logo in difficult lighting conditions. Depending on the essential of gamma transform to enhance the images in complex lighting by converting images to HSV colour images and using the V channel to adjust the lighting, then detecting images using the darknet method which works on a multi-feature recognition network. The resolution of images is $128 \times 128 \times 3$. Each experiment is trained for 100 cycles, a group size of 64, a weight equal to 0.0001, and a learning rate of 0.001. The backbone used for the experiment is resnet 50. Finally, this algorithm applying an optimization method on the gained features from the detected images with high data redundancy. They used Chinese Traffic Sign Detection Benchmark (CCTSDB) dataset, 70% of it as a training set and 10%, 20% as a testing set separately as a result they earned 95.86% rate.

Almost all other works normalize the size of the logo images, while our work deals with different sizes of images. Also, our work deals with a proposed brightness algorithm that works on local areas of the logo image, rather than standard ones which deals with the global image area.

3 The Used Database Description

The used dataset consists of 34 classes of vehicle logo images, each class has 16 samples with a total of 544 images. The samples differ in illumination conditions, rotation angle, and size which ranges from 52 to 670 pixels for the width, while it comes between 32 and 1081 pixels for the height. See Fig. 1. The format of the images is PNG (Portable Network Graphics), and because the BMP (Bitmap) image is the common uncompressed format that is usually used in image processing and machine vision applications, the image samples were converted to BMP using Ashampoo Photo Commander application. More details about the used dataset are described in [24].

Fig. 1. Samples of the used database

4 Layout of the Proposed System

A general layout for the steps of the system can be seen in Fig. 2. Where the necessary steps of training the system are included, and the steps of identifying a logo are shown.

5 Preprocessing Stage

As it is shown in Fig. 1, the used logo images come in deferent variation of size, illumination state and rotation angle, so this intraclass variation will affect the recognition rates. Preprocessing stage is a necessary step to be applied to get better recognition rates. The proposed system includes two main preprocessing steps: the first one is getting the gray images, while the second one is applying of local illumination compensation which leads to raise success recognition rates. The two mentioned steps are explained in details in the following subsections:

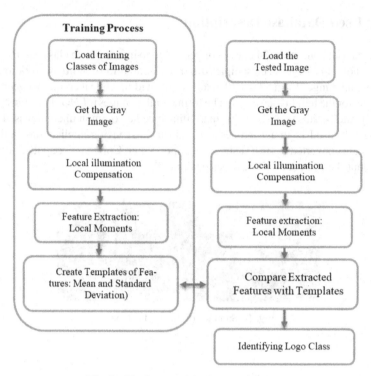

Fig. 2. The layout of the Proposed System

5.1 Converting Images to Grayscale Level

The non-compressed format of Bitmap (BMP) images is a widely used in pattern recognition applications. The true coloured image of this format with 24 bits per pixel; represents a lot of information to be processed at one time by any system, so the three bands of this format: the red band, the green band and the blue band are converted to one band of a gray-scaled image to minimize the amount of information processed by the system as shown in Eq. (1).

$$Gray(x, y) = (Red(x, y) + Green(x, y) + Blue(x, y))/3 \qquad (1)$$

Fig. 3. Original true-coloured vehicle logo images

In Eq. (1). $Gray_{()}$ is gained as the sum of the bands $Red_{()}$, $Green_{()}$ and $Blue_{()}$ divided by their number which is 3, at each pixel location (x, y). See Fig. 3. And Fig. 4.

Fig. 4. Grayscale vehicle logo images

5.2 Local Illumination Compensation

The gray-scaled images shown in Fig. 4 demonstrate different brightness levels that will affect the feature extraction stage negatively. To unify the brightness of images the local illumination compensation method is used. This step is made by subdividing the gray-scaled vehicle logo images into 5x5 areas, and for each of these areas Eq. (2) is applied:

$$Gray_{(x,y)} = 40 * \frac{Gray_{(x,y)} - LocalMean_{(x,y)}}{Localstd_{(x,y)}} + 128 \qquad (2)$$

where 40 is a threshold value that controls the strength of the illumination compensation process, $Gray_{(x,y)}$ is the gray scaled logo image, $LocalMean_{()}$ and $LocalStd_{()}$ are the average and the standard deviation of the sub images. The results of applying the illumination compensation is shown in Fig. 5. One can easily note that the illumination is equalized especially when it is compared to logo images in Fig. 4.

Fig. 5. Preprocessing using local illumination compensation, where brightness almost unified

6 Feature Extraction Using Spatial Moments

Feature extraction stage is the phase where the necessary attributes are extracted from the digital image to identify the logo manufacturer. This is preformed through subdividing the image that is extracted from illumination compensation step into N × N overlapped sub images, after that, from each sub image six local features are extracted: the *Mass* which refers to the local density of a sub image, *MomX, MomY, MomX2, MomY2, MomXY*.

The term Mom refers to the moments extracted where more details are shown in Algorithm (1). The number of features obtained from each logo image is calculated using Eq. (3).

$$T = N^*N^*6 \tag{3}$$

In Eq. (3), the N represents the number of sub images for each image dimension, and 6 refers to the features extracted from each of the sub images. T represent the total number of features extracted from a whole logo image.

To identify a certain image logo, for each class, two main template vectors are created, for each of the 6 the extracted features: a mean template vector and a standard deviation template vector. These two vectors are used in the next phase of classification to perform comparison between the input image and those two vectors. Among the 16 samples that form each class, a vector is created from features extracted from 12 images in that class. Figure 6. Shows the process of subdividing of an illumination compensation image, where features are extracted from each of the overlapped sub images.

Fig. 6. Example of subdividing the illumination compensation image into 4 × 4 areas with 10% rate of overlapping between sub images, No. of extracted features here is 4 × 4 × 6 which equals to 96

Algorithm 1 Calculate Local Moments

```
For Every OverLaped SubImage in the illmunation Comnestaiont
Logo Image
Do
    For Every Pixel X, Y in the SubImage do
        LocaMass ← LocalMass + SubImg(X,Y)
        MomX ← MomX + X × SubImg(X,Y)
        MomY ← MomY + Y × SubImg(X,Y)
    End For
    MomX ← MomX/LocalMass
    MomY ← MomY /LocalMass
    For Every Y in the SubImage do
        YP ← Y - MomY
        For Every X in the SubImage do
            XP ← X - MomX
            MomX2 ← MomX2 + XP^2 × SubImg(X,Y)
            MomY2 ← MomY2 + YP^2 × SubImg(X,Y)
            MomXY ← MomXY + XP × YP × SubImg(X,Y)
        End For
    End For
    MomX2 ← √MomX2/LocalMass
    MomY2 ← √MomY 2/LocalMass
    MomXY ← √MomXy/LocalMass
    LX ← X2 - X1 + 1
    MomX ← MomX/LX
    MomX2 ← MomX2/LX
    LY ← Y2 - Y1 + 1
    MomY ← MomY/LY
    MomY2 ← MomY2/LY
    MomXY ← MomXY/LX/LY
    LoacalMass ← LocalMass/Lx/Ly
End loop
```

7 Classification Process

To decide the input image belongs to which class, then classification process is needed. First, the extracted features from the input image are converted into vector with size of T that appear in Eq. (3)., and after that a comparison is done between that vector and each class vector, i.e., templates' vectors which extracted using the training set to find the most equivalent class. Comparison process is made using: modified Euclidian measurement, shown in Eq. (4), modified city block measurement shown in Eq. (5)., city block measurement shown in Eq. (6)., and Euclidian measurement presented in Eq. (7). The resulted nearest distance refers to the class of that input image.

$$distance = \sum \left(\frac{MeanTamplate - Feature}{StandTamplate} \right)^2 \tag{4}$$

$$distance = \sum \frac{|MeanTamplate - Feature|}{StandTamplate} \tag{5}$$

$$distance = \sum |MeanTamplate - Feature| \tag{6}$$

$$distance = \sum \sqrt{(MeanTamplate - Feature)^2} \tag{7}$$

For the Eqs. 4–7. The **Feature** refers to the feature values extracted from the input logo image, while **MeanTemplate** and **StandTemplate** refer to the values of mean template and standard deviation template vectors for the training sets.

8　Experimental Results and Discussion

The logo recognition accuracy rate of the proposed system is calculated using Eq. (8):

$$Accuracy = \frac{TP}{No.ofSamples} * 100 \tag{8}$$

where **TP** refers to the truly identified digital images, i.e. true posative, and **No.of Samples** denote the total number of tested samples. Three main types of tests are made to evaluate the proposed system. For the three types of tests, the number of tested sub-images ranges from 2×2 to 20×20, and the rate of overlapping between sub-image ranges from 0 to 0.9.

The first evaluation test is the training test. This test includes all classes' samples that are included in the training process and the results are shown in Table 1. Where the highest gained success rate reached to 93.13725%.

The second evaluation test is the hard test. This test includes all classes' samples that are not included in the training process. The results are shown in Table 2., the gained result reached to 77.20588%.

Finally, the third test made to evaluate the performance of the proposed system is the comprehensive test, where are samples are used in the test whether they are included in the training process of creating template vectors or not, and the results test are shown in Table 3. In the comprehensive test the highest gained results reached to 86.94853%. The three mentioned success rates are gained after the preprocessing of local illumination compensation is applied.

For the up mentioned three types of tests, retesting is reapplied without the pre-processing illumination compensation method, and the obtained results reached to: 90.93137%, 65.44118%, 84.00735%, for the training, hard and comprehensive tests respectively. The results are presented in Fig. 7. This leads to the realization of the effec-tiveness of the proposed preprocessing of local illumination compensation method. Note how the success rates are raised for both the training test and for the comprehensive, while it is peaked for the hard test with more than of 12%.

In general, from the highest results of the three tests shown in Table 1, Table 2, and Table 3, as the number of sub-images increased the better recognition rate are obtained, in which the minimum number of sub image of a single dimension is 13.

The rate of overlapping differs, and the equations that present best performance are either Eq. (4) or Eq. (6), and almost always Eq. (7). Gives the lowest discrimination rates.

Fig. 8 Shows the gained results when different combinations of the six extracted features from a sub image are used to get the logo identification, but at the end it is clear that the collection of all of the six features: Local Mass, MomX, MomY, MomX2, MomY2 and MomXY gives the highest recognition rates. All of the single tests shown in Fig. 8 is made without preprocessing.

Note that each single tick shown in Fig. 7 and in Fig. 8. Represents the reached accuracy rates for a single test; when a certain rate of overlapping with a certain number of sub-images is experimented with using a particular distance measure equation. In Table 4. a comprision is shown between the proposed work and other works. One can notice that the proposed work gives a very good results despite the variation of the size images, while this property, i.e. the size of image is unified in reference [4] and in [21], and normalized in [18], and special transformation is used in [19], while the size information is not documented in many other literatures such as [1]. Also, The proposed work presents original local illumination treatment for the logo images; while the other literatures either do not handle this essential issue, such as [17] and [4], or use the traditional global correctness such as [20], and [1] where their methods work with only good illumination conditions.

Table 1. The highest five results obtained from the training test

No. of Samples	No. of Samples Correctly Identified	Success Rate%	Time (in Seconds)	No. of Sub-images	Rate of Overlap ping	Distance Measure Equation
408	380	93.13725	0.054	16	0.5	4
	379	92.89216	0.035	13	0.2	4
	377	92.40196	0.06	17	0.5	4
	377	92.40196	0.069	17	0.9	4
	376	92.15686	0.066	17	0.8	4

Table 2. The highest five results obtained from the hard test

No. of Samples	No. of Samples Correctly Identified	Success Rate%	Time (in Seconds)	No. of Sub-images	Rate of Overlapping	Distance Measure Equation
163	105	77.20588	0.044	15	0.1	4
	105	77.20588	0.054	17	0	4
	105	77.20588	0.09	19	0.4	4
	105	77.20588	0.087	19	0.5	4
	105	77.20588	0.106	19	0.7	4

Table 3. The highest five results obtained from the comprehensive test

No. of Samples	No. of Samples Correctly Identified	Success Rate %	Times in Seconds	No. of Sub-images	Rate of Overlapping	Distance Measure Equation
544	473	86.94853	0.076	20	0	6
	472	86.76471	0.052	17	0	6
	471	86.58088	0.065	19	0	6
	471	86.58088	0.073	20	0.1	6
	470	86.39706	0.058	18	0	6

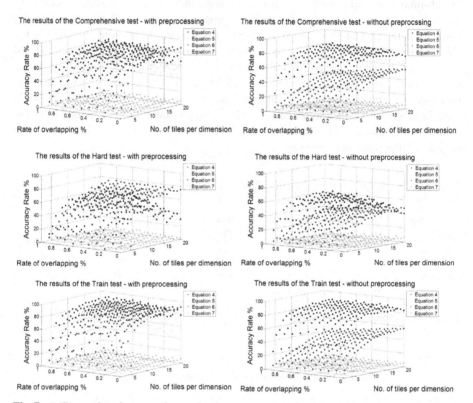

Fig. 7. A Comparison between the result of success rates gained from the three tests: the Comprehensive, the hard and the training test with preprocessing stage (the column to the left) and the same test types without preprocessing stage (the right column).

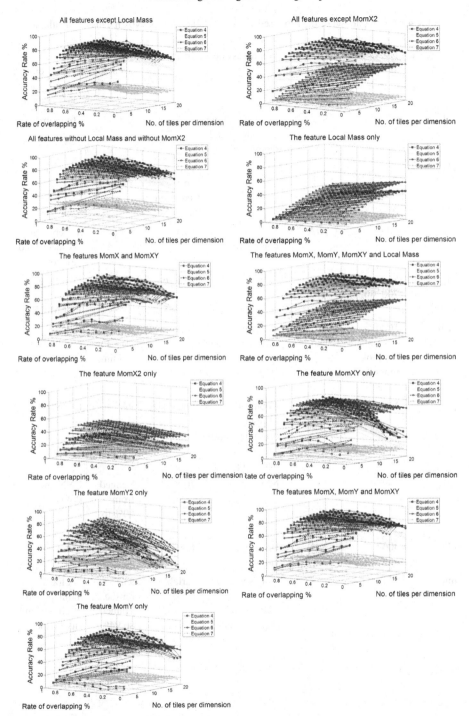

Fig. 8. Different gained recognition results, when different combination of the six features used in feature extraction stage, all that is done without preprocessing.

Table 4. A comparison of the proposed work with other works. N/A means not available

Reference No	No. of total samples	No. of classes	preprocessing	Image Size	Highest accuracy	Feature Extraction and classification
[1]	-	9	Gray stretching, smoothing and denoising	N/A	92%	Hu moment
[17]	-	12	-	Different and assumed to be 34 × 34	92%	Tenseor
[18]	200	N/A	Shape compacting	N/A	92%	Tchebichef Moments and Adaboost
[4]	Seems to be 200*80	80	N/A	64 × 64	97.3%	Overlapping and OE POEM CRC
[19]	1000 Documented but look much less is used	Seem to bee 5	Perspective Transformation	N/A	97% and 99% for training set	R-CNN
[20]	N/A	N/A	Rotation, Flipping, brightness and Contrast and denoising	Small size	93.5%	R-CNN
[21]	14950	13	N/A	70 × 70	Ranged from 60% to 99%	CNN and PSO
[23]	3067	N/A	HSV and gamma	128 × 128	95.86%	YOLOF
ours	544	34	Gray scale and local illumination compensation	variable	93.13% for the Trained 86.94%, overall	Six local moment and Euclidian and city block

9 Conclusion

This proposed work produced a logo image recognition by applying local moments, with six extracted features from overlapped sub-images. The input logo images differ in size, rotation angles and lightness state, and that's why a very effective preprocessing step is done through applying local illumination compensation. The proposed method is tested before performing the illumination compensation process, and results was as following: for the training test 90%, while for the hard test was 65%, and for the comprehensive test was 84%. The effect of local illumination compensation on the obtained results from the proposed method was so obvious, because the results reached to 93%, 77% and 86,9%. From the obtained results from the different types of tests with and without preprocessing and using different combination of the extracted 6 features, one can notice that the CityBlock measurement gives the better recognition ratio, and the number of sub-images starting from 13×13. Also, for the rate of overlapping between the sub-images different values are seen, but depending on the comprehensive test low overlapping rate are affective. In general, in the used dataset, since there is a wide intraclass variation the very good gained results meet the expectations. The created system can be combined with LPR system to identify vehicles logo in many countries. However, this is an offline recognition method which can be developed to an online identification system. So, as future work other preprocessing may be applied to enhance the results such as trying to unify the size and rotation angle of the logo images, or more features may be extracted from a single sub-image, and for classification many machine learning methods can be used.

References

1. Zhao, J., Wang, X.: Vehicle-logo recognition based on modified HU invariant moments and SVM. Multimedia Tools and Applications. **78**, 75–97 (2019). https://doi.org/10.1007/s11042-017-5254-0
2. Llorca, D.F., Arroyo, R., Sotelo, M.A.: Vehicle logo recognition in traffic images using HOG features and SVM. IEEE Conference on Intelligent Transportation Systems, Proceedings, ITSC. 2229–2234 (2013). https://doi.org/10.1109/ITSC.2013.6728559
3. Yu, Y., et al.: A multilayer pyramid network based on learning for vehicle logo recognition. IEEE Trans. Intell. Transp. Syst. **22**, 3123–3134 (2021). https://doi.org/10.1109/TITS.2020.2981737
4. Yu, Y., Wang, J., Lu, J., Xie, Y., Nie, Z.: Vehicle logo recognition based on overlapping enhanced patterns of oriented edge magnitudes. Comput. Electr. Eng. **71**, 273–283 (2018). https://doi.org/10.1016/j.compeleceng.2018.07.045
5. Lu, W., Zhao, H., He, Q., Huang, H., Jin, X.: Category-consistent deep network learning for accurate vehicle logo recognition. Neurocomputing **463**, 623–636 (2021). https://doi.org/10.1016/j.neucom.2021.08.030
6. Yu, S., Zheng, S., Yang, H., Liang, L.: Vehicle logo recognition based on Bag-of-Words. 2013 10th IEEE International Conference on Advanced Video and Signal Based Surveillance, AVSS 2013. 353–358 (2013). https://doi.org/10.1109/AVSS.2013.6636665
7. Pan, H., Zhang, B.: An integrative approach to accurate vehicle logo detection. Journal of Electrical and Computer Engineering. 2013, (2013). https://doi.org/10.1155/2013/391652

8. Huang, Y., Wu, R., Sun, Y., Wang, W., Ding, X.: Vehicle logo recognition system based on convolutional neural networks with a pretraining strategy. In: IEEE Trans. Intell. Transp. Syst. **16**(4), 1951–1960 (2015). https://doi.org/10.1109/TITS.2014.2387069.
9. Huang, Z., Fu, M., Ni, K., Sun, H., Sun, S.: Recognition of vehicle-logo based on faster-RCNN. Springer Singapore (2019). https://doi.org/10.1007/978-981-13-1733-0_10
10. Psyllos, A.P., Anagnostopoulos, C.N.E., Kayafas, E.: Vehicle logo recognition using a sift-based enhanced matching scheme. IEEE Trans. Intell. Transp. Syst. **11**, 322–328 (2010). https://doi.org/10.1109/TITS.2010.2042714
11. NGUYEN, H.: Vehicle logo recognition based on vehicle region and multi-scale feature fusion. Journal of Theoretical and Applied Information Technology. 98, 3327–3337 (2020)
12. Meethongjan, K., Surinwarangkoon, T., Hoang, V.T.: Vehicle logo recognition using histograms of oriented gradient descriptor and sparsity score. Telkomnika (Telecommunication Computing Electronics and Control). 18, 3019–3025 (2020). https://doi.org/10.12928/TEL KOMNIKA.v18i6.16133
13. Ge, P., Hu, Y.: Vehicle Type Classification based on Improved HOG_SVM. 87, 640–647 (2019). https://doi.org/10.2991/icmeit-19.2019.102
14. Hou, S., Li, J., Min, W., Hou, Q., Zhao, Y., Zheng, Y., Jiang, S.: Deep Learning for Logo Detection: A Survey. ACM Transactions on Multimedia Computing, Communications, and Applications. 1–13 (2023). https://doi.org/10.1145/3611309
15. Song, L., Min, W., Zhou, L., Wang, Q., Zhao, H.: Vehicle Logo Recognition Using Spatial Structure Correlation and YOLO-T. Sensors. **23**, 1–15 (2023). https://doi.org/10.3390/s23 094313
16. Chen, R., Jalal, M.A., Mihaylova, L., Moore, R.K.: Learning Capsules for Vehicle Logo Recognition. 2018 21st International Conference on Information Fusion, FUSION 2018. 565–572 (2018). https://doi.org/10.23919/ICIF.2018.8455227
17. Cyganek, B., Woźniak, M.: Vehicle Logo Recognition with an Ensemble of Classifiers. Lecture Notes in Computer Science (including subseries Lecture Notes in Artificial Intelligence and Lecture Notes in Bioinformatics). 8398 LNAI, 117–126 (2014). https://doi.org/10.1007/ 978-3-319-05458-2_13
18. Sam, K., Tian, X.: Vehicle Logo Recognition Using Modest AdaBoost and Radial Tchebichef Moments. Proceedings of 2012 4th International Conference on Machine Learning and Computing IPCSIT. 25, 91–95 (2012)
19. Ansari, I., Lee, Y., Jeong, Y.: Recognition of Car Manufacturers using Faster R-CNN and Perspective Transformation. Journal of Korea Multimedia Society Vol. 21, No. 8, August 2018. 21, 888–896 (2018)
20. Ke, X., Du, P.: Vehicle Logo Recognition with Small Sample Problem in Complex Scene Based on Data Augmentation. Mathematical Problems in Engineering. 2020, (2020). https:// doi.org/10.1155/2020/6591873
21. Soon, F.C., Khaw, H.Y., Chuan, J.H., Kanesan, J.: Hyper-parameters optimisation of deep CNN architecture for vehicle logo recognition, (2018). https://doi.org/10.1049/iet-its.2018. 5127
22. Mustafa S. Kadhm, Yun, L.S.: Propose A Simple and Practical Vehicle Logo Detection and Extraction Framework. International Journal of Emerging Tendrs and Technology in Computer Sciences. 4, 87–90 (2015)
23. Zhao, Q., Guo, W.: Detection of Logos of Moving Vehicles under Complex Lighting Conditions. (2022). https://doi.org/10.3390/app12083835
24. Dataset, https://github.com/GeneralBlockchain/vehicle-logos-dataset

Exploring Enhanced Recognition in Gesture Language Videos Through Unsupervised Learning of Deep Autoencoder

Anwar Mira[(✉)] [ID]

University of Babylon, Hillah, Iraq
Anwar.jaafar@uobabylon.edu.iq

Abstract. The primary objective of a neural network is to achieve generalization, enabling it to recognize previously unseen data from the same category. This concept facilitates the transfer of knowledge between neural networks trained for different purposes. In this study, we investigate the utilization of deep Autoencoder neural networks for unsupervised learning as a regression network to improve the performance of a categorization network. We propose a novel deep Autoencoder network designed to accomplish this goal. Our proposed architecture is applied to video data, specifically focusing on hand gesture recognition using the Chalearn 2014 and IsoGD datasets. Hand gesture recognition plays a crucial role in human-machine interactions, despite challenges posed by temporal information, gesture overlap in videos, and variations in hand gesture orientation. Through our research, we have successfully enhanced the recognition accuracy from 46.1% to 76.4% by employing transfer learning techniques within the same dataset and across different datasets.

Keywords: Deep Neural Network · Deep Autoencoder Network · Transfer Unsupervised Learning · Video Gesture Recognition

1 Introduction

Access to video data for training hand gesture recognition is often limited due to the diverse range of movements and variations in how individuals perform them. To achieve accurate hand movement recognition, it is crucial to have a sufficient dataset for training deep neural networks. This work focuses on the concept of generality in training deep networks and its impact on effectively discriminating hand gesture movements.

Drawing upon previous research, particularly [1], which highlights the benefits of unsupervised pre-training for leveraging recognition on the same dataset, though it did not result in improved accuracy, it demonstrated the effectiveness of the autoencoder network in weight initialization and generalization. Additionally, [2] achieved enhanced classification results by pretraining LSTM networks with predictors for future frames. In another study, [3] presented a model that utilized pre-trained convolutional neural network (CNN) architectures such as Inception V3, ResNet50, VGG-19, VGG-16, and

A. M. Al-Bakry et al. (Eds.): NTICT 2023, CCIS 2096, pp. 179–192, 2024.
https://doi.org/10.1007/978-3-031-62814-6_13

Inception-V2 ResNet to extract features from mammographic images. Furthermore, [4] applied transfer learning-based fine-tuning to state-of-the-art YOLO networks to address the domain-shift problem. In [5], a systematic framework was introduced that combined space decomposition, physics-informed deep learning, and transfer learning to accelerate the multi-objective stochastic optimization of a heat exchanger system. These significant advancements in deep learning design and the wealth of research demonstrating the advantages of pre-trained networks in various domains have served as motivations for our research.

2 Related Works

Deep supervised learning has achieved tremendous success in large datasets and has been widely used in various learning tasks, as demonstrated in [6] and [7]. However, the high cost of labeling data accurately has led researchers to explore alternative approaches. Transfer learning has emerged as a promising solution, where data from different tasks is transferred to a task that may be unrelated [8]. Transfer learning can be implemented through weight initialization and feature extraction [9], allowing the new network to benefit from the design efficiency of the networks it has learned from [10]. Additionally, when dealing with video image data, temporal information becomes crucial as it captures shared events over time [11], in contrast to single images [12]. Autoencoders, on the other hand, possess the ability to train on unlabeled data by extracting underlying features through regression [13]. They can also handle time-shared frames in videos [14]. Notably, autoencoder networks have demonstrated success in building intelligent deep learning-based systems for anomaly detection in industrial machines [15]. Furthermore, [16] introduced multiple input and multiple output spatial division multiplexing using deep learning techniques with open-loop autoencoders.

Given the significance of learning hand gesture movements, which impacts a large segment of the deaf and dumb community, it is crucial to consider the high sensitivity of this type of video to the sequence of movements, with a particular emphasis on hand movement and its spatial relationship to the body. These videos may also involve similar partial movements that overlap with different gesture movements.

In conclusion, this study contributes to the exploration of training deep autoencoder networks and presents an innovative approach based on autoencoders that significantly improves the accuracy of hand gesture recognition. This creative technique pushes the boundaries of human-machine interaction systems, enabling more accurate hand gesture identification.

3 Proposed Methodology

Our work is based on two phases: the first is the training phase through the proposed autoencoder network, and the second phase is the transmission phase of learning from the proposed autoencoder to be able to train on another classification network (Fig. 1). The steps will be explained in detail in the following:

3.1   Data Description

Our work involves the utilization of two distinct datasets for sign language gesture learning, each of which is elaborated upon in the following sections.

3.1.1   ChalearnIsoGD Dataset

Presented in 2016 by Jun Wan and Stan Z. Li [17] as a multi-modal dataset of RGB and depth video images for sign hand gesture recognition of 249 hand gestures that were performed by different 21 signers, the beginning and end of each hand gesture have been labelled manually and given as resources. Each signer has performed hand gestures with clothes and backgrounds that differ from the others. The initial process involved cutting up a continuous video of a single signer into a series of relevant videos, excluding the static movements, such that each video represented a single gestural movement belonging to a certain category. As a result, distinct videos were obtained for each category. As a result, only the training set has 35878 separate videos.

3.1.2   Montalbano2 Dataset

The target dataset used for recognition is called Montalbano2 [18]. It is a dataset specifically designed for the classification neural network task. The dataset consists of a multi-modal gesture detection system that incorporates RGB and Depth videos as well as skeletal data, with an image resolution of 640 * 480. In the dataset, one person is seated in front of the camera while 27 signers perform various motions. The dataset includes labels for the beginning and finishing frames, skeleton data for each signer, diverse backgrounds, and a lexicon of 20 Italian gesture types.

To provide additional information about the signer's movements, a motion tracking sensor is used to collect data about each signer's skeleton. It is worth noting that the signers may alternate between using their left and right hands to symbolize the same action, as the way they use their hands varies for each category.

Initially, the continuous video of each signer was divided into relevant videos, excluding static movements. Each video represents a single gestural movement belonging to a specific category. This process resulted in a training set consisting of 6,400 gesture videos. To accurately fit the size of the signer in each frame, preprocessing was performed using the provided skeleton values from the dataset. This preprocessing effectively removes the background, ensuring that each frame focuses solely on the signer's movements.

3.2   Proposed Autoencoder

An autoencoder neural network usually consists of two parts, one for encoding and the other for decoding. The design of each part varies, as does the target that it predicts, depending on the purpose for which it is trained. Different designs of deep autoencoders have been presented in different publications; in this work, we have uniformed two designs as follows:

The first design is for training from scratch, where the encoder part consists of four blocks of twin convolutional layers with a kernel size of 3×3 in each layer. Each block is followed by a max-pooling layer. The filter sizes for each block are (16, 32, 64, and 128) and there are three fully connected layers with 2048 neurons each. The prediction layer has either 20 or 249 neurons, depending on the dataset being trained.

The second design is for fine-tuning the weights by training the encoder part on a pre-trained network called "VGG_face." This network, proposed by [19], was trained on over two million face images. It consists of 36 layers, divided into five blocks. The first two blocks have two convolutional layers alternating with activation layers, while the remaining three blocks have three convolutional layers alternating with activation layers. Each convolutional layer uses a fixed-length filter of size 3 × 3, and each block is followed by a pooling layer to reduce the overall layer size. The network ends with two flattening layers, which represent a convolutional layer with a 1 × 1 kernel filter, followed by a prediction layer with 20 or 249 neurons.

The decoder part for both types of encoders is symmetric to the encoder part (Fig. 1). Due to the deep nature of the network, there is a skip connection layer between the two parts of the autoencoder. This skip connection helps refine the network's data during training by combining the high-impact layer data from the encoder part with the decoder part. Additionally, this additional layer participates in the gradient process.

Furthermore, we utilized the L2 loss function to minimize the error distance between predictions (P) and ground truth (G) in our neural network, as shown in Eq. (1). The error is then normalized to the image size.

$$loss = (G - P)^2 \tag{1}$$

The backward loss function, as presented in Eq. (2), is calculated as:

$$loss = 2(G - P) \tag{2}$$

Additionally, we replaced the L2 loss, which can be influenced by outliers, with the Huber loss function. The Huber loss is designed to be less sensitive to outliers, as explained in Eqs. (3) and (4). Here, sigma represents a scaling factor applied to the absolute difference between G and P:

$$f(x) = \begin{bmatrix} 0.5 * \text{sigma}^2 * (G - P)^2 & \text{if } |G - P| < 1/sigma^2 \\ |G - P| - 0.5/ \text{sigma}^2 & \text{otherwise} \end{bmatrix} \tag{3}$$

where the drivitive for $f(x)$ is $f\prime(x)$ as in (3)which is used in network backward

$$f'(x) = \begin{bmatrix} \text{sigma}^2 * \text{ x} & \text{if } |G - P| < 1/sigma^2 \\ \text{sign}(G - P) & \text{otherwise} \end{bmatrix} \tag{4}$$

We have chosen the Huber loss to improve the robustness of our network, particularly when dealing with potential outliers in the data.

Skip connection is a technique employed to address feature weakening or loss that may occur during downsampling in a neural network. It involves concatenating corresponding layers in both the encoder and decoder parts and adding an extra layer in the decoder. This concatenation is performed along the channel dimension, allowing the errors to back-propagate through this layer. By stabilizing network training and convergence, the skip connection helps to recover and preserve important features throughout the network [7].

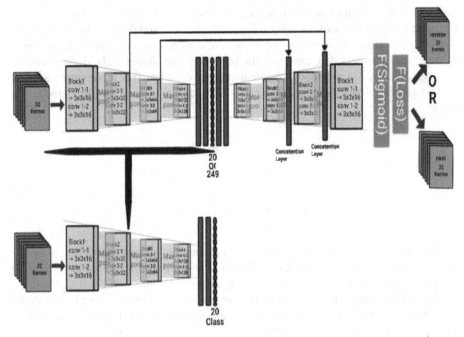

Fig. 1. Layout of the proposed work

3.3 Transfer Learning

Taking advantage of the fixed design of the encoder part in the autoencoder network and the effectiveness of the skip layer in the decoder part, we employ transfer learning. This involves transferring the learned knowledge from a pre-trained deep autoencoder to a new classification network aimed at learning 20 different gestures. The encoder-trained portion of the autoencoder serves as the base weights initialized for the new classification network. When dealing with video images, deep training networks can be applied in two ways, depending on the desired working mechanism and goals. One approach involves training the network on individual image frames, while the other considers frames that share a common time element.

3.3.1 Single Frame Image Deep Autoencoder

While previous studies have demonstrated the superior classification results achieved by deep networks trained on video clips compared to single-frame images, there is still potential to benefit from training deep networks using individual frames. This can be accomplished by incorporating the linking of sequential frames into video clips, as shown in [7] and [8]. In these approaches, features extracted from training a deep network on single frames are combined into a unified clip and inputted into another time-sharing network for further training.

In our case, the videos had relatively short durations, with each video gesture consistently starting and ending with the same frame. The most significant movements

occurred in the middle frames. To account for these specific characteristics, we introduced an autoencoder network designed to train on a sequential, frame-by-frame basis.
This network aims to predict the frame that best represents the central image frame
of each hand gesture video, capturing the key movement. Importantly, during network
training, this frame is excluded from the input images, effectively representing a single
gesture for an individual. Figure 2 provides a visual representation of this process for
clarity.

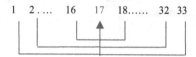

Fig. 2. Predict center of video.

Unfortunately, we encountered difficulties in training the Autoencoder network due
to a bias in predicting the input image. This bias arose from the absence of a sequence
of video frames, which could alter the interpretation of motion. To address this issue,
we adopted the approach of using the dominant image as the future image for the event.
Specifically, we predicted the occurrence of the tenth image based on the previous nine
images. This process is illustrated in Fig. 3.

By implementing this approach, we aimed to overcome the problem of bias in network training, which is a common issue faced by Autoencoders. Additionally, it allowed
us to maintain the cascade of video frames in the training process.

Fig. 3. Predict future image frame.

3.3.2   Time Sharing Image Frames Deep Autoencoder

Dealing with a clip of frames provides a more comprehensive understanding of videos,
considering the sequential nature of events within a single video. To facilitate this, we
constructed an autoencoder capable of training on a clip of video frames. In Sect. 3.2,
we modified the first convolutional layer to accommodate convolving a clip of frames.
The length of the clip was determined to be 32 frames in our work. The autoencoder
was trained with a conditional predictor for the next clip of 32 frames or a predictor for
the same fed frames, which were sequentially inverted in the case of the last clip within
the video. This approach, as suggested by [2], helps avoid bias towards predicting the
same input frames.

For evaluation purposes, both the proposed autoencoder trained on a single frame
and an autoencoder trained on a clip of image frames were compared using root mean

square (RMSE) and accuracy metrics. The accuracy metric (5) measures the accuracy of gesture video predictions by dividing the number of correctly predicted gesture videos (R) by the total number of gesture videos in the test dataset (T) and multiplying the result by 100:

$$\text{Accuracy} = (R/T) \times 100 \tag{5}$$

Furthermore, the average root mean square error (RMSE) for all scanned images or clips of image frames in the training set is calculated using Eq. (6):

$$\text{RMSE} = \sum_{j=1}^{m} \sqrt{\frac{\sum_{i}^{n}(G_{ij}-P_{ij})}{n}} / M \tag{6}$$

In the equations, G and P represent the ground truth and predicted images, respectively. The variables n and m denote the length of each image in G and P, and the number of images or clips of images in the training set, respectively.

4 Experimental Results

The core objective of our practical work is to enhance the recognition performance of Mantalbano2 hand gesture recognition videos by applying the principle of generality. To achieve this, we conducted experiments on various scenarios using this datset, aiming to determine the potential of leveraging this principle for improved results in deeply trained networks. Two deep networks, namely CNN and Autoencoder, were trained for this purpose. It is important to note that the training parameters for both networks remained fixed across all the presented scenarios. These parameters included a learning rate of 0.001, a weight decay of 0.0005, and a momentum of 0.9. We employed SGD as the optimizer and used a mini-batch size of 36. In the following sections, we provide a detailed breakdown of these different scenarios and their corresponding results.

4.1 Single_Frame Deep Autoencoder

In this scenario, the autoencoder network was trained to address the issue of network bias towards the input frame. The training process ensured that each consecutive set of frames within the video accurately predicted a consistent and distinct frame within a given video gesture. This approach aimed to mitigate biases and ensure the occurrence of multiple instances of the same predicted network output, aligning with the typical considerations in classification network design.

However, as depicted in Fig. 4, it is evident that the predicted image may not capture all the intricate details, particularly the subtle nuances of gesture movement and the hand's pose.

Transfer learning was applied to the Montalbano 2 dataset that had previously undergone Autoencoder training. However, the objective of this scenario was significantly different: to classify the 10th image frame from hand gesture videos within the dataset [18]. For this purpose, a network with the same architecture as the encoder segment of

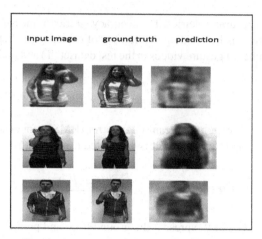

Fig. 4. Autoencoder for a single frame image.

the Autoencoder network was employed. It is important to note that the initial weights of this network were inherited from the Autoencoder's encoder segment, despite the differences in their respective objectives. Table 1 presents a comparative analysis of.

the results obtained by classification networks trained with random initial weights and those trained using weights derived from the pre-trained Autoencoder network. Although the Autoencoder network was initially trained for a different purpose on the same dataset, its design facilitated the process of transfer learning. In this context, we evaluated the efficacy of Autoencoder predictions using Root Mean Square Error (RMSE) as defined in Eq. (6), which involved calculating the average RMSE across all examined images in the training set. Furthermore, our comparison included the precision metric, which measures the accuracy of video predictions as expressed in Eq. (5).

Table 1. Single image frame.

Classification from scratch			Classification By Transferred learning			
Image size	64×64	224×224	64×64		224×224	
Loss function	Softmax	Softmax	Huber	L2	Huber	L2
Acc% from scratch	46.1	**54.8**	51.01	50.9	**59.3**	59.2
Acc% from pre-trained Autoencoder	57.1	**69.1**	61.9	60.3	**70.2**	69.3
RMSE from scratch Autoencoder	–	–	1.40	1.4146	1.38	1.39
RMSE from Pretrained Autoencoder	–	–	1.38	1.39	**1.33**	1.35

The table presents the correlation between the prediction quality of the Autoencoder network and the type of prediction function, subsequently impacting the accuracy of the classification network trained using it. It is worth highlighting that the Huber function exhibited superior performance in comparison to the L2 function, as evidenced by the lower RMSE measurement values.

4.2 Clip of Image Frames Deep Autoencoder

We suppose two strategies to prove generalization in clip of frames on Montalbano2 [18] dataset as follows:

4.2.1 Inside the Same Dataset

The proposed Autoencoder network has applied on the same Montalbano2 dataset, after that the trained weights of encoder part are transferred to another network that has the same design of this part but with different prediction goal which is for classification, so loss function will be softmax loss which was the most convenes for classification, the training on the hosted classification network is started from the first layer after fine tuning the trained weights of autoencoder network because the first layers was giving the best accuracy results. Training autoencoder has considered two different prediction scenarios:

The first is to predict the next clip frames, when network input is the last clip frames in video the prediction will be the inverse frame of the same input frame clips of the video to avoid network biases to the same input as suggested in [2].

The second is to predict the frame of climax movement which has been considered in the middle of clip frames. Figure 5 bellow visualizes the ground truth and prediction from the two proposed network.

As depicted in Fig. 5, the predicted images lack clarity, particularly in the hand pose region, which is the main target for motion recognition. Table 2 explain RMSE metric of predicted clip frames from Autoencoder as a measure of training efficiency as mentioned before, while for the classification network accuracy metric was the measure of number of correct predictive video under the total number of videos as Shawn in the table.

From the table, we observe the potential for transferring learning gained from training the autoencoder network on a complete image clip. It's evident that the efficiency of autoencoder training is influenced by the choice of prediction function, and consequently, this impacts the effectiveness of the classification network that inherits the weights from the pre-trained network.

4.2.2 Outside the Dataset

The same strategy employed previously was applied here but on a different dataset. Specifically, we utilized the Autoencoder trained on the ChalearnIsoGD [17] dataset, which included a larger number of classes and more examples per class compared to the Montalbano [18] dataset.

In this scenario, we followed the identical algorithm described earlier within a different dataset. The network was trained to predict future clip frames, with the additional provision that the same clip frames were predicted in reverse order when dealing with

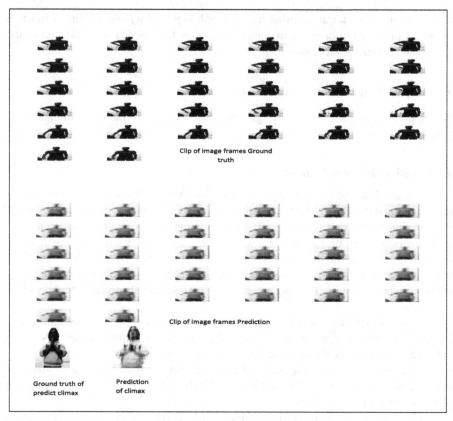

Fig. 5. Prediction image examples from two proposed scenarios of training a clip of image frames.

Table 2. Clip of image frames inside Montalbano2.

Classification from scratch			Classification By Transferred learning			
Image size	64 × 64	224 × 224	64 × 64		224 × 224	
Loss function	Softmax	Softmax	Huber	L2	Huber	L2
Acc% from scratch	69.7	**76.3**	71.9	71.2	**80.6**	79.9
Acc% from pre-trained Autoencoder	73.5	**86.6**	76.2	75.9	**89.1**	89.02
RMSE from scratch Autoencoder	–	–	1.381	1.382	1.372	1.374
RMSE from Pretrained Autoencoder	–	–	1.352	1.355	1.343	1.346

the final clip frames in the video. Figure 6 below provides a visual representation of the ground truth alongside predictions generated by the proposed network.

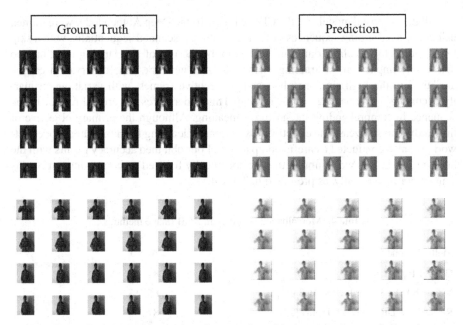

Fig. 6. Prediction image examples of proposed outside Autoencoder transfer learning.

It's evident from Fig. 6, the images predicted by this method still exhibit the issue of blurring in the hand pose region, despite the clarity observed in the body part. In Table 3, we computed the Root Mean Square Error (RMSE) as a metric to gauge the training efficiency of the Autoencoder prediction results. Meanwhile, accuracy served as the measurement for evaluating the training of the network.

Similar to the approach in the first strategy, the training of the pre-trained classification network began from the initial layer without any layer freezing. Consequently, weights were subject to adjustment during the network training process.

Table 3. Clip of frame images outside Montalbano2

Classification from scratch			Classification By Transferred learning			
Image size	64 × 64	224 × 224	64 × 64		224 × 224	
Loss function	Softmax	Softmax	Huber	L2	Huber	L2
Acc% from scratch	69.7	**76.3**	70.5	69.9	**78.6**	77.9
Acc% from pre-trained Autoencoder	73.5	**86.6**	75.3	74.9	**87.1**	87.02
RMSE from scratch Autoencoder	–	–	1.381	1.382	1.372	1.374
RMSE from Pretrained Autoencoder	–	–	1.361	1.363	1.351	1.354

All the proposed models for the Clip of image frame Deep Autoencoder were trained using non-overlapping sequences of frames. Data was scanned sequentially, with a skip length equal to the clip's duration. Interestingly, the use of overlapping clips had no discernible impact on the training of the classifier network. This observation can be attributed to the high similarity between consecutive frames, with variations primarily occurring in the hand gesture's motion. This underscores the crucial role of frame sequence in comprehending the gesture's meaning. Although the primary objective of this study is to showcase the effectiveness of transfer learning from the autoencoder network, we also conducted a careful comparison of our obtained accuracy results with the latest research. It is worth noting that we specifically focused on comparing our results using color images only, as presented in Table 4.

Table 4. Montalbano2 accuracy in RGB image methods.

Method Name	Acc%
GB + LF [20]	88.38%
NP + HOGha 3 × 3 + HOG3D3 [21]	85.5%
Resnet50 + DR-CNN1 + HMM [22]	87.37%
Ours	89.1%

The table demonstrates the potential superiority of the proposed method, specifically highlighting the effectiveness of transferring knowledge through two distinct networks when utilizing color images exclusively.

5   Conclusions and Future Works

We have drawn significant conclusions from our practical application and the results obtained from transfer learning between two entirely distinct networks. Specifically, we found that employing the proposed autoencoder network as a pre-training step has a profound impact on the performance of the classification network. This holds whether the transfer learning is applied within the same dataset or even when dealing with a different type of dataset, particularly one involving diverse body gestures and varying shooting conditions. Notably, this effect was observed despite the considerable size of the videos involved in the latter case.

It is noteworthy that the impact of the proposed autoencoder network was more pronounced when applied within the same dataset, even though the training data for the autoencoder network in this scenario was relatively small (4500 batch samples) compared to the classification network, which had a substantially larger training dataset of 8965 batch samples. We attribute this superiority to the effective training of the proposed autoencoder network and the possibility of sharing crucial decision-making layers when transferring knowledge within the same dataset, which is a vital factor in classification network training. In contrast, during transfer learning between two dissimilar datasets, such sharing becomes challenging due to variations in the number of targets. Despite the Autoencoder network's smaller training dataset size than the corresponding dataset in the classification network for knowledge transfer, the Autoencoder network achieved commendable results in transferring data. This was made possible by its ability to outperform sequential scanning without skipping data, thereby accommodating a larger dataset. This underscores the importance of data-predicted quality in training regression networks. As part of our future work, we recommend focusing on improving the prediction of hand-part features. Through visualization, we observed that the predicted images tend to exhibit blurriness in the hand pose region, which is a critical element in distinguishing hand gestures. Enhancing hand features should result in more effective training of the autoencoder network, which, in turn, will likely have a positive impact on the training of the transferred classifier network.

References

1. Erhan, D., Bengio, Y., Courville, A., Manzagol, P.-A., Vincent, P., Bengio, S.: Why does unsupervised pre-training help deep learning?. J. Mach. Learn. Res. 625–660 (2010)
2. Srivastava, N., Mansimov, E., Salakhudinov, R.: Unsupervised learning of video representations using LSTMs. In: International Conference on Machine Learning, pp. 843–852 (2015)
3. Saber, A., Sakr, M., Abo-Seida, O.M., Keshk, A., Chen, H.: A novel deep-learning model for automatic detection and classification of breast cancer using the transfer-learning technique. IEEE Access **9**, 71194–71209 (2021). https://doi.org/10.1109/ACCESS.2021.3079204
4. Neupane, B., Horanont, T., Aryal, J.: Real-time vehicle classification and tracking using a transfer learning-improved deep learning network. Sensors **22**(10), 3813 (2022). https://doi.org/10.3390/s22103813
5. Wu, Z., et al.: Accelerating heat exchanger design by combining physics-informed deep learning and transfer learning. Chem. Eng. Sci. **282**, 119285 (2023)
6. Nabipour, M., Nayyeri, P., Jabani, H., Mosavi, A., Salwana, E.: Deep learning for stock market prediction. Entropy **22**(8), 840 (2020)
7. Gao, W., Mahajan, S.P., Sulam, J., Gray, J.J.: Deep learning in protein structural modeling and design. Patterns **1**(9) (2020)
8. Long, M., Cao, Y., Wang, J., Jordan, M.: Learning transferable features with deep adaptation networks. In: International Conference on Machine Learning, pp. 97–105 (2015)
9. Long, M., Zhu, H., Wang, J., Jordan, M.I.: Deep transfer learning with joint adaptation networks. In: International Conference on Machine Learning, pp. 2208–2217. PMLR (2017)
10. Pan, S.J., Yang, Q.: A survey on transfer learning. IEEE Trans. Knowl. Data Eng. **22**(10), 1345–1359 (2010)
11. Karpathy, A., et al.: Large scale video classification with convolutional neural networks. In: Proceedings of the IEEE Conference on Computer Vision and Pattern Recognition (CVPR), pp. 1725–1732 (2014)
12. Krizhevsky, A., Sutskever, I., Hinton, G.E.: ImageNet classification with deep convolutional neural networks. Commun. ACM **60**(6), 84–90 (2017)
13. Lore, K.G., So, Akintayo, A., Sarkar, S.: LLNet: a deep autoencoder approach to natural low-light image enhancement. ScienceDirect **61**, 650–662 (2017)
14. Alo, U.R., Nweke, H.F., Teh, Y.W., Murtaza, G.: Smartphone motion sensor-based complex human activity identification using deep stacked autoencoder algorithm for enhanced smart healthcare system. Sensors **20**(21), 6300 (2020)

15. Ahmed, I., Ahmad, M., Chehri, A., Jeon, G.: A smart-anomaly-detection system for industrial machines based on feature autoencoder and deep learning. Micromachines **14**(1), 154 (2023). https://doi.org/10.3390/mi14010154
16. Bui, T.T.T., Tran, X.N., Phan, A.H.: Deep learning based MIMO systems using open-loop autoencoder. AEU-Int. J. Electron. Commun. **168**, 154712 (2023)
17. Wan, J., Zhao, Y., Zhou, S., Guyon, I., Escalera, S., Li, S.Z.: Chalearn looking at people RGB-D isolated and continuous datasets for gesture recognition. In: Proceedings of the IEEE Conference on Computer Vision and Pattern Recognition Workshops, pp. 56–64 (2016)
18. Escalera, S., et al.: ChaLearn looking at people challenge 2014: dataset and results. In: Agapito, L., Bronstein, M.M., Rother, C. (eds.) ECCV 2014. LNCS, vol. 8925, pp. 459–473. Springer, Cham (2015). https://doi.org/10.1007/978-3-319-16178-5_32
19. Parkhi, O., Vedaldi, A., Zisserman, A.: Deep face recognition. In: BMVC 2015- Proceedings of the British Machine Vision Conference 2015. British Machine Vision Association (2015)
20. Escobedo-Cardenas, E., Camara-Chavez, G.: A robust gesture recognition using hand local data and skeleton trajectory. In: IEEE International Conference on Image Processing (ICIP), pp. 1240–1244. IEEE (2015)
21. Xi, C., Koskela, M.: Using appearance-based hand features for dynamic RGB-D gesture recognition. In: 22nd International Conference on Pattern Recognition (ICPR), pp. 411–416. IEEE (2014)
22. Tur, A.O., Keles, H.Y.: Evaluation of hidden Markov models using deep CNN features in isolated sign recognition. Multimed. Tools Appl. **80** (2021)

IOT-Based Water Quality Monitoring for the Tigris River: Addressing Pollution Challenges

Mariam Abdul Jabbar Ali[1]([✉]) [ID], Mahdi Nsaif Jasim[2] [ID], and Saad Najm Al-Saad[1] [ID]

[1] Mustansiriyah University, Baghdad, Iraq
{mariam.a.ali,dr.alsaadcs}@uomustansiriyah.edu.iq
[2] University of Information Technology and Communications (UoITC), Baghdad, Iraq
mahdinsaif@uoitc.edu.iq

Abstract. The Tigris River is a second-largest river in southwestern Asia, flowing through Turkey, Syria, Iraq, and Iran. The river is facing lack of water from the source and increasing pollution from agricultural runoff, industrial wastewater, and sewage. Water quality in the basin is primarily threatened by increased salinity rates caused by intensive irrigated agriculture and high evaporation rates. This paper aims to develop an IoT based and data mining system for monitoring the water quality of the Tigris River. The proposed system achieves real-time monitoring depending on the Internet of Things (IoT) to deter-mine water pH level, total dissolved solids (TDS), temperature, and turbidity or conductivi-ty. The acquired data to be sent online to a hosting server (SQL server) for collecting, storing, and analysis as well as for displaying it on a website The SQL server receives data from the ESP8266 Wi-Fi module (Node MCU). The MUC serves as the central hub, interfacing with sensors and coordinating data transfers to a PC for more analysis. In real-time, quickly compiled data is forwarded to a PHP custom-built website and also displayed on a map using ArcGIS for Power BI to simplify communicate with interested persons. Data mining techniques are used to identify patterns in the data and to detect changes in water quality as well as to predict areas of pollution in order to take corrective action by environment and water resources authorities. Decision Tree, Random Forest, Support Vector Machine, and Support Vector Classifier algorithms are used in this paper. By Using Python code, these methods are compared based on their accuracy rates. It was found that the model using the Random Forest method had the highest accuracy of 0.92, making it the best algorithm for classifying the water quality of the Tigris River in Baghdad, Iraq. Furthermore, due to the nature of these algorithms, Decision Tree has also good classification accuracy due to the nature of these algorithms, where Random Forest is progressively improved from Decision Tree.

Keywords: Arduino UNO · Internet of Things (IoT) · Data mining · water quality · MCU · SQL server · Temperature sensor · pH sensor · TDS sensor · Turbidity sensor

1 Introduction

Water quality monitoring is critical for preserving rivers' ecological balance and guaranteeing the well-being of populations that rely on them. The Tigris River originates in south-eastern Turkey and runs for around 1900 km before entering Iraq. Turkey, Syria, Iraq, and Iran share a river basin of 471,606 km^2 where 54% of which is inside Iraq [1, 2]. The Tigris River, which borders Mesopotamia in the Fertile Crescent, has been a vital source of irrigation, hydroelectric power, and transportation since the oldest known civilizations. Water quality is relatively good in the top half of the basin, while salinity levels rise in the Iraqi part of the basin. Concerns have been raised about rising pollution caused by both residential and industrial sources [3, 4].

Data mining is the process of extracting new information from massive amounts of data in the form of patterns or rules. Statistics, Machine Learning, Artificial Intelligence, Data Science, Database Technology, High-Performance Computing, and Visualization Methods are all involved [5].The main purpose of data mining is to extract hidden information from a data source. Data mining techniques are used to mine a various rules and patterns, including association rules, sequential patterns, classification trees, and so on. It requires data preparation before it can provide relevant information. The retrieved data can be used to make decisions. Currently, there are several common data mining techniques that are effectively used to find predictive information for many different reasons [6, 7].

Many fields has been changed due to the introduction of Internet of Things (IoT) technology in recent years, including environmental monitoring. IoT provides an opportunity for revolutionizing how we manage and perceive water quality by enabling continuous data collection from distant sensors installed across the river system. When combined with powerful data mining techniques, this technology has the potential to enable us to extract valuable insights from the huge datasets collected by these sensors. Such insights can subsequently be used to support informed decision-making and policy formation, ensuring safe and sustainable usage of the Tigris River's waters [8].

The main objective of this research is to study the connection between IoT technology and Data Mining approaches for comprehensive water quality monitoring in the Tigris River. This project aims to provide a complete overview of the modern in monitoring water quality in the Tigris River using IoT and Data Mining through a critical analysis of relevant literature.

In the subsequent sections of this research paper, a comprehensive review of related works, addressing the significant developments achieved through the integration of IoT and Data Mining for water quality monitoring. Moreover, further discussions about the challenges and opportunities involved in developing such a monitoring system. By shedding light on these challenges, this research adds to the developing collection of knowledge and prepares the path for effective management and conservation of the Tigris River's invaluable water resource.

2 Related Work

The traditional water quality monitoring system consists of three steps: water sampling, testing, and inquiry. These steps are done manually by scientists. This technique is not entirely reliable and provides no prior indication of water quality. A remote and automated system for monitoring water quality is offered as a solution to the challenge created by the traditional analytical method. Therefore, several studies have explored the integration of IoT and data mining for water quality monitoring in various contexts. Hussien a. et al. [9] proposed a study deals with using data mining technique to evaluation of water quality of Tigris River in area within Baghdad city. Salwan A. A. et al. [10] highlighted in their study the value of using Multivariate statistical approaches as a beneficial tool for Tigris river management, control, and preservation. Hashim et al. [11] applied artificial neural networks to model the impact of climate change on the streamflow of the Tigris River. These studies underscore the potential of technology-driven solutions in addressing water quality challenges, even though they may not be specific to the Tigris River itself. In reference [12] Salam H. et al. used some statistical techniques such as principal Component Analysis (PCA) and modified Delphi method as well as the Iraq WQI was applied to the Tigris River within Baghdad as a case study.

Iraqi environmental specialists are very interested in the Tigris River because of the severe impact of pollution on it. [13, 14] assessed the quality of Tigris water in a variety of places to estimate pollution consequences. Also in [15], the writers analyze drinking water contamination and its influence on human health in Baghdad, Iraq whereas [16] study the pollution of heavy metals (HMs) in the Euphrates River, which located from Al-Kifl to Kufa city in Iraq. Finally, Sura F. et al. [17] suggested a remote monitoring system for Tigris river water in Baghdad by using different water quality parameters such as Dissolved oxygen (DO), Electrical conductivity (EC), Total Dissolved Solids (TDS), and pH.

3 System Architecture

The architecture of this project is separated into two primary sections: hardware and software.

3.1 Hardware Section

Involves sensors, microcontroller and Wi-Fi module (ESP8266).

3.1.1 Sensors

3.1.1.1. Temperature Sensor: Temperature sensors is used to monitor hotness and coldness of water, which influences various physical, chemical, and biological processes in aquatic ecosystems. It is measured in degree Celsius, with precision of 0.1 steps, which is more accurate than the mercury thermistor and has an operating temperature range between -55 and $+125$ °C with accuracy ± 0.5 °C. Changes in water temperature affect dissolved oxygen levels, nutrient availability, and the metabolic rates of aquatic organisms [8, 18, 19].

3.1.1.2. PH Sensor: The pH sensor measures the hydrogen ion concentration in water and provides insights into water acidity or alkalinity levels. Changes in pH can indicate pollution, eutrophication, or the presence of contaminants [17, 20].

3.1.1.3. TDS Sensor: Total Dissolved Solids (TDS) sensors measure the electrical conductivity of water, which correlates with the concentration of inorganic salts, minerals, and organic matter dissolved in water. High TDS levels can indicate contamination from industrial discharge, agricultural runoff, or natural sources [20].

3.1.1.4. Turbidity Sensor: Turbidity sensors measure the concentration of suspended particles, such as sediment, algae, or organic matter, in water. High turbidity levels can impact light penetration, oxygen availability, and aquatic life.

3.1.1.5. GPS Sensor: GPS (Global Positioning System) sensor provides geographical coordinates like latitude, longitude, altitude, time, etc. from the satellites, allowing for precise location tracking of water quality measurements. It uses signals sent by satellites in orbit and ground stations on Earth to identify its position on Earth [21]. By integrating a GPS sensor with the Arduino Uno and ESP device, the monitoring system can accurately record the geographic location of water samples, facilitating spatial analysis and identification of pollution sources [22].

All the above sensors are connected to the Arduino Uno and ESP device provides real-time monitoring of water. The sensors are shown collectively in Fig. 1.

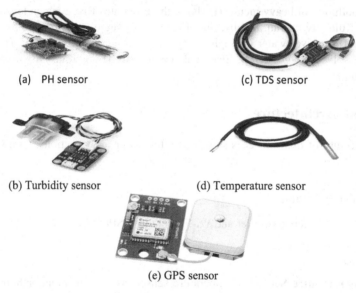

(a) PH sensor   (c) TDS sensor

(b) Turbidity sensor   (d) Temperature sensor

(e) GPS sensor

Fig. 1. Water quality Sensors

3.1.2 Controller

The Arduino Uno microcontroller unit (MCU) is the main processing module that interacts with the environment via sensors and actuators. The microcontroller unit is an essential component for measuring water quality. The reason for this is that the Arduino Uno is the system's heart, which is programmed to read analog input and convert it to digital output. It is responsible for storing and retrieving numerical data from sensors. To minimize communication issues caused by the added-on Wi-Fi module, Arduino Uno is recommended over Raspberry Pi 3 [23]. Raspberry Pi 3 has a process power of 1.2 GHz BCM2836 quad-core ARM Cortex-A7, and memory of 1GB RAM [24] whereas Arduino Uno consumes little power. The Arduino Uno is small, less expensive and has the capacity to monitor sensor readings as well as provide commands [25].

3.1.3 Wi-Fi Module

Wireless module (ESP8266 module) which can provide access to a Wi-Fi network to any microcontroller. The ESP8266 can host an application or offload all Wi-Fi networking functionality to another application processor. Each ES P8266 module comes pre-programmed with an AT command set firmware that can be simply plugged into an Arduino device to provide about as much Wi-Fi functionality as a Wi-Fi Shield [26].

In general, the microcontroller analyses the data collected by the pH, TDS, temperature and turbidity sensors before transferring it to the database via the Wi-Fi module, which acts as a bridge.

3.1.4. A portable laptop provided with relevant software to receive digital data and display it on a screen in an understandable format, in addition to powering the microcontroller.

3.1.5. LCD Screen DISPLAYS the output of sensors [27].

The complete hardware part of the proposed IOT system is demonstrated in Fig. 2.

3.2 Programming (Software) Section

Essentially, the hardware was programmed using Integrated Development Environment (IDE) with C/C++ programing language. The open-source Arduino Software (IDE) makes writing code and uploading it to the board very simple. It is compatible with Windows, Linux, and Mac OS X. The environment is written in Java and depends on processing and other open-source software. This software is also compatible with any Arduino-compatible board such as Arduino UNO.

The program is written, created, debugged, and uploaded to the microcontroller (Arduino uno) for execution. To build the program sketches, there are numerous references available online. Before constructing the system, the sensors should be examined and validated using calibrated instruments. The system was put through a functional test to establish its capabilities. The sensory input was verified, and the total procedure was successful [28–30].

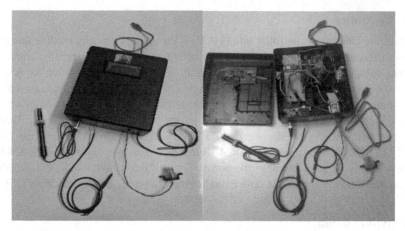

Fig. 2. Sensors node prototype for water monitoring proposed system

4 Methodology

The proposed methodology involves the following steps:

a. Sensor Selection: Choosing suitable pH, TDS, turbidity, GPS, and temperature sensors compatible with Arduino Uno and ESP devices.
b. Hardware Setup: Connecting the sensors to Arduino Uno and ESP devices, ensuring proper wiring and configuration.
c. Data Collection: Collecting water quality data, including pH, TDS, turbidity, GPS coordinates, and temperature, using the integrated sensors.
d. Data Transmission: Transmitting the collected data wirelessly through the ESP device to a hosting server (SQL server) using ESP8266 module.
e. Data Analysis: Analyzing the collected data to assess water quality parameters, identify patterns, and detect potential pollution sources by using data mining techniques.
f. Visualization and Reporting: Presenting the analyzed data through visualizations, maps, and reports to facilitate understanding and decision-making.

In general, the project consists of three phases: data collection, data mining and data visualization (Fig. 3).

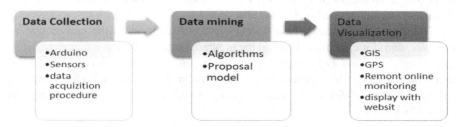

Fig. 3. System stages

The output of the first phase (data collection) will be the input of the next phase (data mining) as well as the output of the second phase will be the input of the last phase (data visualization). In this paper the first phase is achieved through Sensor Selection, Hardware Setup, Data Collection, and Data Transmission steps whereas the second phase includes Data Analysis. Visualization and Reporting and will be the third phase where the data was displayed at PHP-designed website and ArcGIS map.

Figure 4 illustrates the proposed system's diagram. The system monitors and customizes the referred parameters. As a result, new sensors can be simply added and configured via the dashboard of the program. Turbidity, temperature, acidity, and total dissolved solids (TDS) (conductivity) are the current monitoring parameters. Data collected by sensors is uploaded to the SQL server and received by the server over the Internet, where data packets involving information about water quality are included.

The SQL server's responsibilities include storing data uploaded by sensors, which will be displayed on the website, as well as monitoring and controlling the sensors via an application stored on the server. Sensors connected with the Nod MCU (ESP8366) can transfer sensory data to the SQL server. The sensor node can connect to a wireless network and send collected data to a PHP-based web platform implemented on desktop browsers. The user will be able to obtain the uploaded data, allowing for the classification of water quality levels using Python programing code executed by Google Colab platform.

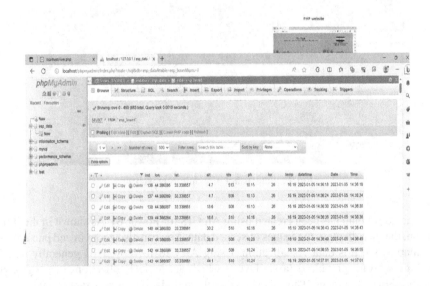

Fig. 4. A block diagram of a complete system

Michael Widenius was used to create the database for the website shown in Fig. 5. Structured MySQL's query language includes features such as a "users table" and a "configuration table". People can access data stored in the database. There are two kinds of users: system administrators, who must have a license and abilities in both writing and content evaluation, and monitoring users, who can only examine data published to a

Fig. 5. Screenshot of MySQL database

PHP-designed website (see Fig. 6) from various locations. Whereas monitoring users are those who can only track information collected by the system from various locations (Fig. 7).

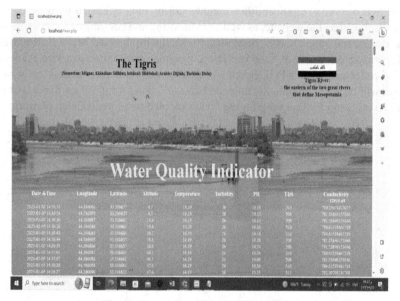

Fig. 6. Screenshot of monitoring website

The above map has been created using ArcGIS for Power BI. As shown in the illustration, the map displays the values of sensed data. You can also zoom in and pan around the map to see this data visualization. Using ArcGIS for Power BI generally entails working with geographic data via layers, which are logical collections of geographic data used to create maps. They also serve as the foundation for geographic analysis.

To test the water, the proposed method was used to collect the five parameters listed in Table 1. The World Health Organization (WHO) specifies these parameters. PH measures acid, whereas turbidity measures haziness. TDS measures the amount of salt concentration in water. We may also characterize water's ability to carry electricity via the conductivity parameter using the following relationship between total dissolved solids (TDS) in mg/L and conductivity.

$$TDS = Conductivity * 0.65 \tag{1}$$

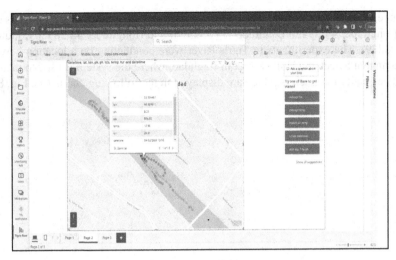

Fig. 7. ArcGIS Map

Table 1. Limits of Water Health Organization for safe drinking water

Parameters	WHO standards	Units
Turbidity	5–10	NTU
pH	6.5–8.5	pH
Temperature	10–22	°C
TDS	500–1000	mg/L
Conductivity	300–800	microS/cm

5   Results and Discussion

As previously mentioned, in this research the system examined tracked specific sets of water quality parameters. IOT and WSN have been used for remote real-time monitoring technology. The data from sensors provides information of some water samples taken from the Tigris Iraqi' river in Baghdad governorate.

The system was designed to be operated in remote locations using a portable laptop to power the MCU [18]. The Arduino IDE code was uploaded to the sensors node in order to connect the wireless network and send measured data to the website. The software module for the proposed website is developed which shows the values of the water quantity indicator on the website of the area of interest. Furthermore, the system was tested on the Tigris River in Baghdad. The dataset includes a real-time water quality indicator within two hours duration. There are 833 records in the dataset, each with the following attributes: longitude, latitude, altitude, pH, TDS, turbidity, temperature, date, and time. The proposed system uses GPS data to determine location information such as longitude, latitude, and altitude. The GPS coordinates of recorded data for the Tigris river were (longitude: 44.394806, latitude: 33.332596) and (longitude: 44.383869, latitude: 33.3408839). The experiments in Tigris River were conducting on a location between Al-Mutanabbi Street and Haifa Street in Baghdad, where the system is used to measure the water quality parameters which are displayed in the following charts (see Fig. 8).

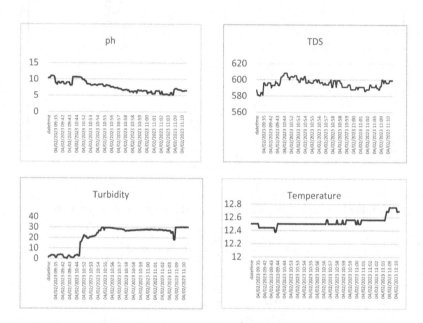

Fig. 8. The water quality monitoring of pH, TDS, Turbidity and Temperature

Table 2. The statistical summary of parameters

	Average	Maximum	Minimum	Standard Deviation
pH	7.592	11.03	5.18	1.6064
TDS	595.126	608	580	5.8306
Conductivity	915.58	935.38	892.31	8.9702
Turbidity	21.064	30.04	0.01	10.6262
Temperature	12.527	12.75	12.38	0.07365

The degree to which data is correct and error-free is referred to as accuracy. Data accuracy is essential for machine learning algorithms to perform properly. In other words, it aims to define the degree to which data accurately represents real-world occurrences. To make correct predictions and judgments, machine learning systems require accurate data for training and validation. Inaccurate data can lead to biased or erroneous outcomes, reducing dependability and utility of these systems [31].

The experiments are performed on a computer with the following specification: Window 10, 11th Gen Intel(R) Core(TM) i7-11370H @ 3.30 GHz 3.30 GHz processor and 16 GB memory. Four machine learning algorithms are used for classification: Decision Tree, Random Forest, Support Vector Machine and Support Vector Classifier. The dataset is divided into two subsets: training set which is used to train model and test set that is used to test the train model. All the above machine learning algorithms are assessed based on their performance feature accuracy rate and are compared to find the best algorithm that has a higher accuracy rate in classifying the data. The dataset are classified depending on the pH values (water acidity or alkalinity level). Depending on the accuracy rates of the mentioned algorithms shown in Table 3, the Random Forest Algorithm and Decision Tree algorithm achieved higher performance comparing with other algorithms.

Table 3. Performance of machine learning algorithms for classifying Tigris River water quality

	Algorithms			
	Decision Tree (%)	Random Forest (%)	Support Vector Machine (%)	Support Vector Classifier (%)
Accuracy Rate	0.90	0.92	0.68	0.35

According to the study's results, the Tigris River has poor drinking water quality. Except for temperature, the measured pH, TDS, and turbidity values are outside the range of the WHO water quality requirements listed in Table 1. Because of the high levels of pH, turbidity, TDS, and conductivity, the water is considered very acidic and salty. Table 2 includes certain statistical data such as average, maximum, minimum, and standard deviation.

The water quality at the aforementioned locations may not be adequate for most individuals. However, in order to establish a successful plan for improving contaminated water quality, a functional system that can swiftly assess and provide real-time data is required [32, 33].

6 Conclusion

The project aims to develop a smart IOT system to monitor the water quality of Tigris River in real time. The research focused on integration of IOT, Data Mining, and GIS. The study utilized various sensors such as: pH, TDS, turbidity, temperature and GPS to obtain sensors readings related to water quality. The collect data was then used to developed classifier models and algorithms for predicting and classifying pH level and water purity. After the experiments, it shows that the model using Random Forest algorithm had the most outstanding accuracy of 0.92 which is the best algorithm that can used to classify the water quality of Tigris River, Baghdad, Iraq. Moreover, Decision Tree also have good classification accuracy due to the nature of these algorithms where Random Forest is progressively improved from decision Tree. The ensemble of decision trees has high accuracy because it applies randomness on two levels. In the first level, the algorithm chooses a subset of features randomly to use as candidates at each split. This avoids a large number of decision trees from relying on the same set of features. When generating splits in the second level, each tree uses a random sample of data from the training dataset. This offers an additional element of unpredictability, preventing the individual trees from overfitting the data. They cannot overfit since they cannot traverse all of the data. By combining decision trees, we may compensate for the weaknesses of each tree individually.

The accuracy of results allows decision makers of water resource management to make precise procedure according to the precise predicated results taken from the proposed system.

Presenting the analyzed data in the form of visualizations, maps, and reports which have been created using Power BI that helps with understanding and decision making.

While the proposed system seems promising, it is limited by various limits and impediments including the obtaining the essential materials for developing the envisioned prototype and the cost of the entire system rises due to limited access to electronic components as a result of severe constraints implemented in Iraq due to suspicion of terrorist activities, affecting the affordability of the proposed system.

In the future, more sensors will be needed to increase the system's capabilities and collect data from all parts of Tigris River. Moreover, the device will detect the problem by using on-device intelligence. Finally, the system will predict future pollution conditions based on current and historical data from water sources, and the system will be deployed in most of Iraqi wetlands and marshes.

References

1. Rahi, K.A., Halihan, T.: Salinity evolution of the Tigris River. Reg. Environ. Change **18**(7), 2117–2127 (2018). https://doi.org/10.1007/s10113-018-1344-4
2. Hussein, H.A., Alshami, A.H., Al-Awadi, A.T., Ibrahim, M.A.: Hydrological characteristics of the Tigris River at the Baghdad Sarai station. Ain Shams Eng. J. **14**(2), 101846 (2023). https://doi.org/10.1016/j.asej.2022.101846
3. https://education.nationalgeographic.org/resource/tigris-river/
4. Hashim, M.H.: Elemental analysis of river, marshes and ground water in Thi Qar region, Iraq. Al-Mustansiriyah J. Sci. **29**(2), 182–187 (2018). https://doi.org/10.23851/mjs.v29i2.394
5. Sarker, I.H.: Data science and analytics: an overview from data-driven smart computing, decision-making and applications perspective, vol. 2, no. 5 (2021). https://doi.org/10.1007/s42979-021-00765-8
6. Al-Refaie, A., Abu Hamdieh, B., Lepkova, N.: Prediction of maintenance activities using generalized sequential pattern and association rules in data mining. Buildings **13**(4), 946 (2023). https://doi.org/10.3390/buildings13040946
7. Shah weli, Z.N.: Covid-19 prediction model using data mining algorithms. Al-Mustansiriyah J. Sci. **33**(1), 45–50 (2022). https://doi.org/10.23851/mjs.v33i1.1076
8. AlMetwally, S.A.H., Hassan, M.K., Mourad, M.H.: Real time internet of things (IoT) based water quality management system. Procedia CIRP **91**, 478–485 (2020). https://www.sciencedirect.com/science/article/pii/S2212827120308532
9. Hussien, A., Mariana, M., Adina, F.: Analysis of data mining tools used for water resources management in Tigris River. Adv. Manage. Sci. **3**(2) (2014). https://doi.org/10.7508/AMS-V3-N2-1-10

10. Abed, S.A., Hussein, E.S., Al-Ansar, N.: Evaluation of water quality in the Tigris River within Baghdad, Iraq using multivariate statistical techniques. J. Phys.: Conf. Ser. **1294**(7) (2019). https://doi.org/10.1088/1742-6596/1294/7/072025

11. Hashim, M., Al-Ansari, N., Alsamanawi, M.: Modeling the impact of climate change on Tigris River's streamflow using artificial neural network. J. Hydrol. **570**, 444–455 (2019)

12. Salam, H., Salwan, A., Nadhir, A., Riyadh, M.: Development and evaluation of water quality index for the iraqi rivers. Hydrology **7**, 67 (2020). https://doi.org/10.3390/hydrology7030067

13. Farhan, A.F., Al-Ahmady, K.K., Al-Masry, N.A.A.: Assessment of Tigris River water quality in Mosul for drinking and domestic use by applying CCME water quality index. In: IOP Conference Series: Materials Science and Engineering, vol. 737, no. 1, p. 012204. IOP Publishing (2020)

14. Chabuk, A., Al-Madhlom, Q., Al-Maliki, A., et al.: Water quality assessment along Tigris River (Iraq) using water quality index (WQI) and GIS software. Arab. J. Geosci. **13**, 654 (2020). https://doi.org/10.1007/s12517-020-05575-5A.M

15. AL-Dulaimi, G.A., Younes, M.K.: Assessment of potable water quality in Baghdad City, Iraq. Air Soil Water Res. **10** (2017). https://doi.org/10.1177/1178622117733441

16. Kamel, L.H., Al-Zurfi, S.K.L., Mahmood, M.B.: Investigation of heavy metals pollution in Euphrates River (Iraq) by using heavy metal pollution index model. In: IOP Conference Series: Earth and Environmental Science, vol. 1029, no. 1, p. 012034. IOP Publishing (2022)

17. Sura, F., Hussein, A.: Int. J. Eng. Technol. **7**(4), 2784–2788 (2018). https://doi.org/10.14419/ijet.v7i4.16699

18. Hong, W.J., et al.: Water quality monitoring with Arduino based sensors. Environments **8**, 6 (2021). https://doi.org/10.3390/environments8010006

19. Chowdury, M.S.U., et al.: IoT based real-time river water quality monitoring system. Procedia Comput. Sci. **155**, 161–168 (2019). https://doi.org/10.1016/j.procs.2019.08.025

20. Ahmed, A.F., Mohamed, I.S.: IOP Conference Series: Materials Science and Engineering, 2nd International Scientific Conference of Al-Ayen University (ISCAU-2020), 15–16 July 2020, Thi-Qar, Iraq, vol. 928 (2020) https://iopscience.iop.org/article/10.1088/1757-899X/928/3/032054

21. https://www.electronicwings.com/arduino/gps-module-interfacing-with-arduino-uno

22. Yigit Avdan, Z., Kaplan, G., Goncu, S., Avdan, U.: Monitoring the water quality of small water bodies using high-resolution remote sensing data. ISPRS Int. J. Geo-Inf. **8**(12), 553 (2019). https://doi.org/10.3390/ijgi8120553

23. Erboz, G.: How to define industry 4.0: the main pillars of industry 4.0, no. July (2018)

24. Paper, C.: An IoT-based water supply monitoring and controlling system, vol. 9, no. 3, pp. 202–206 (2018). www.ijarcs.info

25. Jain, A., Malhotra, A., Rohilla, A., Kaushik, P.: Water quality monitoring and management system for residents. Int. J. Eng. Adv. Technol. **9**(2), 567–570 (2019). https://doi.org/10.35940/ijeat.b3521.129219

26. Geetha, S., Gouthami, S.: Internet of things enabled real time water quality monitoring system. Smart Water **2**, 1 (2016). https://doi.org/10.1186/s40713-017-0005-y

27. Ibrahim M.K., Hussien N.M., Alsaad S.N.: Smart system for monitoring ammonium nitrate storage warehouse, vol. 23, no. 1 (2021). https://doi.org/10.11591/ijeecs.v23.i1.pp583-589

28. Samsudin, S.I., Salim, S.I.M., Osman, K., Sulaiman, S.F., Sabri Cent, M.I.A.: Indon. J. Electr. Eng. Comput. Sci. **10**(3), 951–958 (2018). https://doi.org/10.11591/ijeecs.v10.i3.pp951-958. ISSN: 2502-4752

29. Jasim, M.: A GIS assessment of water quality in euphrates river/Iraq. J. Univ. Babylon Eng. Sci. **23**(2) (2015)

30. Talib, A.M., Jasim, M.N.: Geolocation based air pollution mobile monitoring system. Indon. J. Electr. Eng. Comput. Sci. **23**(1), 162–170 (2021). https://doi.org/10.11591/ijeecs.v23.i1.pp162-170

31. Aldoseri, A., Al-Khalifa, K.N., Hamouda, A.M.: Re-thinking data strategy and integration for artificial intelligence: concepts, opportunities, and challenges. Appl. Sci. **13**(12), 7082 (2023). https://doi.org/10.3390/app13127082
32. Malche, T., Tharewal, S., Bhatt, D.P.: A portable water pollution monitoring device for smart city based on internet of things (IoT). In: IOP Conference Series: Earth and Environmental Science, vol. 795, p. 012014 (2021). https://iopscience.iop.org/article/10.1088/1755-1315/795/1/012014
33. Ramadhan, A.J.: Smart water-quality monitoring system based on enabled real-time internet of things. J. Eng. Sci. Technol. **15**(6), 3514–3527 (2020). https://jestec.taylors.edu.my/Vol%2015%20issue%206%20December%202020/15_6_1.pdf

Crimes Tweet Detection Based on CNN Hyperparameter Optimization Using Snake Optimizer

Zainab Khyioon Abdalrdha[1]([envelope]) [iD], Abbas Mohsin Al-Bakry[2] [iD],
and Alaa K. Farhan[3] [iD]

[1] Informatics Institute of Postgraduate Studies, Iraqi Commission for Computers and
Informatics, Baghdad, Iraq
phd202120695@iips.edu.iq
[2] University of Information Technology and Communications, Baghdad, Iraq
abbasm.albakry@uoitc.edu.iq
[3] Department of Computer Sciences, University of Technology, Baghdad, Iraq
110030@uotechnology.edu.iq

Abstract. The increased use of social media has transformed how people share information. In the voluminous social media content, tweets involving criminal activity have increased. The detection and tracking of crime-related tweets helps law enforcement, researchers, and legislators protect the public and identify trends. This study uses Natural Language Processing (NLP) and machine learning (ML) algorithms to identify illicit tweets related to theft, violence, cybercrime, and drug use. Deep convolutional neural networks (CNNs) are used to classify these tweets, but their ideal structure requires further study. The study uses CNN hyperparameters to select relevant features from a dataset, improving model performance. The embedding dimension, dense unit count, learning rate, and batch size are considered. Snake optimization algorithms, inspired by natural processes, are used to provide optimal solutions for complex situations. The paper introduces CNN-SO, a framework that uses a snake optimization algorithm for CNN feature selection of hyperparameter optimization and improving Arabic Twitter crime identification. The model detects crime-related tweets and the results show that CNN-snake optimization outperformed the Conventional CNN used, as the accuracy in Dataset 1 was 99.57% after optimization, and 99.65% with CNN-snake optimization when using Dataset 2, which was built for this model after optimization. However, the accuracy is 88% when using Dataset 1 before optimization CNN, and 87.63% when using Dataset 2, aiming to reduce the burden of these tweets and improve public safety.

Keywords: Snake Optimizer · Crime Tweet · Selection Schemes · Aho-Corasick Algorithm · Deep Learning

A. M. Al-Bakry et al. (Eds.): NTICT 2023, CCIS 2096, pp. 207–222, 2024.
https://doi.org/10.1007/978-3-031-62814-6_15

1 Introduction

The rapid growth of social media platforms, particularly Twitter's name current (X), is driving a global information explosion, enabling knowledge discovery scholars to explore innovative methodologies and hidden discoveries [1]. Technology has significantly impacted modern life, altering communication and daily tasks. However, on the negative side, the digital revolution has also led to increased cyberbullying, theft, and cybercrime due to pseudonyms [2]. While social media offers numerous benefits, it also brings to the fore dishonest and fraudulent people who harm others and ruin their reputations [3]. Culture-specific crime impacts individuals and groups, necessitating studies to understand behavior, identify individuals, and detect, anticipate, and prosecute such crimes [4]. While communication and technology can enhance task performance, they can also increase the risk of bugs in data transfer. [5]. Social media sites like Facebook, blogs, wikis, and Twitter transform [6]. This study focuses on rising global crime rates, particularly in the Arab world and Iraq, and uses the Aho-Corasick algorithm to collect accurate criminal activity data. The study employs natural optimization approaches like the snake method with the CNN Model to improve Arabic Twitter crime detection.

The following are this work's main contributions:

1. The study creates a novel dataset of Arabic crime tweets, enriched with features for crime analysis and classification using the Python library and Aho-Corasick algorithm.
2. This study is the first to combine behavioral-inspired and deep learning (DL) techniques for crime detection on Arabic Twitter, evaluating their performance on Arabic language crime tweet tasks. The paper is organized as follows: Sect. 2 addresses related work, Sect. 3 details the proposed method, and Sect. 4 describes the experimental result and discusses it, while Sect. 5 provides the conclusion.

2 Related Work

Social media platforms have revolutionized communication, especially in the Arabic language, enabling news dissemination and emotional expression, and other subjects across diverse topics [7]. This study used DL and ML to analyze Arabic-language Twitter criminal activity and compare it to English-language cases. It [8] used BERT to analyze Twitter tweets for criminal activity, comparing SVM, ANN, and TF-IDF. Results showed that logistic regression has the highest precision and recall, with low detection rates. Classifier median accuracy was 90.57%. The detection rates (81.39%) and execution times (0.014 s and 2.5287 s, respectively) of Multinomial NB and Multinomial RF were low. In this paper, the authors [9] used DL models, specifically CNN and Bi-GRU, to identify abusive Arabic language and hate speech on Twitter, achieving an optimal F1-score of 85.9% for foul language recognition and 75% for hate speech detection. The authors [10] suggested using ML, DL, feature extraction, N-gram models, and performance measurements to identify Arabic tweets, with the neural network method outperforming others. The classical paradigm uses support vector machines, neural networks, logistic regression, and naive Bayes. GloVe and fast Text models are used in DL. The authors [11] suggested using Naive Bayesian, Random Forest, J48, and ZeroR classifiers to accurately identify illegal terms in tweets. Classifiers performed well in this task with good accuracy, precision, and recall. Machine learning techniques help discover illegal behavior. The approach had a 94.92 F1 score and 94.91% classification accuracy. Strategy loss was 16.26%, strategy precision 94.94%, and recall accuracy 93.91%. Nan Sun and colleagues proposed real-time Twitter spam detection using account and content attributes [12]. The study used nine supervised ML algorithms; DL performed best in accuracy, TPR, FPR, and F-measure parameters. The DL algorithm outperformed the Random Forest and C5.0 algorithms with 80% accuracy in 200,000 instances. Zaheer et al. [13] proposed a three-stage crime prediction framework using MNB, KNN, and SVM models, achieving a high precision exceeding 0.9, with SVM outperforming network-based feature selection approaches. The study's authors [14] used CNN-LSTM models in "fast and simple" approaches to evaluate 8,000 Arabic tweets. Results showed that ML models could not match brain learning, with the CNN-LSTM hybrid model scoring the highest at 73%. Finally, Muhammad et al. [15] developed a method to detect cyberbullying in Indonesian tweets using a hybrid CNN-BiLSTM algorithm and Fast Text. The study found that the CNN-BiLSTM model had the highest accuracy (80.55%) in Fast Text.

3 Methodology

This strategy proposes using an improved CNN model to find the best CNN model hyperparameters and input data features for crime-related tweet detection. Figure 1 shows the block diagram of the proposed CNN hyperparameter optimization method.

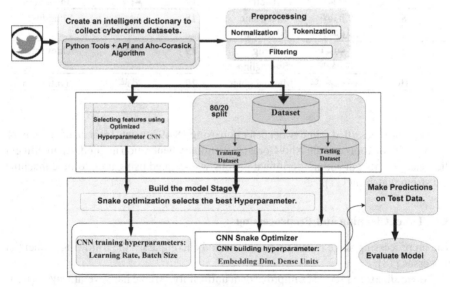

Fig. 1. The framework for CNN Hyperparameter Optimization.

Describe the proposed method's main framework in each paragraph in Fig. 1, and to complete the schematic explanation, we move on to the rest of the methods.

3.1   Tweet Collection

This section shows how to build the dictionary to gather Twitter datasets, although the Twitter development team also offers an official Twitter API [16]. Python has many library utilities, like Tweepy and Scrape. In this research, the snscrape library employed the Aho-Corasick algorithm to scrape the dataset using crime and non-crime keywords, an intelligent method for graph and metadata analysis that classifies crime-related tweets. Some of the terms in Table 1 are intelligent dictionary keywords.

Table 1. Intelligent dictionary keywords.

No. keywords	Keywords Arabic	Keywords English	No. keywords	Keywords Arabic	Keywords English
1	'تنمر'	Bullying	11	'حقوق الإنسان'	Human rights
2	'المخدرات'	Drugs	12	'السلم '	'The peace'
3	'ابتزاز'	Blackmail	13	'الأمن'	Security
4	'إرهاب'	Terrorism	14	التسامح ' والاحترام'	Tolerance and respect'
5	الاحتيال ' الالكتروني'	Electronic fraud'	15	'التعاون'	Cooperation
6	التزوير ' الالكتروني'	Electronic forgery'	16	العدالة ' والمساواة'	Justice and equality
7	'سرقة'	Theft'	17	'الإلهام'	Inspiration
8	الابتزاز ' الإلكتروني'	'Electronic Blackmail'	18	'التحفيز'	Stimulus
9	تهريب ' المخدرات'	Drug Smuggling	19	'الإنجاز'	Achievement
10	'قتل غير متعمد'	Manslaughter	20	'التفاؤل'	Optimism

The study classified crime tweet detection using 18,493 tweets from diverse domains through Python tools and the Aho-Corasick dictionary generation method. An intelligent dictionary for illegal and non-criminal actions was created using a finite state machine and string matching [17].

3.2   Tweet Cleaning and Preprocessing

The preprocessing stage in our proposed method encompasses the following sequential processes [18]:

Normalization processes improve data uniformity. Arabic has several ways to transcribe identical words. For example, "ابتزاز"", which denotes extortion, might be written as "إبتزاز.". Spelling differences may reduce categorization accuracy. Normalizing methods are suggested to fix this, like adjusting the user's writing to academic style. The Arabic characters "إ" "أ," and "آ" will be substituted by "ا." Additionally, "ه" will become "ة", and "ي" will become "ى".

Tokenization divides the text into tokens and removes unnecessary spaces, leaving one space between each word. This phase is crucial for text data because it converts unstructured data into a processing-friendly format.

Filtering: The approach optimizes text classification by removing non-alphabetic characters, focusing on Twitter symbols like # and @, and removing unnecessary stop words by comparing them to Arabic stop words.

3.3   Feature Selection and Hyperparameter Optimization

Feature selection (FS) methods enhance criminal tweet detection by identifying crucial features, eliminating irrelevant traits, reducing data size, and improving performance [19]. Hyperparameter optimization enhances feature selection by determining function evaluations per phase, ensuring efficient input selection for maximum output in dataset classification [20]. Hyperparameters help ML models evaluate a dataset objectively and choose the best subset for performance [21]. Hyperparameter optimization is the process of selecting the optimal hyperparameter combination on an independent dataset to provide the best model based on a loss function [22]. DL models require extensive data training to achieve precision and generalizability, especially in crime tweet detection, which may be scarce due to its extensive parameterization [23]. Machine algorithms configure model architectures in DL, with hyperparameters modified during training. Network parameters like filters, kernel size, hidden layers, and training parameters like epochs and learning rate [24]. This study used CNN-snake optimization methods to find the best CNN hyperparameters and also identified the most relevant input data criteria for crime tweet detection to ensure representativeness. This paper discusses CNN-snake optimization, which helps construct a precise Arabic tweet crime detection system by acquiring appropriate model parameters. This study improved the CNN model classification accuracy using embedding dimension, dense units, learning rate, and batch size. This section details the six CNN optimization approaches.

3.3.1   Snake Optimizer (SO)

This section explains the inspiration and mathematical formulation of the recently published snake optimization [25].

A. **Pairing behavior of snakes**
    Snake breeding in late spring and early summer needs plenty of food and low temperatures. Males compete for mating, while females lay eggs and leave after hatching.

B. **Motivational source**
    Snake optimization (SO) is regulated by mating behavior, where snakes engage in food exploration or consumption under unfavorable conditions. The search strategy encompasses both exploration and exploitation, with transitional methods enhancing the overall efficiency of global search. Mating entails a process of conflict and reproduction, wherein males and females engage in competition to secure the most desirable mate.

C. **Mathematical model of Snake Algorithm**

Stepwise details of the mathematical model for SO are described in this section.

Using Eq. (1), initializing solutions provides a collection of random solutions in the search space that serve as the population of the snake for optimization in later stages.

$$N_i = N_{min} + r \times (N_{max} - N_{min}) \tag{1}$$

The notation N_i represents the location of individuals in the population at time i. The term r represents a random value between 0 and 1 at time i. The upper and lower limits of the population are denoted by N_{max} and N_{min}, respectively.

Division of solutions is done using Eq. (2). The population is 50% male and 50% female.

$$S_m = S_f = \frac{S_{all}}{2} \tag{2}$$

In this context, S_{all} represents the overall population size, whereas S_m and S_f refer to the specific quantities of male and female snakes, respectively. The solutions will be evaluated by selecting the optimal one from the male ($N_{best, male}$) and female groups ($N_{best, female}$) and determining the position of the food (L_{food}). Two additional notions are defined, namely temperature (Temperature) and quantity of food (quantity), as represented by Eqs. (3) and (4) respectively.

$$T = exp(\frac{-t}{T}) \tag{3}$$

$$FQ = c_1 \times exp^{(t- T)/T} \tag{4}$$

The variables "t" and "T" denote the current and total iterations, respectively, and "c1" is a constant with a value of 0.5. Search space exploration (food not found) relies on applying a given threshold value. When the Quantity is less than 0.25, the search space's solutions update their locations to a predetermined random place to search globally. Equations (6)–(9) model this.

When the food quotient (FQ) is less than 0.25 during the exploration phase, snakes use random selection to identify and update their food-foraging positions. Equations (5) and (6) can simulate exploration.

$$N_i^m(t+1) = N_r^m(t) \pm C_2 \times exp^{\frac{-fmr}{fmi}} \times N_i = \begin{Bmatrix} N_r^m(t) + C_2 \times e^{\frac{-fmr}{fmi}} \times N_i \\ N_r^m(t) - C_2 \times e^{\frac{-fmr}{fmi}} \times N_i \end{Bmatrix} \tag{5}$$

$$N_i^f(t+1) = N_r^f(t) \pm C_2 \times exp^{\frac{-ffr}{ffi}} \times N_i = \begin{Bmatrix} N_r^f(t) + C_2 \times e^{\frac{-ffr}{ffi}} \times N_i \\ N_r^f(t) - C_2 \times e^{\frac{-ffr}{ffi}} \times N_i \end{Bmatrix} \tag{6}$$

where $N_i{}^m$ denotes the male snake's position at the i-th time, while $N_r{}^m$ represents its random position. The fitness of male snake $N_r{}^m$ is represented by $f^m{}_r$, while the fitness of each male snake at time i is represented by $f^m{}_i$. The female snake's position at the i-th time is indicated by the symbol $N^f i$, while its random position is indicated by $N^f r$. The fitness of the i-th unique female snake $N^f r$ is represented by $f^f r$, where c2 is a fixed constant equal to 0.05. In the exploratory phase, the snake does not engage in mating behavior and only consumes food when FQ > 0.25 and Temp > 0.6. Equation (7) can be used to represent the process.

$$N_i^{m,f}(t+1) = N_{food} \pm C_3 \times Temp \times rand_i \times (N_{food} - N_i^{m,f}(t)$$

$$= \begin{cases} N_{food} + C_3 \times Temp \times rand_i \times (N_{food} - N_i^{m,f}(t) \\ N_{food} - C_3 \times Temp \times rand_i \times (N_{food} - N_i^{m,f}(t) \end{cases} \tag{7}$$

In the present context, the representation of the male snake's positional value in the i-th generation is indicated as N_{food}. Additionally, a fixed constant of 2 is represented as c3.

Female quality (FQ) > 0.25 and temperature (Temp) $\leq$ 0.6 0.6 indicate mating behavior in snakes. As both males and females want to mate with the best heterosexual partner, competition may ensue. The individual who emerges as the victor has the privilege of selecting its mating partner before others. Consequently, there are two distinct modes in the mating process: a fighting mode and a mating mode. These modes are denoted by Eqs. (8) and (9) to represent the fighting and mating modes, respectively.

$$N_i^m(t+1) = N_r^m(t) + C_3 \times exp^{\frac{-ff_{best}}{f_i}} \times rand_i \times \left(FQ \times N_{best}^f - N_i^m(t) \right) \tag{8}$$

$$N_i^f(t+1) = N_r^f(t) + C_3 \times exp^{\frac{-fm_{best}}{f_i}} \times rand_i \times (FQ \times N_{best}^m - N_i^m(t)) \tag{9}$$

The male snake's position in the i-th generation is denoted by N_i^m. Similarly, N_{best}^f stands for the positioning value of the snake's most ideal individual, which is the female. N_i^f represents the female snake's place in the i-th generation, while N_{best}^m represents the male snake's ideal individual. In addition, the fitness value of the most ideal female snake during battle mode is ff_{best}, whereas the male snake's is fm_{best}. Finally, f_i is an individual's fitness. Equations (10) and (11) show the mating patterns.

$$N_i^m(t+1) = N_r^m(t) + C_3 \times exp^{\frac{-m_i^f}{f_i^m}} \times rand_i \times (FQ \times N_i^f(t) - N_i^m(t) \tag{10}$$

$$N_i^f(t+1) = N_r^f(t) + C_3 \times exp^{\frac{-m_i^m}{f_i}} \times rand_i \times (FQ \times N_i^m(t) - N_i^f(t) \tag{11}$$

The variable N_i^m represents the positional value of the i-th individual within the male snakes, while N_i^f represents the positional value of the i-th individual within the female snakes. The variable "i" represents the fitness of the male snake in the mating pattern, whereas "m_i^f" represents that of the female snake in the same mating pattern. Upon completion of mating, female snakes engage in egg laying and incubation. This reproductive strategy serves to generate offspring, which subsequently have the potential to replace the least favorable male or female individuals within the initial population, contingent upon the gender of the newly hatched snake. The replacement process is mathematically represented by Eqs. (12) and (13).

In the mating pattern, "i" indicates male snake fitness, while m_i^f indicates female snake fitness. After mating, female snakes lay and incubate eggs. The gender of the freshly hatched snake determines whether the progeny can replace the least desirable male or female in the initial population. Equations (12) and (13) represent replacement.

$$N_{\text{worst}}^m = N_{min} + rand_i \times (N_{max} - N_{min}) \tag{12}$$

$$N_{\text{worst}}^f = N_{min} + rand_i \times (N_{max} - N_{min}) \tag{13}$$

The terms "N_{worst}^m" and "N_{worst}^f" refer respectively to the male and female individuals with the lowest performance or characteristics.

3.4   Model Construction

DL uses multi-layered and neural networks to extract complex features from dimensional data for abstract computing [26]. DL techniques improve accuracy and reduce training time in complex situations, advancing research, engineering, data analysis, pattern detection, and prediction in various fields [27]. DL algorithms are used in various fields like autonomous vehicles, robotics, intelligent systems, community-related applications, and social network analysis to identify user behavior patterns and trends [28]. Predictions by supervised and unsupervised DL algorithms require labeled training data [29, 30]. DL has revolutionized data analysis, pattern detection, decision-making, and innovation. A CNN model consists of eight layers, each with a different learning goal. The first layer maps input words to continuous-valued vectors, with word embedding as a hyperparameter using a snake optimizer. The second layer, Conv1D, has three 1D convoluted layers. In this model, each layer has 128 filters and filters of sizes 3, 4, and 5. These layers use ReLU activation. After convolution layers, GlobalMaxPooling1D is used. The model uses Dense Layers to extract important text fragments, and the layers perform complex cognitive processing on convolutional reconstructed features. Hyperparameters are optimized using snake optimizer and ReLU activation for nonlinearity, L2 regularization for overfitting reduction, and ReLU activation for dropout layers. Hyperparameters like dense layers and neurons per layer are used in optimization. The model's dropout layers are not linearized by each dense layer's ReLU activation function. Dense layers are added after flowing layers to avoid over-processing. The two-unit output layer uses softmax activation for two-class binary classification. The CNN architecture hyperparameter optimization method considers learning rate and batch size, which affect convergence speed and efficiency based on the snake optimizer. The model is dynamic and depends on hyperparameters, enabling the architecture to adapt to the best structure for the dataset and task. The output layer of the network delivers a probability score for binary classification, indicating whether a Tweet is a crime. The CNN Architecture Hyperparameter Optimization method is shown in Fig. 2. According to the architectural configuration, Algorithm 1 shows the pseudocode for the snake optimizer for the CNN model's best features and hyperparameters, as shown in Algorithm 1.

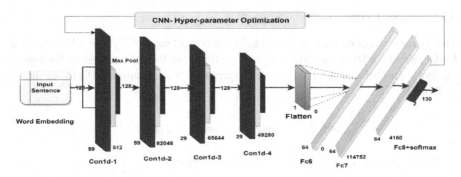

Fig. 2. The CNN Architecture Hyperparameter Optimization.

Algorithm 1: Pseudocode for Snake Optimizer for CNN Model's Best Features and Hyperparameters

Input: Dim=n_features, UB=[1] * dim, LB=[0] * dim , and Nall_symbolic = Symbol('Nall'), T_symbolic = Symbol('T'), t_symbolic = Symbol('t')

Output: Best Snake (features and hyperparameters).

1: Initialize the Snakes Randomly X, hyperparameters, fitness, Xfood, hyperparameters_food, Gbest = initialize_snake (Nall, Dim, LB, UB)

2: N_i **(i = 1, 2, ..., Nall_symbolic)** using Eqs. (2)

3: **while (t_symbolic < T_symbolic) do**

4: **Evaluate fitness_snake** ($N_{features}$, $N_{hyperparameters}$, fitness, Dim, LB, UB)

5: Find the best snake (features and hyperparameters)

6: best_snake = find_best_snake ($N_{features}$, fitness)

7: Define *Temperature_symbolic Phase* using **Eq.** (3).

8: Define Food_Quantity_symbolic **FQ using Eq.** (4).

9: **if** (FQ < 0.25) **then**

10: Perform the exploration Phase using **Eq.** (5) **and Eq.** (6)

11: $N_{features\ new}$, $N_{hyperparameters\ new}$ = perform_exploration ($N_{features}$, $N_{hyperparameters}$, Dim, LB, UB)

12: **else if** (*Temperature_symbolic* > 0.6) **then** Perform Exploitation phase **Eq.** (7)

13: $N_{features\ new}$, $N_{hyperparameters\ new}$ = perform_exploitation ($N_{features}$, $N_{hyperparameters}$, Dim, LB, UB)

14: **else if** (rand_symbolic > 0.6) **then**

15: Snakes in Fight Mode **Eq.** (8) **and Eq.** (9)

16: $N_{features\ new}$, $N_{hyperparameters\ new}$ = snakes_in_fight_mode ($N_{features}$, $N_{hyperparameters}$, fitness)

17: **else**

18: Snakes in Mating Mode **Eq.** (10) **and Eq.** (11)

19: $N_{features\ new}$, $N_{hyperparameters\ new}$ = snakes_in_mating_mode ($N_{features}$, $N_{hyperparameters}$, fitness)

20: Replace the worst snake with the new snake **Eqs.** (12) **and** (13)

21: replace_worst_snake ($N_{features}$, $N_{hyperparameters}$, fitness, $N_{features\ new}$, $N_{hyperparameters}$)

22: **t_symbolic += 1 end if**

23: **end if**

24: **end if**

25: **end while**

26: Return the best Snake (features and hyperparameters)

27: best_snake_features = X [Gbest]

28: best_snake_hyperparameters = hyperparameters [Gbest].

3.5 Training Conditions

The research was conducted using an Intel Core i7 machine with Python 3.11.4, Google Collab, Keras for DL, Matplotlib for graph plotting, and Pandas and Numpy for dataset reading and array handling. The proposed hyperparameter and ranges for CNN are shown in Table 2.

Table 2. Hyperparameters and ranges for CNN

CNN Hyperparameter	Ranges
Embedding _dim	[32, 128]
dense units	[16, 128]
learning rate	[-4, -1]
Batch Size	[16, 128]

3.6 Model Evaluation

The methodology analyzes classifier performance using accuracy, precision, recall, and f1-measure. Superior measurement values determine the best classifier. Equations 14–17 consider precision, recall, accuracy, and F1-score [31].

$$Precision = \frac{TP}{TP + FP} \tag{14}$$

$$Recall = \frac{TP}{TP + TN} \tag{15}$$

$$Accuracy = \frac{TP + TN}{TP + TN + FP + FN} \tag{16}$$

$$F1score = \frac{2 \times Precision \times Recall}{Precision + Recall} \tag{17}$$

4   Experimental Results

Our study tested CNN-snake optimization. This section will discuss and present many scenarios of the outcomes.

4.1 Experiment Compares Results Using Dataset1 and Dataset2

As described in the previous section, the model was trained and the results are presented in this section The study focuses on the efficiency and performance of classification methods using evaluation measurements. The dataset includes keywords for crimes, usernames, tweets, and other features. The proposed method categorizes the dataset using the DL model using CNN before and after optimization to find the optimal CNN hyperparameters and select the most relevant features from the input data for crime tweet detection based on snake optimizer is described in Sect. 3.3 in this paper to create an extremely accurate detection of crimes in Arabic tweets, enabling the obtaining of the optimal parameters for the model. When we compare the results before and after the optimized CNN based on two datasets, dataset1 [32] of (13240) tweets and Dataset2 which was built for this model, the outcomes of optimizing the CNN model with the snake optimizer used in the proposed model achieved excellent accuracy of 99.65% for the CNN snake optimizer algorithm. This compares very favorably with other optimizers used in terms of accuracy after feature selection based on optimal hyperparameters for the CNN model, while before optimization, the CNN model outcomes of measuring the classifier obtained with computation of total accuracy is 87.63%. The total classifier precision, recall, and f1-measure are 86.80%, 87.63%, and 87% respectively, for Dataset2. While Datset1 [32] achieved an accuracy of 99.57% for the CNN snake optimizer after feature selection based on optimal hyperparameters for optimizing the CNN model, before optimizing the CNN model, the outcomes of measuring the classifier obtained with computation of overall accuracy, were, for total classifier precision, recall, and f1-measure, 86.51, 85.26, 86.51 and 85.65 respectively. Table 3 illustrates the performance measurements for optimizing the CNN model for Dataset1 [32] and Dataset2 before and after optimization. The optimal values of the hyperparameters obtained are summarized in Table 4. However, optimizing the CNN model using a snake optimizer outperformed most classification algorithms, compared to other optimizers used in terms of accuracy.

Table 3. CNN before and after Optimization results for Dataset 1 [32] and Dataset2.

Model with dataset	Precision	Recall	F1-score	Accuracy
Dataset 1(13240) [32] based on Conventional CNN	88%	86.81%	88%	87.17%
Dataset 1(13240) [32] based on CNN- Snake optimizer	**99.57%**	**99.57%**	**99.57%**	**99.57%**
A dataset 2 of Arabic crime tweets built(18493) for Conventional CNN	87.63%	86.80%	87.63%	87%
A dataset 2 of Arabic crime tweets built(18493) for CNN- Snake optimizer	**99.65%**	**99.60%**	**99.65%**	**99.60%**

4.2 Discussion

The findings of Dataset 1, as reported by CNN, are depicted in Fig. 3. The optimization outcome of the CNN Algorithm indicates that the optimized model exhibits superior performance, compared to the non-optimized model. Concerning performance evaluation,

Table 4. CNN Hyperparameter Values – Optimized for CNN-snake optimizer.

CNN Hyperparameter	Optimal Values Dataset1 [32]	Optimal Values Dataset2
Embedding _dim	42	108
dense units	63	125
learning rate	0.017	0.035
Batch Size	64	78

the CNN snake optimization technique performed the best for Dataset 1. Test accuracy differences were 11.57%, precision differences 12.76%, recall differences 11.57%, and F1 score differences 12.4%, while the results of CNN optimization of the Arabic criminal Twitter dataset (18493) created using the suggested method are illustrated in Fig. 3. The results of the CNN Algorithm optimization indicate that the optimized model exhibits superior performance compared to the unoptimized model. Concerning performance evaluation, the CNN snake optimization technique performed the best for Dataset 2. Test accuracy differences were 12.02%, precision differences 12.8%, recall differences 12.02%, and F1 score differences 12.6%.

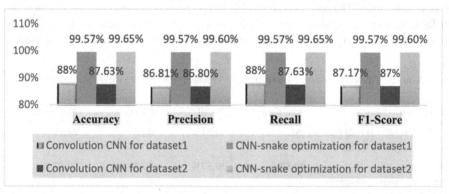

Fig. 3. The performance of Conventional CNN before and after optimization for Dataset 1 [32] and Dataset 2 (18,493).

A comparison in Table 5 displays the best results of the proposed algorithms, compared to the related work.

Table 5. Comparison with related work current published work on crime detection.

Paper authors	Year of Publication	Data source/Language	The best performance of the algorithm	F1 score or Accuracy
S.P.C.W Sandagiri, and others [8]	2020	TW/EN	LR	90.57%
B. Haddad, Z. Orabe et al. [9]	2020	Tw/Arabic	Bi-GRU	75%
Kaddoura, Sanaa; Henno, Safaa [10] and dataset 1 [32]	2023	Tw/Arabic	Fast Text + LSTM, GloVe + LSTM	95.1%, 97%
Vijendra Singh and others [11]	2020	Tw/EN	RF	98.1%
Nan; Lin, Guanjun and others [12]	2020	Tw/EN	Evaluate various algorithms' performance using kNN, GBM, C5.0, and DL	80%
Zaheer Abbass and others. [13]	2020	Tw/EN	SVM, MNB, and K-Nearest Neighbors KNN	93%
Abuzayed et al. [14]	2020	Tw/Arabic	CNN-LSTM	73%
Muhammad et al. [15]	2023	Tw/Indonesian	CNN-BiLSTM hybrid	80.55%
Conventional CNN model for the dataset 1[32]	2023	Tw/Arabic	**Before optimization CNN Model**	88%
Conventional CNN model for the dataset 2	2023	Tw/Arabic	**Before optimization CNN Model**	87.63%
Method Proposed based on CNN-snake optimization for dataset 1[32]	2023	Tw/Arabic	**After optimization CNN Model**	**99.57%**
Method Proposed based on CNN-snake optimization for dataset 2	2023	Tw/Arabic	**After optimization CNN Model**	**99.65%**

5 Conclusion

This study aims to develop optimized models for detecting criminal tweets using a hyperparameter approach to address the issue of digital crime through DL models. It selects the most relevant characteristics from the input data for Arabic crime tweet detection and utilizes a snake optimizer to determine the optimal hyperparameters for a CNN model. Using systematic research and optimization it enhances the algorithm's ability to identify criminal activity in Arabic tweets. Based on the CNN Model, the CNN-snake optimization algorithm outperformed other methods on Datasets 1 and 2. By distinguishing between crimes related to tweets and those not, the models' performance was enhanced, proving the need to adjust hyperparameters to improve the precision and efficacy of Twitter crime detection systems. In summary, the proposed intelligent dictionary model gathers datasets, including Arabic tweets, using the Aho-Corasick algorithm to identify tweets about crimes on Twitter. This intelligent dictionary increases its efficacy by comparing features against a vast dictionary of keywords and contextual cues related to criminal activities. This combination of techniques makes it possible to accurately and quickly identify criminal conduct in Arabic tweets. In future research, the model's accuracy can be enhanced by increasing the number of hyperparameters. Only the statement or text was used in this study to make predictions. To determine whether or not a tweet constitutes a crime, the suggested technique can also be investigated using criminal behavior analysis, etc.

Acknowledgment. The authors would like to thank Informatics Institute of Postgraduate Studies, Iraqi Commission for Computers & Informatics (https://iips.edu.iq/), Baghdad-Iraq, for its support of the present work.

References

1. Thaher, T., Saheb, M., Turabieh, H., Chantar, H.: Intelligent detection of false information in Arabic tweets utilizing hybrid Harris hawks based feature selection and machine learning models. **13**, 556 (2021). https://doi.org/10.3390/sym13040556
2. Al-Ajlan, M.A., Ykhlef, M.: Optimized Twitter Cyberbullying Detection based on Deep Learning, pp. 978–1. IEEE (2018). https://doi.org/10.1109/NCG.2018.8593146
3. Islam, M.M., Uddin, M.A., Islam, L., Akter, A., Sharmin, S., Acharjee, U.K.: Cyberbullying detection on social networks using machine learning approaches. In: IEEE Asia-Pacific Conference on Computer Science and Data Engineering (CSDE), Asia-Pacific (2020). https://doi.org/10.1109/CSDE50874.2020.9411601
4. Yang, D., Heaney, T., Tonon, A., et al.: Crime telescope: crime hotspot prediction based on urban and social media data fusion. World Wide Web **21**, 1323–1347 (2018). https://doi.org/10.1007/s11280-017-0515-4
5. Mansoor Al-Amri, R., Hamood, D.N., Farhan, A.K.: Generation Initial key of the AES algorithm based on randomized and chaotic systems. Al-Salam J. Eng. Technol. **2**(1), 53–68 (2022). https://doi.org/10.55145/ajest.2023.01.01.007
6. Abbass, Z., Ali, Z., Ali, M., Akbar, B., Saleem, A.: A Framework to Predict Social Crime through Twitter Tweets By Using Machine Learning. IEEE (2020). https://doi.org/10.1109/ICSC.2020.00073

7. Guellil, I., Adeel, A., Azouaou, F., et al.: A semi-supervised approach for sentiment analysis of Arab(ic+izi) messages: application to the algerian dialect. SN Comput. Sci. **2**, 118 (2021). https://doi.org/10.1007/s42979-021-00510-1

8. Sandagiri, S., Kumara, B.O.: Deep Neural Network-Based Approach to Identify the Crime-Related Twitter Posts. IEEE (2020). https://doi.org/10.1109/DASA51403.2020.9317098

9. Haddad, B., Orabe, Z., Al-Abood, A., Ghneim, N.: Arabic offensive language detection with attention-based deep neural networks. In: Proceedings of the 4th Workshop on Open-Source Arabic Corpora and Processing Tools, with a Shared Task on Offensive Language Detection, pp. 76–81 (2020). https://aclanthology.org/2020.osact-1.12

10. Kaddoura, S., Alex, S.A., Itani, M., et al.: Arabic spam tweets classification using deep learning. Neural Comput. Appl. **35**, 17233–17246 (2023). https://doi.org/10.1007/s00521-023-08614-w

11. Singh, V., Asari, V.K., Li, K.C.: Analysis and classification of crime tweets. Procedia Comput. Sci. **167**, 1911–1919 (2020). https://doi.org/10.1016/j.procs.2020.03.211

12. Sun, N., Lin, G., Qiu, J., Rimba, P.: Near real-time Twitter spam detection with machine learning techniques. Int. J. Comput. Appl. **44**(4), 338–348 (2020). https://doi.org/10.1080/1206212X.2020.1751387

13. Abbass, Z., Ali, Z., Ali, M., Akbar, B., Saleem, A.: A framework to predict social crime through twitter tweets by using machine learning. In: IEEE 14th International Conference on Semantic Computing (ICSC), San Diego, CA, USA (2020). 3–5 F. https://doi.org/10.1109/ICSC.2020.00073

14. Abuzayed, A., Elsayed, T.: Quick and simple approach for detecting hate speech in Arabic tweets. In: Proceedings of the 4th Workshop on Open-Source Arabic Corpora and Processing Tools, with a Shared Task on Offensive Language Detection, pp. 109–114 (2020). https://aclanthology.org/2020.osact-1.18

15. Nasution, M.A.S., Setiawan, E.B.: Enhancing cyberbullying detection on Indonesian twitter: leveraging fast text for feature expansion and hybrid approach applying CNN and BiLSTM. Revue d'Intelligence Artificielle **37**(4), 929–936 (2023). https://doi.org/10.18280/ria.370413

16. "Twitter Developer. https://developer.twitter.com/en/docs/twitter-api/v1/rate-limits. Accessed 1 Nov 2021

17. Ourlis Lazhar, B.D.: SIMD implementation of the Aho-Corasick algorithm using Intel AVX2. Scalable Comput. Pract. Exp. (SCPE) **20**, 563–576 (2019). https://doi.org/10.12694/scpe.v20i3.1572

18. Alzanin, S.M., Azmi, A.M., Aboalsamh, H.A.: Short text classification for Arabic social media tweets. J. King Saud Univ. Comput. Inf. Sci. **34**(9), 6595–6604 (2022). https://doi.org/10.1016/j.jksuci.2022.03.020

19. Bolón-Canedo, V., Remeseiro, B.: Feature selection in image analysis: a survey. Artif. Intell. Rev. **53**(4), 2905–2931 (2020). https://doi.org/10.1007/s10462-019-09750-3

20. Yang, H., Liu, J., Sun, H., Zhang, H.: PACL: piecewise arc cotangent decay learning rate for deep neural network training. IEEE (2020). https://doi.org/10.1109/ACCESS.2020.3002884

21. Muzakir, A., Adi, K., Kusumaningrum, R.: Advancements in semantic expansion techniques for short text classification and hate speech detectio. Ing. Syst. Inf. **28**(3), 545–556 (2023). https://doi.org/10.18280/isi.280302

22. Feurer, M., Hutter, F.: Hyperparameter Optimization. In: Hutter, F., Kotthoff, L., Vanschoren, J. (eds.), Automated Machine Learning. The Springer Series on Challenges in Machine Learning. Springer, Cham (2019). https://doi.org/10.1007/978-3-030-05318-5_1

23. Sarker, I.H.: Deep learning: a comprehensive overview on techniques, taxonomy, applications and research directions. SN Comput. Sci. **2**, 420 (2021). https://doi.org/10.1007/s42979-021-00815-1

24. Yousaf, I., Anwar, F., Imtiaz, S., Almadhor, A.S., Ishmanov, F., Kim, S.W.: An optimized hyperparameter of convolutional neural network algorithm for bug severity prediction in Alzheimer's-based IoT system. Hindawi, Computational Intelligence and Neuroscience, Article ID 7210928, p. 14 (2022). https://doi.org/10.1155/2022/7210928
25. Hashim, F.A., Hussien, A.G.: Snake optimizer: a novel meta-heuristic optimization algorithm. Knowl.-Based Syst. **242**, 108320 (2022). https://doi.org/10.1016/j.knosys.2022.108320
26. Alsaedi, E.M., Farhan, A.K.: A comparative study of combining deep learning and homomorphic encryption techniques. Al-Qadisiyah J. Pure Sci. 17–33. (2022). https://doi.org/10.4018/IJCAC.309936
27. Liu, F., et al.: Deep learning for community detection: progress, challenges, and opportunities. In: Proceedings of the Twenty-Ninth International Joint Conference on Artificial Intelligence (IJCAI-20), vol. 693, pp. 4981–4987 (2021). https://doi.org/10.24963/ijcai.2020/693
28. Ravi, D., et al.: Deep learning for health informatics. IEEE J. Biomed. Heal. Inform. **21**(1), 4–21 (2017). https://doi.org/10.1109/JBHI.2016.2636665
29. Mosavi, A., Ardabili, S., Várkonyi-Kóczy, A.R.: List of deep learning models. In: Várkonyi-Kóczy, A. (eds.) Engineering for Sustainable Future. INTER-ACADEMIA 2019. LNNS, vol. 101, pp. 202–214. Springer, Cham (2020). https://doi.org/10.1007/978-3-030-36841-8_20
30. Kamath, U., Liu, J., Whitaker, J.: Deep Learning for NLP and Speech, p. 621. Springer International Publishing, New York (2019). (978-3-030145-95-8). https://doi.org/10.1007/978-3-030-14596-5
31. Zhang, Z., Sabuncu, M.R.: Generalized cross entropy loss for training deep neural networks with noisy labels. In: Proceedings of the 32nd International Conference on Neural Information Processing Systems, pp. 8778–8788 (2018). https://doi.org/10.48550/arXiv.1805.07836
32. Kaddoura, S., Henno, S.: Dataset of Arabic Spam and Ham Tweets. Mendeley Data, V1 (2023). https://doi.org/10.17632/86x733xkb8.1

Optimized Intelligent PID Controller for Propofol Dosing in General Anesthesia Using Coati Optimization Algorithm

Ammar T. Namel[1]([✉]) [iD] and Mouayad A. Sahib[2] [iD]

[1] Informatics Institute for Postgraduate Studies, Iraqi Commission for Computers and Informatics, Baghdad, Iraq
phd202110690@iips.edu.iq
[2] University of Information Technology and Communications College of Engineering, Baghdad, Iraq
mouayad.sahib@uoitc.edu.iq

Abstract. General anesthesia is an essential component of any surgical procedure, requiring careful monitoring and precise control. Administering the correct dosage is crucial to ensure the patient remains adequately anesthetized. The pharmacokinetic/pharmacodynamic (PK/PD) model that clarifies the interactions between the administered drug and patient's response was used. In this paper, we introduce an optimized PID controller to automate the control of general anesthesia and adjust the infusion rate of the drug using the bispectral index (BIS) as the process variable. We tuned one PID controller for each individual patient and tested it on a group of 8 virtual patients. The controller utilizes the Coati Optimization Algorithm (COA) to optimize the PID controller parameters. Simulation period was 250 s and results were obtained using MATLAB software. The results demonstrate the effectiveness and robustness of the controller in accurately assessing and regulating the Depth of Hypnosis (DOH).

Keywords: General anesthesia · PK/PD model · DOH · PID control · coati optimization algorithm · CLAD

1 Introduction

Anesthesia is the state of being insensitive to sensory and emotional stimuli, and it is used to reduce a patient's response to pain during surgical procedures. Anesthesia can be achieved while retaining some degree of sensation or by inducing complete insensitivity. There are two main types of anesthesia: general anesthesia, which affects the entire body and results in complete insensitivity, and local anesthesia, which involves the administration of substances to temporarily block pain perception in a specific body region while allowing the patient to remain aware of the surgical process [1].

A. M. Al-Bakry et al. (Eds.): NTICT 2023, CCIS 2096, pp. 223–237, 2024.
https://doi.org/10.1007/978-3-031-62814-6_16

In the operating room, it is crucial to achieve a complete absence of responsiveness in the patient. One of the challenges during surgical procedures is the anesthetist's skill in maintaining the patient's consciousness at an optimal level throughout the operation. The initial dosage of medication sets the baseline level of anesthesia, which is then adjusted based on factors such as the patient's age, weight, height, and gender. Since the required anesthesia level can vary and the administered drug quantities may not always be accurate, the concept of closed-loop anesthesia control has emerged. However, designing such a controller must take into account the individual variations in patient characteristics to ensure the desired dosage [2]. Using closed-loop control in anesthesia has several advantages in determining dosage. This approach improves patient safety by preventing excessive administration and unwanted sensations during the surgical procedure. Additionally, an optimal feedback controller can reduce healthcare costs by avoiding dosage-related issues and shortening the recovery phase. The notable advantage of closed-loop techniques is their resilience and reliability, even in the presence of uncertainties within the system.

The primary source of uncertainty stems from substantial variances among patients; Some individuals showed limited responsiveness to dosage effects, while others exhibited increased sensitivity to dosage variations [3]. Various intelligent techniques have been introduced to ensure the effectiveness of Depth of Anesthesia (DOA) regulation. In [4], a model predictive control system for the depth of hypnosis is proposed and analyzed. The control scheme utilizes a nonlinear multiple-input–single-output (MISO) model to predict the influence of remifentanil on the hypnotic effect of propofol. It then employs a generalized model predictive control algorithm and a ratio between the two drugs to determine the optimal dosage for achieving the desired BIS level. A PID-based control system was developed in [2] to regulate the depth of hypnosis through the coadministration of propofol and remifentanil. This controller was designed to fulfill specific clinical requirements that are relevant to anesthesiologists. A nonlinear control technique called backstepping was developed in [5] to track the desired level of hypnosis in patients during surgery. In [6], a genetically optimized PID control was introduced to regulate the depth of hypnosis. A fuzzy logic control system, developed in [7], was used to regulate the depth of general anesthesia during surgery under the influence of propofol. This controller effectively controlled various parameters that impact the patient's condition during the procedure.

In [8], a proposed machine learning model for predicting the impact of drug dosage was introduced, while [9] utilized a reinforcement learning model for drug infusion control. Unfortunately, the practical applicability of these approaches is still limited. This is because they do not take into account the full range of population characteristics, due to the limited number of subjects. Additionally, they fail to consider the inheritance relationships between anesthetic and analgesic drugs. In [10], a method for explainable RL was proposed to stabilize the BIS and predict BIS levels. This method helps physicians understand how the RL agent makes optimal action decisions. On the other hand, [11] presents a nonlinear design method that uses the SMC method to calculate the infusion rate. In all cases, the values obtained are better than the desired value within the acceptable range for surgery. Also, in [12], the extended prediction, self-adaptive control (EPSAC) method has been utilized. Subsequently, a closed-loop method has been implemented with an appropriate sampling time and an optimal value for the prediction horizon in the EPSAC algorithm. Finally, in [13], a positive state observer is designed to implement a control scheme proposed for automatic administration, aiming to track a desired level for the BIS. All of these methodologies elaborate on their respective capabilities in upholding DOA levels effectively.

Within this paper, we use an optimized PID controller to regulate the Depth of Anesthesia (DOA). This controller effectively manages the nonlinearity in the patient's mathematical model. Additionally, we employ the COA tuning technique to determine the optimal drug dosage. The paper is structured as follows: Sect. 2 explains the mathematical model of the patient. In Sect. 3, we elaborate on the COA tuning method. Section 4 outlines the DOA controller scheme. We discuss the simulation results in Sect. 5 and provide concluding remarks in Sect. 6.

2 Mathematical model for the Patient

Anesthesia dosage administration is based on a mathematical model that incorporates both pharmacokinetic (PK) and pharmacodynamic (PD) components. The PK model determines the direction and rate of dosage, considering patient characteristics like age, weight, height, and gender. On the other hand, the PD model establishes the relationship between the dose and the resulting response recorded by the monitoring device, indicating how the dose affects the patient. Figure 1 illustrates the PK/PD model [14]:

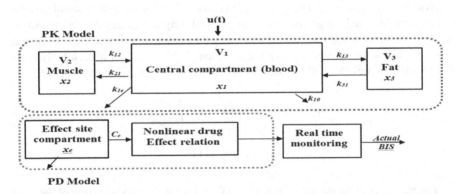

Fig. 1. PK/PD model.

The mathematical expressions of the PK/PD model are developed with a focus on the use of the propofol drug [15]:

$$x_1' = \frac{u(t)}{V_1} - k_{12}x_1 - k_{13}x_1 - k_{10}x_1 + k_{21}x_2 + k_{31}x_3 \tag{1}$$

$$x_2' = k_{12}x_1 - k_{21}x_2$$

$$x_3' = k_{13}x_1 - k_{31}x_3$$

In this context, x_1 delineates the dosage quantity in the central compartment, which is deemed crucial (blood). The remaining two components, x_2 and x_3, represent the dosage quantity in the muscle and fat compartments respectively. The coefficients k_{ij}, where $j \neq i$, signify the distribution of dosage rate from the j_{th} compartment to the i_{th} compartment. Meanwhile, the variable k_{10} encapsulates the dosage metabolism, while $u(t)$ denotes the propofol drug rate into the primary compartment (blood). Among the assortment of models that have emerged, the Schnider model stands out as the most suitable for three distinct bodily compartments [16]. This distinction arises from its capacity to delineate the requisite variables within relationships that accurately mirror the patient's characteristics, as illustrated in Eqs. (2, 3, 4):

$$V_1 = 4.27. V_2 = 18.9 - 0.391(\text{age-53}). V_3 = 238.$$

$$k_{10} = CL_1/V_1. \; k_{12} = CL_2/V_1. \; k_{13} = CL_3/V_1. k_{31} = CL_3/V_3 \min^{-1} \qquad (2)$$

$$k_{1e} = 0.456, \; k_{e0} = 0.456 \min^{-1}$$

$$CL_1 = 1.89 + 0.0456(\text{weight} - 77) - 0.0681(lbm - 59) + 0.0264(\text{height} - 177).$$

$$CL_2 = 1.29 - 0.024(\text{age} - 53), \; CL_3 = 0.836 \min^{-1}.$$

Evidently, the parameters k_{10}, k_{12}, k_{13}, k_{21}, and k_{31} are derived from actual patient-specific attributes including gender, age, height, and weight. CL_1 denotes the extraction rate from the body, while CL_2 and CL_3 quantify the dosage transferred from the central to peripheral compartments during administration. The calculation of lean body mass (lbm) for both males (M) and females (F) is detailed in the following passage [17]:

$$\text{lbm}\_m = 1.1 \, \text{weight} - 128 \frac{\text{weight}^2}{\text{height}^2} \qquad (3)$$

$$\text{lbm}\_f = 1.07 \, \text{weight} - 148 \frac{\text{weight}^2}{\text{height}^2} \qquad (4)$$

The pharmacodynamic segment relies on C_p which is the concentration of the drug in the blood of the central compartment, as depicted in the following context (5):

$$x_e'(t) = k_{1e}x_1'(t) - k_{e0}x_e(t) \qquad (5)$$

K_{e0} and k_{1e} possess predefined values, whereas x_e signifies the rate of dosing within the pharmacodynamic component, elucidating the effect segment. Because k_{1e} is a significantly smaller value compared to k_{e0}, it can be considered negligible. The concentration in the effect site compartment can be calculated using the Eq. (6) provided below:

$$C_e'(t) = k_{e0}\big(c_p(t) - c_e(t)\big) \qquad (6)$$

C_e signifies the concentration within the effect site compartment of the body. Various medical devices can be employed to compute C_e during surgery. Among these, the BIS monitoring index is the favored monitoring device chosen for monitoring level of unconsciousness (LOU) sometimes called Depth of Anesthesia (DOA) or Depth of Hypnosis (DOH). This index directly mirrors the patient's sensory state, with values ranging from 0 to 100. A value of zero signifies a lack of brain activity within the patient's body, while a value of 100 indicates that the patient is free from medication influence and fully awake. Upon the commencement of surgery, the targeted BIS value is set at 50, ideally maintaining it within the range of 40 to 60. This range indicates the achievement of a desirable level of hypnosis as intended by clinicians. The connection between the BIS value and the concentration of the dose effect (C_e) in the effect site compartment, is established by the relation outlined in (7). This connection underscores the time-varying and nonlinear characteristics of the BIS value, aligning with the attributes of the Sigmoid Hill Equation [18]:

$$BIS(t) = E_0 - E_{max} \frac{C_e^\gamma}{C_e^\gamma + EC_{50}^\gamma} \tag{7}$$

Here, E_0 represents the effect when no dosage is administered and is set at 100. E_{max} denotes the maximum value reached as a result of the dosage rate, Meanwhile, EC_{50} represents the infusion rate value at which the midpoint of the maximum hypnosis effect is achieved, revealing the body's reaction to the dose. The parameter γ indicates the slope of the equation i.e. patient response to medication. Equation (8), depicting the reciprocal of the Hill equation [19]:

$$C_e(t) = EC_{50}(\frac{E0 - BIS(t)}{Emax - E0 + BIS(t)})^{\frac{1}{\gamma}} \tag{8}$$

3 Coatis Optimization Algorithm

Optimization techniques have been applied to address intricate engineering systems. The COA methodology is formulated by emulating the innate behaviors of coatis, particularly their interactions with predators, which involve intelligent processes. This approach is inspired by the way coatis engage with iguanas during these interactions.

3.1 Initializing the Algorithm

The COA (Coati Optimization Algorithm) methodology, developed by Mohammad Dehghani et al. [20], adopts a population-based metaheuristic approach, where coatis are considered integral members of the algorithm's population. Within the search space, the position of each coati determines the values assigned to the decision variables. Therefore, in this algorithm, the position of a coati represents a potential solution to the problem. To initialize the COA, the positions of coatis in the search space are randomly established using Eq. (9) [20].

$$X_i : x_{i,j} = lb_j + r. (ub_j - lb_j), i = 1, 2, \ldots, N, j = 1, 2, \ldots, m \tag{9}$$

In this equation, Xi denotes the position of the ith coati within the search space, xi, j represents the value of the jth decision variable, N stands for the total count of coatis, m signifies the number of decision variables, r corresponds to a random real number within the range [0, 1], and lbj and ubj denote the lower and upper bounds of the jth decision variable respectively. The population of coatis at the COA is represented mathematically by the matrix X, also referred to as the population matrix [20]. When potential solutions are assigned to decision variables, it results in the calculation of various values for the objective function of the problem. These computed values are expressed using Eq. (11). In this equation, F represents the vector of objective function values attained, with Fi indicating the specific objective function value obtained from the ith coati. In this equation, F represents the vector of achieved objective function values, with Fi representing the specific objective function value obtained from the ith coati. In metaheuristic algorithms such as COA, the evaluation of a candidate solution's quality depends on the value of the corresponding objective function. Therefore, the best individual in the population is the member that produces the most favorable objective function. As candidate solutions undergo updates during the algorithm's iterations, the best individual in the population also gets updated in each iteration [20].

$$X = \begin{bmatrix} X_1 \\ \vdots \\ X_i \\ \vdots \\ X_N \end{bmatrix} = \begin{bmatrix} x_{1,1} & \cdots & x_{1,j} & \cdots & x_{1,m} \\ \vdots & \ddots & \vdots & \cdot & \vdots \\ x_{i,1} & \cdots & x_{i,j} & \cdots & x_{i,m} \\ \vdots & \cdot & \vdots & \ddots & \vdots \\ x_{N,1} & \cdots & x_{N,j} & \cdots & x_{N,m} \end{bmatrix} \tag{10}$$

$$F = \begin{bmatrix} F_1 \\ \vdots \\ F_i \\ \vdots \\ F_N \end{bmatrix} = \begin{bmatrix} F(X_1) \\ \vdots \\ F(X_i) \\ \vdots \\ F(X_N) \end{bmatrix} \tag{11}$$

3.2 COA Mathematical Model

In COA, the updating of the positions of coatis that represent candidate solutions relies on emulating two natural inherent behaviors observed in coatis. These behaviors include: (1) attacking strategy, and (2) escaping strategy. Therefore, the COA population undergoes updates through two distinct phases:

a. Phase 1: attacking iguana strategy (exploration phase)

Simulating coati behavior during iguana attacks is the first step in updating the coati population within the search space. In this approach, a group of coatis climbs a tree to provoke an iguana. Meanwhile, other coatis stay beneath the tree, waiting for the iguana to come down to the ground. Once the iguana descends, the coatis attack and capture it. This strategy encourages the coatis to explore different areas within the search space, demonstrating the COA's ability to engage in global exploration in problem-solving [20].

In the COA framework, the position of the finest population member is analogized to the position of the iguana. The scenario is conceived in such a way that the coati population is divided into two equal parts. One part of the coatis ascend the tree, while the second part remains stationed, awaiting the iguana's descent. Consequently, the mathematical simulation of the positions of coatis climbing the tree is expressed using Eq. (13).

$$X_i^{p1} : X_{ij}^{p1} = x_{i,j} + r. = (Iguana - I.x_{i,j}), \text{ for } i = 1, 2, \ldots, \left[\frac{N}{2}\right]$$

$$\text{and } j = 1, 2, \ldots, m. \tag{12}$$

The iguana is randomly positioned within the search space once it lands on the ground. In response to this newly assigned position, the coatis situated on the ground adjust their positions within the search space. This adjustment is simulated through Eqs. (13) and (14).

$$Iguana^G : Iguana_j^G = lb_j + r.(ub_j - lb_j), \quad j = 1, 2, \ldots m, \tag{13}$$

$$X_i^{p1} : X_{ij}^{p1} = \begin{cases} x_{i,j} + r. \left(Iguana - I.x_{i,j}\right), & F_{Iguana^G} < F_i, \\ x_{i,j} + r. \left(x_{i,j} - Iguana_j^G\right), & else, \end{cases}$$

$$\text{for } I = \left\lfloor\frac{N}{2}\right\rfloor + 1, = \left\lfloor\frac{N}{2}\right\rfloor + 2, \ldots, N \text{ and } j = 1, 2, \ldots, m. \tag{14}$$

If the value of the objective function is improved by the recalculated position for the coati, then this recalculated position is considered suitable for the update procedure. Otherwise, the coati maintains its previous location. This updating procedure is applied for values of i ranging from 1 to N, as illustrated in Eq. (15).

$$Xi = \begin{cases} X_i^{p1, F_i^{p1} < F_i} \\ X_i, else \end{cases} \tag{15}$$

In this context, X_i^{p1} refers to the updated position of the *ith* coati, where X_{ij}^{p1} denotes its position along the *jth* dimension, F_i^{p1} represents its objective function amount. The variable 'r' denotes a randomly selected real number from the interval [0, 1]. We use 'Iguana' to denote the best member's position in the search space. 'Iguana_j' signifies its position along the *jth* dimension. The variable 'I' is a randomly chosen integer from the set {1, 2}. 'Iguana^G' corresponds to the position of the iguana on the ground, generated randomly, with 'Iguana$_j^G$' representing its position along the *jth* dimension. Finally, 'F$_{Iguana}$G' signifies the objective function value associated with this position, and the notation $\lfloor \cdot \rfloor$ denotes the floor function, sometimes referred to as the greatest integer function [20].

b. Phase 2: Escaping strategy (exploitation phase)

Escaping from predators is the second step in updating coati positions within the search space. This step is represented mathematically by drawing inspiration from the coati behavior when they encounter and evade predators. The creature instinctively flees its current location when a predator threatens it. In this escaping strategy, the movements of coati in this strategy result in securing a safe spot near its current position, showcasing the COA's proficiency in local search and exploitation. A random position in close proximity to the current location of each coati was generated to mimic this behavior, as illustrated in the following equations provided in (16) and (17) [20].

$$lb_j^{local} = \frac{lb_j}{t}, \; ub_j^{local} = \frac{ub_j}{t} \; \text{where} \; t = 1, \, 2, \ldots, T. \tag{16}$$

$$X_i^{p2} = X_{i,j}^{p2} = x_{i,j} + (1 - 2r) \cdot \left(lb_j^{local} + r \cdot \left(ub_j^{local} - lb_j^{local}\right)\right),$$
$$i = 1, \, 2, \, \ldots, N, \; j = 1, \, 2, \, \ldots, m, \tag{17}$$

The newly computed position is considered valid if it results in an enhancement of the objective function's value, a criterion established and modeled in (18).

$$Xi = \begin{cases} X_i^{p2, \; F_i^{p2} < F_i} \\ X_i, \, else \end{cases} \tag{18}$$

In this context, X_i^{p2} represents the newly computed position for the *ith* coati, as determined in the second phase of the algorithm. X_{ij}^{p1} signifies its position along the *jth* dimension, while F_i^{p2} stands for the value of its corresponding objective function. The variable r represents a random real number drawn from the period [0, 1], t denotes the current loop count, lb_j^{local} and ub_j^{local} refer to the local lower and upper bounds for the *jth* decision variable, respectively. Additionally, lb_j and ub_j represent the lower and upper bounds for the same *jth* decision variable, respectively [20].

The flowchart in Fig. 2 presents the different stages of the COA implementation.

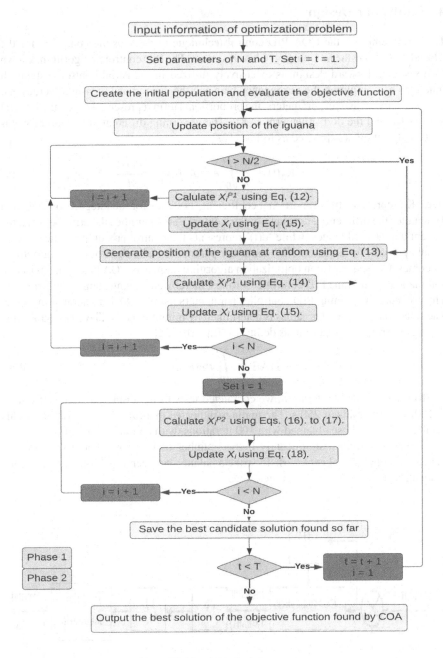

Fig. 2. COA flowchart

4 Controller Design

This study employs the COA-PID control technique to work as the controller for the Closed-Loop Anesthesia Delivery (CLAD) system. The PID controller algorithm, known for its straightforward design, is extensively utilized in industrial control systems. It functions by computing the discrepancy between the desired and measured values and consists of three distinct parameters: the proportional term represented as K_p, the integral term as K_i, and the derivative term as K_d. The governing differential equation for this control algorithm is expressed as follows [21]:

$$d(t) = K_p(t) + \frac{1}{T} \int_0^i e(t)dt + T_d + \frac{de(t)}{dt} \tag{19}$$

Here, Kp represents the proportional coefficient, Ti stands for the integral constant, and Td denotes the differential constant. In this context, 'e(t)' signifies the error between the current and desired values, while 'd(t)' represents the controller's output. The conventional approach for adjusting PID controller parameters falls short of ensuring optimal outcomes. Consequently, an optimization algorithm known as COA has been devised to automatically fine-tune PID controller settings, thereby enhancing control performance. The procedure for tuning PID controller parameters using COA is illustrated in Fig. 3. The selection of parameter values is determined by minimizing the ITAE (Integral Time Absolute Error) cost function, as defined in Eq. (20) [22].

$$\text{ITAE} = \int_0^t t|e(t)| \, dt \tag{20}$$

To select a suitable iteration, we conducted tests using a range of values, from 30 to 100, with intervals of 250 s. The findings indicated that running 35 iterations yields results similar to those obtained with 100 iterations, with an interval of 250 s.

However, setting a significantly higher number of iterations would substantially increase computational expenses. To strike a balance between reliability and efficiency in identification, we opted for 50 iterations.

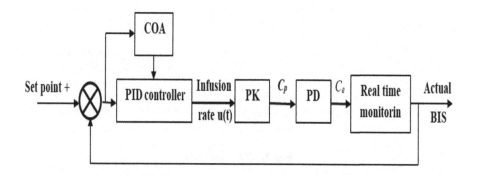

Fig. 3. CLAD system block diagram

5   Results and Discussion

We utilized MATLAB software to simulate the controller model, focusing on a cohort of eight patients arranged according to the specifications detailed in Table 1 [23]. These patients have been categorized into three distinct groups: sensitive, insensitive, and nominal, based on their unique attributes, as outlined in Table 1. The anesthesia drug infusion rate is directly influenced by the attributes associated with these classes.

Table 1. Patients characteristics [23]

Patient ID	Age (year)	Height (cm)	Weight (kg)	Gender	E_{C50}	E_0	E_{max}	γ
1	40	163	54	Female	6.33	98.8	94.1	2.24
2	36	163	50	Female	6.76	98.6	86	4.29
3	28	164	52	Female	8.44	91.2	80.7	4.1
4	50	163	83	Female	6.44	95.9	102	2.18
5	28	164	60	Male	4.93	94.7	85.3	2.46
6	43	163	59	Female	12.1	90.2	147	2.42
7	37	187	75	Male	8.02	92	151	2.1
8	38	174	80	Female	6.56	95.5	76.4	4.12

The table labeled Table 2 provides a listing of the variables employed within the Coati Optimization Algorithm.

Table 2. Variables employed within the Coati Optimization Algorithm.

COA variables	3
No. of Search Agents	30
No. of iterations	100

When applying the proposed control scheme to the patient model, the resulting response is illustrated in Fig. 4. Upon careful analysis of this figure, we observe that the induction phase, in accordance with the anesthesia stages (induction, maintenance, recovery), facilitates the transition of the patient's condition from a state of wakefulness to a level suitable for surgery commencement. For sensitive patients, the settling time begins at around 53 s, or approximately 1 min, indicating the speed at which the system found the desired BIS 50. Whereas for insensitive patients, it starts at 240 s, or roughly 4 min later. While this may not seem significantly different, it does impact the infusion rate, which increases every second, as shown in Fig. 5. The maintenance stage, on the other hand, begins when the BIS value appears to stabilize and remains constant throughout the remainder of the surgery. According to Fig. 4, the settling time lasts for approximately 1 min for sensitive patients and 4 min for insensitive patients. After this period, the system maintains its stability without any undershoot or overshoot, eventually reaching a steady-state condition where the value remains constant. Ultimately, the error is computed by assessing the variance between the desired BIS value and the actual BIS value for all patients, as illustrated in Fig. 6.

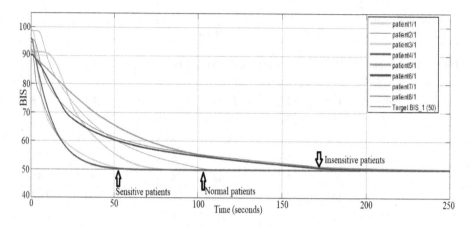

Fig. 4. BIS reading for patients.

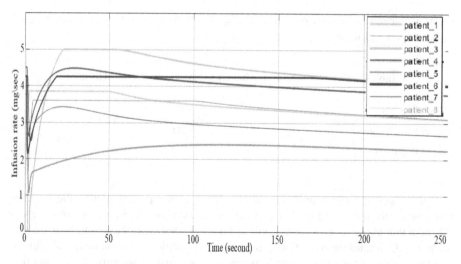

Fig. 5. Propofol infusion rate (control signal) for 8 patients.

The PID controller parameters, fine-tuned through the COA optimization process, are consolidated in Table 3.

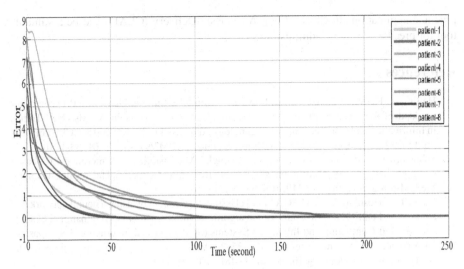

Fig. 6. Tracking error for all patients

Table 3. Optimized PID parameters by COA.

Patient	Kp	Ki	Kd	Patient	Kp	Ki	Kd
1	4.2419	0.81841	3.9378	5	7.1683	0.19757	3.2901
2	4.3928	0.72176	5.4421	6	9.788	0.5172	7.6889
3	2.8368	0.31442	5.0336	7	3.195	0.28224	0.87226
4	8.7226	0.75251	5.7964	8	3.9822	0.92923	7.4085

6 Conclusion

This paper presents an optimized PID controller that has been designed to regulate the administration of propofol for managing the depth of hypnosis during general anesthesia using Bispectral index (BIS) as a process variable. The PID parameters were adjusted using the coati optimization algorithm, with the goal of minimizing the Integrated Time Absolute Error (ITAE) across a group of eight patients. In this approach, the nonlinear behavior of the hill function was approximated using linear functions. This enabled the tuned controller, initially identified in response to patient 1, to be applied across different patient behaviors. The introduced controller achieved the BIS interval within approximately 53 s for sensitive patients and 240 s for insensitive patients. Despite variations in patient characteristics and the simplicity of the PID controller, the proposed control scheme consistently produces a stable response. However, additional studies are needed to ensure safety before proceeding with clinical trials. In future research, we plan to explore the potential of COA-PID in closed-loop control for inhalation anesthesia and combined anesthesia. We expect that the combination of BIS with COA-PID can have practical applications in Closed-Loop Anesthesia Delivery (CLAD) and be a valuable tool for precise anesthetic administration.

References

1. Hattim, L., Karam, E.H., Issa, A.H.: Implementation of selftune single neuron PID controller for depth of anesthesia by FPGA. In: Al-mamory, S., Alwan, J., Hussein, A. (eds.) New Trends in Information and Communications Technology Applications. NTICT 2018. CCIS, vol. 938, pp. 159–170. Springer, Cham (2018). https://doi.org/10.1007/978-3-030-01653-1_10
2. Schiavo, M., Padula, F., Latronico, N., Paltenghi, M., Visioli, A.: A modified PID-based control scheme for depth-of-hypnosis control: Design and experimental results. Comput. Methods Programs Biomed. **219**, 106763(2022)
3. Dang, T.L., Hoshino, Y.: An-FPGA based classification system by using a neural network and an improved particle swarm optimization algorithm. In: 2016 Joint 8th International Conference on Soft Computing and Intelligent Systems (SCIS) and 17th International Symposium on Advanced Intelligent Systems (ISIS), 25–28 August 2016, Sapporo, Japan, pp. 97–102. IEEE (2016). https://doi.org/10.1109/SCIS-ISIS.2016.0033
4. Pawłowski, A., Schiavo, M., Latronico, N., Paltenghi, M., Visioli, A.: Model predictive control using MISO approach for drug co-administration in anesthesia. J. Process Control **117**, 98–111 (2022)
5. Khaqan, A., Bilal, M., Ilyas, M., Ijaz, B., Ali Riaz, R.: Control law design for propofol infusion to regulate depth of hypnosis: a nonlinear control strategy. J. Comput. Math. Methods Med. **2016**, 1810303 (2016)
6. Padula, F., Ionescu, C., Latronico, N., Paltenghi, M., Visioli, A., Vivacqua, G.: Optimized PID control of depth of hypnosis in anesthesia. J. Comput. Methods Programs Biom. **144**, 21–35 (2017)
7. Samira, B., Hanane, Z.: Fuzzy logic control system in medical field. In: International Conference on Industrial Engineering and Operations Management, pp. 26–27, © IEOM Society International, Paris, France (2018)
8. Lee, H.C., Ryu, H.G., Chung, E.J., Jung, C.W.: Prediction of bispectral index during target-controlled infusion of propofol and remifentanil a deep learning approach. Anesthesiology **128**(3), 492–501 (2018)
9. Moore, B.L., Pyeatt, L.D., Kulkarni, V., Panousis, P., Padrez, K., Doufas, A.G.: Reinforcement learning for closed-loop propofol anesthesia: a study in human volunteers. J. Mach. Learn. Res. **15**(1), 655–696 (2014)
10. Yun, W.J., Shin, M., Jung, S., Ko, J., Lee, H.C., Kim, J.: Deep reinforcement learning-based propofol infusion control for anesthesia: a feasibility study with a 3000-subject dataset. Comput. Biol. Med. **156**, 106739 (2023)
11. Khaqan, A., Riaz, R.A.: Depth of hypnosis regulation using nonlinear control approach. In: 2016 IEEE International Conference on Electro Information Technology (EIT), pp. 100–104. IEEE, Budapest, Hungary (2016)
12. Ionescu, C.M., Copot, D., De Keyser, R.: Anesthesiologist in the loop and predictive algorithm to maintain hypnosis while mimicking surgical disturbance. IFAC-PapersOnLine **50**(1), 5080–15085 (2017)
13. Nogueira, F.N., Mendonça, T., Rocha, P.: Positive state observer for the automatic control of the depth of anesthesia - clinical results. Comput. Methods Programs Biomed. **171**, 99–108 (2019)
14. Caiado, D.V., Lemos, J.M., Costa, B.A., Silva, M.M., Mendonça, T.F.: Design of depth of anesthesia controllers in the presence of model uncertainty. In: 21st Mediterranean Conference on Control and Automation (MED), pp. 213–218. IEEE, Platanias, Greece (2013)

15. Naşcu, I., Oberdieck, R., Pistikopoulos, E.N.: An explicit hybrid model predictive control strategy for intravenous anaesthesia. IFAC-PapersOnLine 28(20), 58–63 (2015)
16. Ionescu, C.M., De, K.R., Torrico, B.C., De, S.T., Struys, M.M.R.F., Normey-Rico, J.E.: Robust predictive control strategy applied for propofol dosing using BIS as a controlled variable during anesthesia. IEEE Trans. Biomed. Eng. 55(9), 2161–2170 (2008)
17. Schnider, T.W., Minto, C.F., Shafer, S.L.: The influence of age on propofol pharmacodynamics. Anesthesiology 90(6), 1502–1516 (1999)
18. Schiavo, M., Consolini, L., Laurini, M., Latronico, N., Paltenghi, M., Visioli, A.: Optimized robust combined feedforward/feedback control of propofol for induction of hypnosis in general anesthesia. In: IEEE International Conference on Systems, Man and Cybernetics, pp. 1266–1271. IEEE, Melbourne, Australia (2021)
19. Schamberg, G., Badgeley, M., Meschede-Krasa, B., Kwon, O., Brown, E.N.: Continuous action deep reinforcement learning for propofol dosing during general anesthesia. Artif. Intell. Med. 123, 102227 (2022)
20. Dehghani, M., Montazeri, Z., Trojovská, E., Trojovský, P.: Coati optimization algorithm: a new bioinspired metaheuristic algorithm for solving optimization problems. Knowl.-Based Syst. 259, 110011 (2023)
21. Kagami, R.M., Franco, R.M., Reynoso-Meza, G., Freire, R.Z.: PID control of hypnotic induction in anaesthesia employing multiobjective optimization design procedures. IFAC-PapersOnLine 54(15), 31–36 (2021)
22. Abood, L.H.: Optimal modified PID controller for automatic voltage regulation system. In: AIP Conference Proceedings, vol. 2415, no. 1. AIP Publishing, Baghdad, Iraq (2022)
23. Liang, Z., Fu, L., Li, X., Feng, Z., Sleigh, J.W., Lam, H.K.: Ant colony optimization PID control of hypnosis with propofol using renyi permutation entropy as controlled variable. IEEE Access 7, 97689–97703 (2019)

Eye Movement Recognition: Exploring Trade-Offs in Deep Learning Approaches with Development

Ali A. Masaoodi[1]([✉]) [iD], Haider I. Shahadi[2], and Hawraa H. Abbas[1]

[1] College of Information Technology Engineering, Al-Zahraa University for Women, Karbala, Iraq
{Ali.m,Hawraa.h}@alzahraa.edu.iq
[2] Department of Electrical and Electronic Engineering, University of Kerbala, Karbala, Iraq
haider_almayaly@uokerbala.edu.iq

Abstract. Eye movement recognition has garnered substantial attention in recent years across diverse disciplines such as Human-Computer Interaction (HCI), medical diagnostics, and assistive technologies. This technology offers transformative possibilities, especially for individuals with paralysis and disabilities. Yet, the deployment of deep learning models for eye movement classification using non-intrusive head-free cameras like webcams remains fraught with challenges. These challenges include the lack of comparative and benchmarking studies that guide researchers and practitioners in choosing appropriate deep learning models that match such complex tasks. To address these challenges, we conducted a meticulous comparative analysis of selected deep learning architectures, including customized and fine-tuned versions of ResNet-18, EfficientNet-B0, and AlexNet. Our analysis primarily aims to evaluate the performance and generalizability for each model across a complex dataset encompassing various conditions. The n-fold cross-validation is employed to assess the robustness of our findings. Our empirical assessments reveal a nuanced landscape. For instance, ResNet-18 excels in terms of accuracy with 99.5% and acquires a competitive small model size of 43MB, while AlexNet acquires around 222MB. While this advantage comes with slightly higher computational and memory overhead compared to models like EfficientNet-B0. This study offers critical insights into the trade-offs involved in selecting an optimal deep learning model for eye movement recognition under real-world conditions.

Keywords: Eye Movement Recognition · Eye Gesture Recognition · Eye Tracking · Deep Learning · Human-Computer Interaction

1 Introduction

The study of eye movement recognition spans a broad array of fields, from HCI and psychology to medical diagnostics. The importance of this research is underlined by its interdisciplinary applicability in various areas such as assistive technologies [1], gaze-oriented user authentication [2, 3], and early diagnosis of both psychological and neurological disorders [4–7] including diagnosis of autism spectrum disorder (ASD), Reading Impairment, and Parkinson's disease. Recently eye movement recognition used to augment stroke diagnosis specifically for posterior circulatory stroke (PCS) diagnosis [8]. Despite its significance, the endeavor is fraught with challenges, primarily because eye movements are subtle and highly susceptible to external variables like fluctuating light conditions and physical obstructions. Current approaches often depend on specialized hardware ranging from infrared cameras to wearable devices [9], sensor-based solutions [10], and mobile phone cameras [11]. These methods are not only costly but also complex, thereby restricting the widespread adoption of eye movement recognition technologies. Previous research has predominantly employed machine learning techniques based on electrooculography (EOG) for eye movement detection, often sidelining the application of deep learning models. Furthermore, there has been a wide lack of comparative evaluations carried out under consistent conditions, which is the gap that our study aims to fill. Our fine-tuned models are trained on data collected in various environmental conditions and using two different cameras including both low and high resolutions contributing to more robust models.

1.1 Objectives and Contributions

This paper aims to achieve the below objectives:

- Evaluating the performance of fine-tuned deep learning models on a collected eye movement dataset, aiming to address a significant gap in eye movement research characterized by the absence of deep learning models benchmarking that targets this important field. The models ResNet-18, EfficientNet-B0, and AlexNet are customized and fine-tuned for the eye movement recognition task.
- Conduct a systematic comparative study of these deep learning models, focusing on an array of performance metrics such as accuracy, loss, inference time, and model size.
- Implementing an n-fold cross-validation to mitigate the risk of overfitting and to confirm the models' robustness across different scenarios.

This study rigorously compares the models across multiple aspects, encountering practical considerations such as model generalizability and disk size consumption, in addition to accuracy and inference time. Hence, it offers a nuanced view of the strengths and limitations of each approach, thereby contributing significantly to the advancement of eye movement recognition research.

1.2 Eye's Anatomy, Movements, and Physiology

Before exploring the techniques for recognizing eye movements, it is essential to gain a comprehension of the eye's anatomical and physiological complexities. Key components such as the cornea, pupil, iris, and retina not only serve as integral parts of the visual system but also as salient features for the task of eye movement recognition in this research endeavor. Figure 1 shows the external anatomy of the human eye.

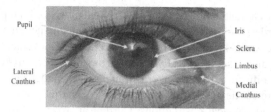

Fig. 1. Eye's external anatomy

1.3 Categories of Eye Movements

The diversity of human eye movements can be primarily sorted into three major categories, saccades, smooth pursuits, and fixations. Each category exhibits unique attributes and motion patterns, rendering the challenge of identifying them intricate and multi-dimensional. The following are the principal types of eye movements.

- Saccades: These are abrupt eye movements designed to swiftly shift the visual focus to a new point of interest. Saccades are particularly relevant in activities that necessitate quick changes in visual attention.
- Smooth Pursuits: This type of eye movement facilitates the tracking of a moving object seamlessly, maintaining the object's image at the center of the visual field. They are crucial for effective object tracking.
- Fixations: In this mode, the eyes remain stationary to enable in-depth visual processing of a specific object. Fixations are frequently observed in tasks that require attention to detail, such as reading.

2 Related Work

Eye movement recognition has evolved through a variety of research paradigms. Traditional methodologies have commonly relied on electrooculography (EOG) [10] or infrared oculography (IROG) to capture eye movement patterns. However, the burgeoning advancements in deep learning have led to a significant rise in the application of Convolutional Neural Networks (CNNs) for eye movement recognition using raw images [12]. Despite this, a comprehensive study comparing these diverse deep learning techniques across a range of performance metrics in the domain of eye movement recognition remains conspicuously absent. While there are studies that approached this task using traditional machine learning models, although, these models are not as reliable as deep learning models [13].

Real-time recognition of eye movements has been successfully implemented to steer wheelchairs, thereby enhancing mobility options for individuals with physical disabilities [15, 16]. In the realm of eye-controlled assistive devices, Xu et al. [17] present an eye-gaze controlled wheelchair for ALS patients, integrating a monocular camera and deep learning models, the system achieves a 98.49% accuracy in eye-movement recognition. A recent study employed eye movement recognition for assessing surgical skills showed that the highest accuracy accomplished for GBDT is 76.64% and for XGBoost is 76.63% [14]. Ezzat et al. [11] introduce "Blink-To-Live", an affordable and accessible eye-based communication system for individuals with speech impairments. Leveraging a modified Blink-To-Speak language, this system utilizes a mobile phone camera to track eye movements, translating four movements (Left, Right, Up, Blink) into over 60 daily life commands, although the study is lacking in mentioning the recognition accuracies and models comparison. Algorithms designed for eye movement recognition have extended their utility to the medical realm, aiding in the early detection of neurological conditions such as Alzheimer's and Parkinson's diseases [7].

3 Methodologies

The primary objective of this research is to pinpoint the most accurate, yet computationally efficient multi-class image classification model tailored for eye movement recognition. Specifically, this study concentrates on the identification of five principal eye movement directions: center, down, left, right, and up. Furthermore, the methodology encompasses four integral phases, data acquisition, preprocessing steps, model implementation, and performance evaluation criteria. By methodically structuring the research into these phases, the study aims to provide a nuanced and comprehensive analysis of the applicability and efficacy of various models in the context of eye movement recognition.

The dataset is collected under various conditions and consists of 3,380 images varying in age, gender, and environmental settings. It is divided into five classes, left, right, up, down, and center. The image format is RGB and sized 150 pixels x 300 pixels, and the overall disk size of the dataset was 169 MB.

The region of interest (ROI) for the eye is extracted from each image within both the training and testing datasets. Subsequently, the intensity values of the images are normalized to a range between 0 and 1. This standardization is critical as it stabilizes the process for deep learning models. To normalize each pixel within its Red, Green, and Blue (R, G, B) components in the image, the following Eqs. (1), (2), and (3), show how the normalization is performed on each pixel in the input image.

$$R_{normalized} = \frac{R}{255} \tag{1}$$

$$G_{normalized} = \frac{G}{255} \tag{2}$$

$$B_{normalized} = \frac{B}{255} \tag{3}$$

3.1 K-Fold Cross-Validation

In order to evaluate deep learning models, two values of k-fold cross-validation are employed. The final metrics are obtained for the best fold, in addition, the curves of the loss and accuracy to epochs are used to show the model performance over epochs. The confusion matrix was recorded for every model for the best fold. With this strategy of accumulating results and analyzing them, we are providing a process to identify the best model for eye movement classification. A 3-fold cross-validation process illustrated in Fig. 2 shows the possibilities of choosing the training split of the dataset.

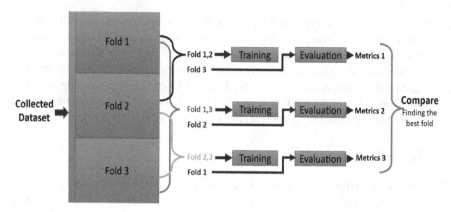

Fig. 2. Illustration of 3-fold cross-validation

3.2 Metrics Evaluation

The following are the metrics that are used to compare the performances of each model on eye movement recognition.

- Model Size: refers to the amount of storage required to store a model. This size is crucial as it impacts the model's deployability and efficiency, especially in environments with limited storage or computational resources. The model size was measured in MB.
- Accuracy: the proportion of true results (both true positives and true negatives) in the total number of cases examined as shown in Eq. (4).

$$Accuracy = \frac{TP+TN}{TP+TN+FP+FN} \tag{4}$$

- Precision: It measures the accuracy of positive predictions. It is the ratio of correctly predicted positive observations to the total predicted positive observations. The Precision equation is shown in Eq. (5).

$$Precision = \frac{TP}{TP+FP} \tag{5}$$

- Recall: It measures the ability of a model to find all the relevant cases within a dataset, and it represents the ratio of correctly predicted positive observations to all the observations in the actual class. The equation for recall is shown in Eq. (6).

$$Recall = \frac{TP}{TP+FN} \tag{6}$$

Where TP is the true positive, TN is the true negative, FP is the false positive, and FN is the false negative.

- F1 Score: It is the harmonic mean of precision and recall, and it is particularly useful when the class distribution is imbalanced. The equation for the F1 Score is shown in Eq. (7).

$$F1Score = 2 \times \frac{Precision \times Recall}{Precision + Recall} \tag{7}$$

- Loss: The loss metric often used is Cross-Entropy loss, which measures the difference between the predicted probabilities and the actual labels as in Eq. (8).

$$Cross - EntropyLoss = H(y, p) = \sum_i \left[y_i \log(p_i) + (1 - y_i) \log(1 - p_i) \right] \quad (8)$$

where y_i is the actual label and p_i is the predicted probability for each class in a multi-class classification problem.

- Model Generalization: how the model generalizes on the unseen images, where unseen images are used to estimate each model generalization. The batch-testing is used to evaluate each model on different unseen images.

3.3 Model Implementation

CNNs are inherently structured to excel at discerning spatially hierarchical features [18], which are imperative for challenges such as image classification. Nevertheless, the computational demands for training deep-layered CNNs are significant and often necessitate specialized computing resources. Additionally, techniques such as batch normalization and residual linkages are incorporated to mitigate issues related to vanishing or exploding gradient phenomena.

To experiment each model capability on new features of eye movements and to examine its performance for classifying eye movements into five classes (center, down, left, right, and up), this study utilized transfer learning and fine-tuning in the implementation of each model, with the freezing of convolutional layers and using trainable custom classifier layers (see Fig. 4).

The proposed architecture is shown in Fig. 3, where it shows the preprocessing block and augmentation. The machine that is used for computing is powered by CUDA GPU which is Nvidia GeForce RTX 3060 4 GB, 16 GB RAM, and Intel Core i7-11800H processor.

ResNet-18. The ResNet-18 architecture was selected for its proven effectiveness in various image classification tasks. Introduced by Kaiming He et al. in 2015 [19], this architecture employs shortcut connections to enable the training of deeper networks while mitigating and vanishing gradient problems. Fine-tuning parameter values were two-fold cross-validation, 10 epochs, 0.001 learning rate, and 32 batch size. The Validation accuracy and validation loss for the fine-tuned ResNet-18 is shown in Fig. 6, while its architecture is shown in Fig. 5.

244   A. A. Masaoodi et al.

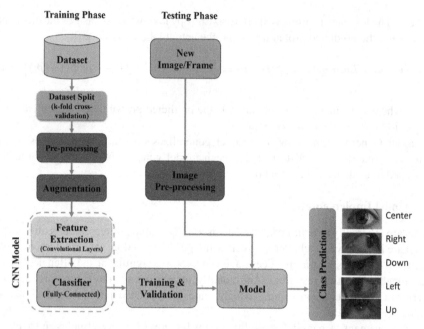

Fig. 3. Proposed architecture for training and testing for each fine-tuned model.

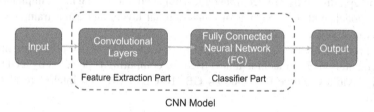

Fig. 4. The two components of each CNN model are the feature extractor and the classifier.

The key innovation of ResNet-18 is the shortcut connections (skip connections) that were introduced to allow the network to skip one or more layers. So, it effectively simplifies the network during training, making it easier to optimize and reducing the risk of vanishing gradients, a challenge in deep networks where gradients become too small to propagate useful learning information. The architecture of ResNet-18, while being shallower than other ResNet variants like ResNet34 or ResNet50, still demonstrates remarkable efficiency and accuracy, making it a popular choice for tasks requiring a balance between performance and computational resource usage. Its ability to perform well with fewer parameters also makes it a practical option for applications where model size and speed are critical.

EfficientNet-B0. EfficientNet-B0 is a CNN architecture, part of the EfficientNet family, developed by Tan and Le in 2019 [20]. Its core innovation lies in a methodical strategy for scaling the network, achieving optimization in not only depth but also width and

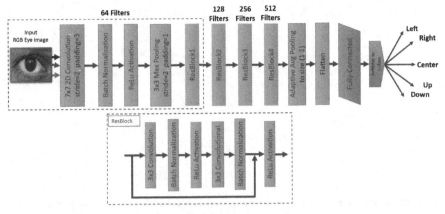

Fig. 5. ResNet-18 Architecture

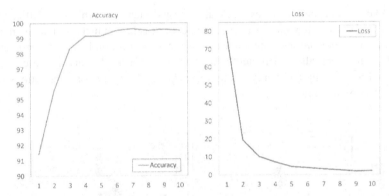

Fig. 6. The validation accuracy and loss metrics for the best-performing fold during the fine-tuning of ResNet-18 were evaluated using the collected diverse dataset.

resolution. This is accomplished through a compound scaling method, which uniformly increases all dimensions of the network based on a consistent set of scaling coefficients. EfficientNet-B0 demonstrates improved transfer learning capabilities, making it highly effective for tasks beyond its initial training scope. The architecture's efficiency in parameter usage translates to lower energy consumption during training and inference, making it an environmentally friendly option in the growing field of AI, where energy efficiency is increasingly important. The Validation accuracy and validation loss for the fine-tuned EfficientNet-B0 are shown in Fig. 7, while its architecture is shown in Fig. 8. Fine-tuning parameters values were two-fold cross-validation, 32 batch size, 10 epochs, and 0.001 learning rate.

EfficientNet-B0's unique contribution to neural network architecture lies in its balanced scaling of all three dimensions, depth, width, and resolution. This balance is achieved through the compound scaling method which systematically determines the optimal relationship between these dimensions for a given resource constraint.

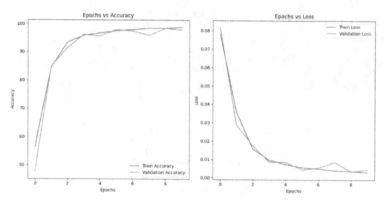

Fig. 7. The validation and training accuracy, as well as loss metrics, for the best-performing fold during the fine-tuning of EfficientNet-B0 were evaluated using the collected diverse dataset.

Unlike previous architectures that scaled these dimensions arbitrarily, EfficientNet-B0's method ensures that each aspect of the network is scaled in a harmonized manner, leading to better performance without excessive increase in complexity or computational cost. This approach allows EfficientNet-B0 to outperform other architectures that are larger and more computationally intensive, making it a choice for applications where efficiency is paramount.

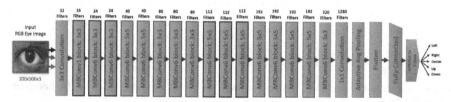

Fig. 8. The EffiecientNet-B0 architecture, is fundamentally built on the mobile inverted bottleneck convolution (MBConv), which is designed for efficiency, featuring inverted residuals, bottleneck, and channel expansion for optimal performance.

AlexNet. AlexNet is a CNN architecture proposed by Alex Krizhevsky, Ilya Sutskever, and Geoffrey Hinton in 2012 [21]. It considerably outperformed the architectures in the ImageNet Large Scale Visual Recognition Challenge (ILSVRC), by initializing the deep learning revolution toward more robust computer vision tasks. The network consists of eight layers, divided into five convolutional layers followed by three classifier layers (fully connected layers). One of its prominent features in this network is the employment of the ReLU (Rectified Linear Unit) activation function, which led to its higher training speeds.

AlexNet's innovation extended to its use of data augmentation and dropout techniques, which were crucial for preventing overfitting in its large network. Data augmentation involved artificially enlarging the dataset through transformations like flipping and cropping, enhancing the network's ability to generalize from its training data.

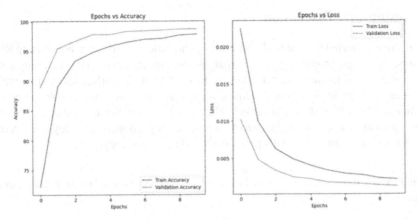

Fig. 9. The validation and training accuracy, as well as loss metrics, for the best-performing fold during the fine-tuning of AlexNet were evaluated using the collected diverse dataset.

The dropout technique that is used is applied in the first two fully connected layers, randomly disabling a fraction of neurons during training. This means that this approach ensured that the network did not become overly reliant on any specific neuron path, further bolstering its generalization capabilities.

Additionally, AlexNet was one of the first models to use GPU computing, specifically NVIDIA GTX 580 GPUs, which was a significant factor in its training efficiency and performance improvement over previous architectures. Fine-tuning parameter values were two-fold cross-validation, 10 epochs, 0.001 learning rate, and 32 batch size.

The validation accuracy and loss for the implemented fine-tuned AlexNet are shown in Fig. 9, while a simplified architecture of AlexNet model is shown in Fig. 10.

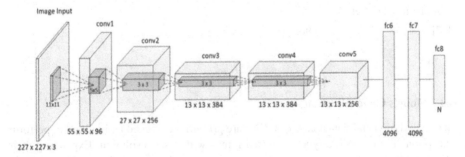

Fig. 10. Simplified diagram of AlexNet architecture [22].

4 Results and Discussion

The result is collected for best fold after training and validation for each model using RGB images, data augmentations, and hyperparameters tuning. A performance comparison across different fine-tuned models, namely ResNet-18, EfficientNet-B0, and AlexNet, is conducted with respect to key metrics, recognition accuracy, model size, precision, recall, and F1-score as shown in Table 1. The inclusion of precision, recall, and F1-score allows for a more nuanced evaluation of each model beyond mere accuracy, considering both the relevance of the model's predictions and its consistency.

Table 1. Comparison between proposed models (fine-tuned) that trained on the eye movements dataset.

Model	Validation Accuracy	Precision	Recall	F1-Score	Model Size (Model disk size)
ResNet-18	99.5%	0.9981	0.9975	0.9978	43 MB
AlexNet	98.8%	0.9887	0.9970	0.9928	222 MB
EfficientNet-B0	98.6%	0.9669	0.9968	0.9817	15 MB

These tests proved that our trained and fine-tuned model has promising results compared to other works; we obtained 99.5% accuracy while [17] achieved 98.49%. As shown in Table 2.

Table 2. Comparison of our results with related works

	Accuracy	Number of Eye Movements Categories
Our fine-tuned model	99.5%	5
[17]	98.5%	3

4.1 Models Generalization

In the field of visual data analysis, CNNs are primarily designed and refined to perform exceptionally on specific types of visual inputs with a set resolution. Expanding their effectiveness to new unseen images continues to be a challenging endeavor. Despite these difficulties, our adjusted models show a strong ability for generalization, as evidenced not only by high accuracy but also by excellent precision and recall rates. This indicates that the models are not only accurate overall but are also reliable in their positive predictions (precision) and comprehensive in identifying all relevant instances (recall). Sustaining their performance even under circumstances where the subject is located far from the camera, or when the camera's resolution is reduced. Figure 11 shows the result of testing an unseen image input using the EfficientNet-B0 model.

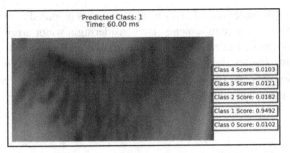

Fig. 11. Batch-Testing for fine-tuned EfficientNet-B0 on unseen eye image, where class 1 represents the down eye movement class.

4.2 Trade-Offs

Accuracy is an essential metric in evaluating the effectiveness of our fine-tuned model in classifying the correct label for a given input. As indicated in Table 1, ResNet-18 stands out with perfect accuracy scores, complemented by its high precision and recall, indicating fewer misclassifications both in terms of false positives and false negatives. Its F1 score closely mirrors its accuracy, suggesting a balanced classification capability. While EfficientNet-B0 and AlexNet lag slightly in accuracy, their precision, recall, and F1 scores are indicative of their robustness. The F1 score reflects the trade-off between precision and recall, which is crucial in applications where the cost of false positives and false negatives is significant. EfficientNet-B0 has the smallest model size which suggests it is more suitable for real-time and constrained environments applications compared to AlexNet, providing a balance between compactness and performance as reflected in its precision, recall, and F1-scores. While the near-perfect metrics of ResNet-18 suggest robust generalization capabilities, its model size also represents a balance compared to other models.

5 Conclusion

The evaluation of ResNet-18, EfficientNet-B0, and AlexNet on diverse eye movement dataset reveals an array of trade-offs. For real-time applications with limited computational resources, the EfficientNet-B0 is ideal due to its smaller model size compared to its competitive F1-score, suggesting a good balance despite its slightly lower accuracy. ResNet-18 presents a good solution, since its high accuracy, precision, recall, indicating its effectiveness in correct classification with minimal false positives and negatives, and its F1-score further underscores its high performance. The balanced model size of ResNet-18 with an almost neglected increase in inference time compared to EfficientNet-B0 makes it suitable for scenarios where reliability is paramount, and resources are less constrained. The choice of a deep learning architecture for eye movement recognition has profound implications, extending beyond mere accuracy to include operational costs and computational complexity. Real-time applications might necessitate investments in optimization for speed, whereas edge computing solutions could demand compromising accuracy to maintain cost efficiency. Therefore, the optimal model is highly contextual and dependent on the specific trade-offs an application is prepared to make among cost, complexity, and real-time requirements. By considering a wider array of performance metrics, can make more informed decisions that align with operational objectives and constraints. Future studies should aim to explore the performance of these models with various image formats and different resolutions. And optimizing the models for more energy efficiency and less computational cost.

References

1. Rani, V.U., Poojasree, S.: An IOT driven eyeball and gesture-controlled smart wheelchair system for disabled person. In: 2022 8th International Conference on Advanced Computing and Communication Systems (ICACCS), pp. 1287–1291. IEEE (2022). https://doi.org/10.1109/ICACCS54159.2022.9785180
2. Cheng, S., Wang, J., Sheng, D., Chen, Y.: Identification with your mind: a hybrid BCI-based authentication approach for anti-shoulder-surfing attacks using EEG and eye movement data. IEEE Trans. Instrum. Meas. 72, 1–14 (2023). https://doi.org/10.1109/TIM.2023.3241081
3. Rajanna, V., Malla, A., Bhagat, R., Hammond, T.: DyGazePass: a gaze gesture-based dynamic authentication system to counter shoulder surfing and video analysis attacks. In: Proceedings of the 2018 IEEE International Symposium on Biometric and Security Technologies (ISBA), pp. 1–8. IEEE (2018). https://doi.org/10.1109/ISBA.2018.8311458
4. Duymaz, E., et al.: Early diagnosis of autistic children with eye tracker and artificial intelligence approach. In: 2022 Medical Technologies Congress (TIPTEKNO), pp. 1–4. IEEE (2022). https://doi.org/10.1109/TIPTEKNO56568.2022.9960148
5. Nagarajan, H., Inakollu, V.S., Vancha, P., Amudha, J.: Detection of reading impairment from eye-gaze behaviour using reinforcement learning. Procedia Comput. Sci. 218, 2734–2743 (2023). https://doi.org/10.1016/j.procs.2023.01.245
6. Budarapu, A., Kalyani, N., Maddala, S.: Early screening of autism among children using ensemble classification method. In: 2021 3rd International Conference on Advances in Computing, Communication Control and Networking (ICAC3N), pp. 162–169. IEEE (2021). https://doi.org/10.1109/ICAC3N53548.2021.9725586
7. Brien, D.C., et al.: Classification and staging of Parkinson's disease using video-based eye tracking. Parkinsonism Relat. Disord. 110, 105316 (2023). https://doi.org/10.1016/j.parkreldis.2023.105316
8. Hassan, M.A., et al.: Approach to quantify eye movements to augment stroke diagnosis with a non-calibrated eye-tracker. IEEE Trans. Biomed. Eng. 70(6), 1750–1757 (2023). https://doi.org/10.1109/TBME.2022.3227015
9. Zhang, S., et al.: An EMG-based wearable multifunctional Eye-control glass to control home appliances and communicate by voluntary blinks. Biomed. Signal Process. Control 86 (2023). https://doi.org/10.1016/j.bspc.2023.105175
10. Lin, C.-T., et al.: EOG-based eye movement classification and application on HCI baseball game. IEEE Access 7, 96166–96176 (2019). https://doi.org/10.1109/ACCESS.2019.2927755
11. Ezzat, M., Maged, M., Gamal, Y., Adel, M., Alrahmawy, M., El-Metwally, S.: Blink-To-Live eye-based communication system for users with speech impairments. Sci. Rep. 13(1) (2023). https://doi.org/10.1038/s41598-023-34310-9
12. Belaiche, R., Liu, Y., Migniot, C., Ginhac, D., Yang, F.: Cost-effective CNNs for real-time micro-expression recognition. Appl. Sci. 10(14), 4959 (2020). https://doi.org/10.3390/app10144959
13. Rasmussen, S.H.R., Ludeke, S.G., Klemmensen, R.: Using deep learning to predict ideology from facial photographs: expressions, beauty, and extra-facial information. Sci. Rep. 13(1), 5257 (2023). https://doi.org/10.1038/s41598-023-31796-1

14. Kuo, R.J., Chen, H.-J., Kuo, Y.-H.: The development of an eye movement-based deep learning system for laparoscopic surgical skills assessment. Sci. Rep. **12**(1), 11036 (2022). https://doi.org/10.1038/s41598-022-15053-5

15. Utaminingrum, F., Somawirata, I.K., Pengestu, G., Thaipisutikul, T., Shih, T.K.: Selecting control menu on electric wheelchair using eyeball movement for difable person. JOIV: Int. J. Inf. Visualizat. **7**(1), 37 (2023). https://doi.org/10.30630/joiv.7.1.1011

16. Tharwat, M., Shalabi, G., Saleh, L., Badawoud, N., Alfalati, R.: Eye-controlled wheelchair. In: 2022 5th International Conference on Computing and Informatics (ICCI), pp. 097–101. IEEE (2022). https://doi.org/10.1109/ICCI54321.2022.9756116

17. Xu, J., Huang, Z., Liu, L., Li, X., Wei, K.: Eye-gaze controlled wheelchair based on deep learning. Sensors **23**(13), 6239 (2023). https://doi.org/10.3390/s23136239

18. Eitel, F., Albrecht, J.P., Weygandt, M., Paul, F., Ritter, K.: Patch individual filter layers in CNNs to harness the spatial homogeneity of neuroimaging data. Sci. Rep. **11**(1), 24447 (2021). https://doi.org/10.1038/s41598-021-03785-9

19. He, K., Zhang, X., Ren, S., Sun, J.: Deep residual learning for image recognition. In: 2016 IEEE Conference on Computer Vision and Pattern Recognition (CVPR), Las Vegas, NV, USA, 27–30 June 2016, pp. 770–778 (2016). https://doi.org/10.1109/CVPR.2016.90

20. Tan, M., Le, Q.V.: EfficientNet: rethinking model scaling for convolutional neural networks. In: Proceedings of the International Conference on Machine Learning (ICML), pp. 6105–6114. PMLR (2019). https://doi.org/10.48550/arXiv.1905.11946

21. Krizhevsky, A., Sutskever, I., Hinton, G.E.: ImageNet classification with deep convolutional neural networks. Commun. ACM **60**(6), 84–90 (2017). https://doi.org/10.1145/3065386

22. Hemmer, M., Khang, H.V., Robbersmyr, K., Waag, T., Meyer, T.: Fault classification of axial and radial roller bearings using transfer learning through a pretrained convolutional neural network. Designs **2**(4), 56 (2018). https://doi.org/10.3390/designs2040056

Enhancing the Early Prediction of Learners Performance in a Virtual Learning Environment

Safa Ridha Albo Abdullah[✉] and Ahmed Al-Azawei

Department of Software, College of Information Technology, University of Babylon, Babil, Iraq
safaruda@uobabylon.edu.iq

Abstract. Educational institutions have widely adopted virtual learning environments (VLEs) in contemporary education. The limits of students' location are no longer a problem because they can learn from anywhere and anytime based on this approach. Hence, by forecasting students' performance in VLEs, educational institutions can enhance their online offerings and provide quality online learning content. This is not possible without taking into account various features that may have a great influence on students' academic accomplishments. The present paper intends to predict students' performance in an online platform. Four classifiers are utilized in the proposed model. This research integrates the whole dataset for science and social science modules. Moreover, to improve the model's prediction accuracy several steps are followed. First, new features are generated based on the available features namely, the total number of clicks before, the total number of clicks after, engagement, and average. Second, the model's hyperparameters are adjusted using the random search optimizer, whereas a feature selection approach is performed to choose the maximum influential features. The experimental results showed that the prediction accuracy is significantly enhanced based on the procedure proposed in this research. The suggested model successfully provides an early prediction of students' performance with an average accuracy of 84%. The outcomes of this research are discussed further to highlight its possible implications on theory and practice.

Keywords: Machine learning · random search optimizer · student performance · feature selection

1 Introduction

Improving learners' engagement with technological platforms can help enhance their learning experience [1]. Universities today strive to raise the level of instruction and learning while also boosting students' achievement [2]. Currently, a variety of educational formats, including online learning, blended learning, and face-to-face (F2F) learning are available [3]. However, a significant number of students in online learning environments fail or dropout of their classes. Additionally, it was shown that students who enrol in online learning courses leave them far more frequently than those who enrol in traditional education courses [4, 5]. As a result, educational institutions must devote greater attention to comprehending features that may affect learners' performance in online courses.

A. M. Al-Bakry et al. (Eds.): NTICT 2023, CCIS 2096, pp. 252–266, 2024.
https://doi.org/10.1007/978-3-031-62814-6_18

In general, the features could be classified into strongly relevant, weakly relevant, or irrelevant [6]. As such, strategies of feature selection could determine the maximum influential features on a target class [7]. As a result, the implementation of such feature selection techniques can lead to reducing a model's complexity and improving its accuracy [8]. Wrapper, filter, and embedding strategies are the most adopted feature selection methods [9]. The most successful elements have been frequently utilized to examine students' performance in different online learning environments [4].

This research aims at achieving four objectives. Firstly, a classification model is built to forecast student performance in online courses. Secondly, the research adopts an approach of feature selection to include the most important features in the prediction process. Thirdly, it predicts learners' academic achievement using newly generated features and the original features in the dataset. Finally, the research enhances the prediction accuracy of the implemented classifiers by using an optimization technique. In comparison to previous literature, this research has many contributions, including 1) The accuracy of the classification model is enhanced by applying the random search optimizer, 2) An approach to feature selection is utilized to identify the maximum important features, 3) Integrating the whole dataset and proposing a general prediction model regardless of a particular course or its specific features, and 4) comparing the findings of the research in different machine learning techniques based on suggesting an early prediction model.

In Sect. 2, the previous studies are explained. The research methodology is presented in Sect. 3. The research findings are reported and discussed in Sect. 4. Finally, Sect. 5 concludes this research and highlights possible future directions.

2 The Literature Review

2.1 Machine Learning Techniques

Different kinds of machine learning techniques have been previously utilized as predictive models. According to earlier literature [10], machine learning can successfully and accurately predict students' performance on online platforms. In this research, four machine learning techniques were used in predicting students at risk which have been widely used in previous research. Such techniques are explained here.

2.1.1 Random Forest (RF)

It is one of the machine learning techniques. RF is employed for estimating and forecasting probability. By mixing various classifiers, ensemble methods seek to increase classification accuracy. Random forest does this by building several decision trees in which each of these is trained using a different subset of data and features [11]. The results of all decision trees are then combined to create a final prediction model which is frequently more accurate than the outcomes of each single decision tree alone [12].

2.1.2 Decision-Tree (DT)

A decision tree is a hierarchical structure that can handle categorical as well as numerical data. It has nodes and edges that are directed [12]. A decision tree's leaf nodes stand for a certain class. The properties and test conditions that are used to categorize the data are represented by the root and various internal nodes. Based on Hunt's algorithm many decision tree algorithms, including ID3, C4.5, CART, and REPTree are built [13].

2.1.3 Support Vector Machine (SVM)

SVM is a sort of supervised machine learning which could be utilized to address problems such as classification and regression [13]. A useful classification method that can handle both linear and non-linear issues is SVM. Statistical learning theory serves as the foundation for SVM because it is also a practical learning method [14]. Equation 1 [15] can be used to compute it.

$$W = \sum_{i=1}^{n} \measuredangle iXiYi \tag{1}$$

where: N refers to the number of support vectors; Xi refers to the input vector of data points closest to the hyperplane; $\measuredangle$irefers to a Lagrange multiplier of Xi; Yi refers to the class label.

2.1.4 Logistic Regression (LR)

A statistical paradigm that pertains to linear regression is called logistic regression (LR). It enables the description of a binomial variable in terms of a group of random variables, whether categorical or numerical. It is used to forecast probabilities using extra information regarding variable values that can be connected to or explained by that event. In LR, several anticipated variables which may be categorical or numerical are employed. LR is also known as the Logit model or the Entropy-based general classifier [16]. LR can be calculated based on Eq. 2 [17].

$$p = \frac{e^{a+bx}}{1 + e^{a+bx}} \tag{2}$$

where: p refers to the predicted value; e refers to the base of the natural logarithm (about 2.72); x refers to the input value; a: the bias or intercept term; b refers to the coefficient for input (x).

2.1.5 Random Search Algorithm (RS)

An example of a Monte Carlo approach is the random search optimizer [18]. Random search is an optimization algorithm that is utilized to find the ideal collection of hyperparameters for a particular machine-learning model. The arbitrary search algorithm is utilized for classification and regression issues across a variety of areas to select the best hyperparameters that can lead to increasing prediction accuracy [18]. By using the random search optimizer, this research strives to maximize the performance of the four classifiers [18]. Random search begins with the random initialization of a set of hyperparameters in each model [19].

2.2 Previous Studies

Daud et al. [20] constructed a model to guess if learners will complete their degrees or not. The proposed model was examined using demographic features. Support Vector Machine (SVM) performed better than other techniques with an F1 score of 0.867. The study's outcomes also showed that geography, self-employment, natural gas consumption, and electricity expenditure were the best predictors. Umer et al. [21] compared the performance of four predictive models namely, Naive Bayes (NB), Linear Discriminant Analysis (LDA), K-Nearest Neighbor (KNN), and Random Forest (RF) and to predict students' results. Based on VLE's engagement data and assignment scores, the forecast was run every week. The results demonstrated that the scores of the assignment were the strongest discriminative variable, with a prediction accuracy of 70% after the first week. Additionally, the outcomes demonstrated that Random Forest consistently, in all weeks, outperformed other classifiers. Hussain et al. [22] compared the performance of J48, gradient-boosted trees, decision trees, JRIP, CART, and Nave Bayes in predicting students' engagement at the Open University (OU) social science course. It was found that the J48 algorithm obtained the highest accuracy of 88.52%.

Soni et al. [23]prepared a model which analyzed the performance of pupils from their last output using NB, DT, and SVM. For the extraction process, 20 out of 48 features were selected. Support vector machine was better than Naïve Bayes and Decision Tree with 83.33% accuracy. An early prediction model proposed by Al-Azawei and Al-Masoudy [3] was set on four time periods of the examined online course. Three features were generated in the process of improving the prediction accuracy in which data of only 1938 students who joined the science module were used from the Open University (OU) dataset. The ensemble method and the decision tree approaches were used. The key findings proved that the accuracy was well enhanced after utilizing the features generated from 70.4334% to 91.3829%. Jawad et al. [24] based on different activities taken through their profiles on VLEs of the Open University (OU) dataset, a model to estimate learners' academic achievement and engagement was presented. The SMOTE technique was used with the RF classifier where 84.2% was the highest obtained accuracy.

Such works, however, are not without limitations. First, most of the studies used a particular course to predict students' performance for that course. Second, most of the studies have not generated new features. Unlike previous works, this research predicts learners' performance on VLEs using the Open University dataset and for all courses. Previous literature such as the research study conducted by Al-Azawei and Al-Masoudy [3], however, relied solely on a science course. Moreover, the prediction of this present research is based on three quarters which are before the start of the course as well as for the first and second quarters. In every quarter prediction, twenty features are used. This study also applies the random search optimizer to select the best set of hyperparameters. Furthermore, four features are generated to enhance the prediction accuracy. An approach of feature selection is utilized to distinguish the best features for the process of prediction. Finally, the very impressive features that could affect learners' academic standards are highlighted. As the current study seeks to enrich the performance of the four prediction models applied, its overall findings extend earlier literature.

3 Research Methodology

This research aims to predict learners' academic achievement in three quarters to high-light the best features that could help learners pass their courses successfully. To achieve this aim, the proposed methodology adopts a specific procedure to enhance the prediction accuracy of the adopted classifiers and for all courses in the research dataset. This requires data preprocessing, generating new features based on the available features, using an optimization technique to select the best hyperparameters, and finally, applying an approach of feature selection to shed light on the superior feature that could affect learners' performance. The proposed methodology shows its effectiveness in enhancing the models' performance.

3.1 Dataset

The Open University (OU) published the dataset on November 28th, 2017 [25]. It includes information on 32,593 students during nine months in 2014–2015. This encompasses their demographics, clickstream behaviour, and assessment results. It consists of seven courses which were offered at least two times and at various times throughout the year. The dataset includes details on four Science modules and three Social Science modules [26].

The dataset consists of seven files [26, 25]. The student information table contains demographic data about students and the outcomes of each course. Details about the courses that students are registered for are available in the course table. The table of registration shows the recorded learner timestamps and dates of course joining. The evaluation table contains the evaluation data. The evaluation outcomes for various students are listed in the student assessment table. The student-VLE table keeps track of how a diverse range of students interact with diverse materials and activities. The VLE interaction data are made by the amount of clicks that learner made while reading the materials of the course. The activity of each course is specified by a label (activity type) like dataplus, folder, forumng, dualpane, externalquiz, glossary, oucontent, ouelluminate, homepage, htmlactivity, oucollaborate, quiz, repeatactivity,ouwiki, page, sharedsubpage, subpage, questionnaire, resource, and url, in the VLE table all these activities are stored.

In the present study, the provided students' performance was changed into a binary classification by setting two classes 'pass' and 'fail'. To create this category, the pass and distinction class labels are mixed into one label 'pass', and the class labels fail and withdrawal are mixed into one label 'fail'. This is performed to avoid the class imbalance problem. Each module is divided into quartiles [1], an early intervention provided for the learners who are at risk of failure. Nine months is the duration of a module [26]. Each module has three quartiles to make an early prediction. The clickstream data of each activity is computed separately in quartiles. In the online learning platform, the clickstream data indicate the interaction behavor of the students. Temporal features that are updated are concerning for each quarter. In each quarter The number of features equals the number of activities in the VLE. However, the clickstream information varies in each quarter. The assessment scores' pattern is deemed to be another parameter that captures the behavior of learners and a positive impact is induced on their performance.

Moreover, in this research, new features are generated from the available data to improve the prediction accuracy. Figure 1 displays the proposed system of the present study.

3.2 Data Preprocessing

The dataset was produced in a new framework with some data files [27]. The data has passed through a preprocessing phase before being used in the machine learning algorithms. Preprocessing consists of several different strategies and techniques [3]. Data preprocessing includes techniques like omitting noise, handling missing values, coding categorical data, normalization, and feature extraction [28].

Missing Values. The assessment scores and the deprivation band (imd band) features in the used dataset have a few missing values. Instead of missing assessments, all -1s were issued following the Open University's allegation of negligence for all assessment values that students failed to complete. The goal of choosing -1 rather than zero is to distinguish between students who received a zero score and those who did not examine at all. Furthermore, the greatest recurring value in the imd band characteristic of the derivative band was investigated as a substitute for missing data.

Coding the Categorical Data. Encoding categorical data involves converting categorical variables into numerical representations that machine learning algorithms can use. Common encoding techniques that are used in this dataset are ordinal encoding for the highest education, age band, code module, and code presentation as well as nominal encoding for gender, region, and disability.

Normalization. For all numeric feature values that would serve as the input for machine learning methods, normalization was carried out. This included many features, such as the number of previous attempts, dataplus, the forum, glossary, collaboration, content, resources, subpage, homepage, and URL. To make sure that all feature values fall inside a single range, this step was carried out based on Eq. 3 [28].

$$\widehat{V} = \frac{V - min_A}{max_A - min_A} \tag{3}$$

If the variable V represents the feature value, the variables min_A and max_A represent the lowest and maximum original values, respectively, for any given feature.

3.3 Feature Generation, Feature Generation

The process of generating new features from the original ones is called feature generation so that the new predictors may capture more important information effectively [19]. The most popular methods used to generate features are 1) Mapping the data to a new space, 2) Feature extraction, and 3) Feature construction [19, 29].

Features Extraction. There are three categories of attributes: behavioral, demographic, and performance. The student VLE table was used to extract the behavioral attributes. This was accomplished by adding the sum of the interactions of the students to the sum of their interactions with each site with the sort to which this site belongs. The type of site

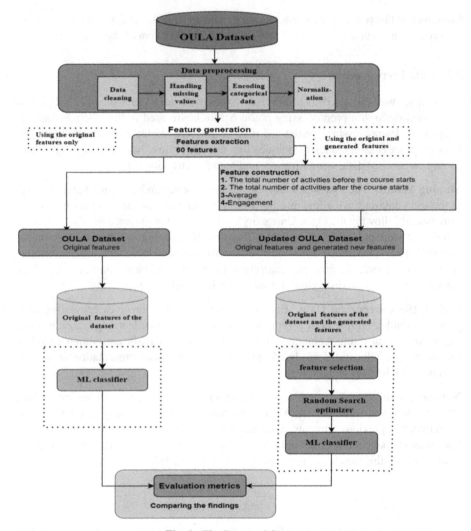

Fig. 1. The Proposed System.

is realized by using the VLE table, that has all ID sites and their types. For each course, twenty different sorts of ID sites were discovered. As a result, twenty new characteristics were obtained. The interaction of students with these activities was calculated at three separate points. The first quarter occurred before the official start of the course. The last two quarters occurred after the formal commencement of the course.

Feature Construction. In this study to improve the prediction accuracy, four new features are generated:

1) The total number of activities before the course begins: This feature is generated based on using twenty features in the original dataset. This feature is calculated from

the students' behavioral features before starting the courses. Before the course began, the number of clicks for each action type was regarded as another set of features [1].

2) The total number of activities after the course begins: This feature is generated based on using twenty features in the original dataset from four different quarters. This feature is calculated from the students' behavioral features after starting the courses for four quarters.

3) Average: This feature is derived based on the results of the tests completed by students in each subject up until the prediction day. Initially, each exam's results were collected individually for the students. This is because, in the raw dataset, grades were randomly assigned to a single column. Next, the averages of the students for each period were determined. Each evaluation is assigned a weight based on its weight in relation to the other assessments in the course [3].

4) Engagement: This feature is generated based on the score of assignment one and the entire number of actions. The entire number of clicks on student's VLE actions was taken to be a sign of involvement [22]. Understanding student engagement is crucial for web-based education. It influences retention in e-learning courses, dropout rates, and student achievement. Students who were most engaged generally performed better [22]. The score of learners' engagement in the web-based system is associated with the educational level of students. Earlier literature found that engagement had a positive and significant effect on student performance [22].

3.4   Feature Selection

Feature selection (FS) is the procedure of choosing the most relevant and most affected set of features on the problem from the original features. The role of FS techniques is to know the strongly relevant features of the target class [7]. Feature selection is responsible for increasing and decreasing the number of attributes and this may affect the accuracy of a particular model. The Wrapper technique is used in this study as a standard feature selection approach. Wrapper techniques are categorized into sequential selection and heuristic search algorithms [30]. The nature of adding features is what gives sequential selection algorithms their name. These methods use strategy search to identify the relevant features [30]. The sequential feature selection is an example of the wrapper method, and this method begin with an empty set. In each phase, features are independently added to a previous subset. The procedure is continued until the appropriate amount of features is added and the best accuracy is acquired [29].

3.5   Evaluation

It's critical to assess how well machine learning methods work to obtain an overall picture of the model's accuracy. The performance of the suggested model is assessed in this study using the assessment metrics f1-measure, accuracy, recall, and precision. Since the accuracy metric is the most widely used metric for model evaluation in classification, it is used in this research. When tested with unknown data, the trained model is assessed on the total number of accurately predicted instances [31]. To define the goal function of RS, this study used a K-fold cross-validation procedure with K = 5. The dataset is divided into five mutually exclusive groups using the cross-validation method. To assess the model, one set is utilized in each iteration, while the remaining four sets are used as training data [31]. A confusion matrix includes four measures which are abbreviated as FN, TN, TP, and FP as shown in Table 1 [6].

Table 1. A two-dimensional confusion matrix.

Actual	Predicted	
	Positive	Negative
Positive	TP	FN
Negative	FP	TN

1. True Positive (TP): is the true positive and predicted positive. 2. False Negative (FN): is the false negative and predicted negative. 3. False Positive (FP): is the false positive when it is actually negative and predicted positive. 4. True Negative (TN): is the true negative when it is positive and predicted negative.

These measures can be calculated based on Eqs. 4, 5, 6, and 7 [19]. The accuracy measure is computed based on the number of right predictions divided by the total number of predictions as presented in Eq. 2 [19].

$$Accuracy = (TP + TN) / TP + TN + \sum FP + \sum FN \qquad (4)$$

$$Recall = TP/TP + \Sigma FN \qquad (5)$$

$$Precision = TP/TP + \Sigma FP \qquad (6)$$

$$F1 - measure = (2 * TP)/2 * TP + \Sigma FN + \Sigma FP) \qquad (7)$$

4 Results and Discussions

In the present study, four prediction models were created over three quartiles. It was carried out over such a wide time span to offer a continuous indication of students' final outcomes if they maintained their current academic standing. Figure 2 depicts the actual number of students who either passed or failed in their courses. The accuracy obtained for the period before the course started and for Q1 and Q2 after the official start of the module is summarized in Table 2.

The following attributes were entered into the machine before generating the new features and without the use of the random search optimizer: folder, forum, region, highest_education, dataplus, dualpane, imd_band, age_band, number_of_previous_attempts, externalquiz, glossary, homepage, studied_credits, disability, htmlactivity, ouwiki, page, aquestionnaire, oucollaborate, oucontent, ouelluminate, and quiz. Table 2 shows that the RF and LR classifiers outperformed others in the period before the course started. In

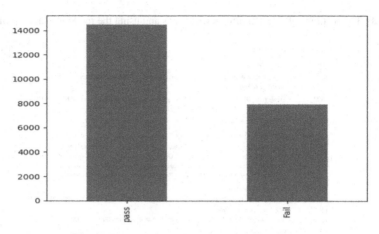

Fig. 2. The actual number of pass and fail classes.

Table 2. The accuracy with original features only and without integrating the random search optimizer.

Quartiles	Categories	Classifier	Feature used	Accuracy
Before	Fail/Pass	DT	Original features	0.67
		RF		0.68
		SVM		0.67
		LR		0.68
Q1		DT		0.69
		RF		0.71
		SVM		0.72
		LR		0.71
Q2		DT		0.75
		RF		0.78
		SVM		0.80
		LR		0.82

Q1, SVM classifiers performed significantly better than the other classifiers, whereas in Q2, LR classifiers achieved higher accuracy than others. Table 2 presents the model's accuracy without generating new features or integrating the random search optimizer.

Based on the results shown above, it is obvious that, despite the relevance of demographic, behavioral, and environmental factors, and assessment score features, there is a need to improve the early prediction of learners' performance. This may indicate that there is a substantial need to employ the random search optimizer and generate new attributes. Table 3 shows the model's accuracy after employing the features that are generated in the three quartile. After including the random search optimizer and using the generated attributes, the accuracy increased considerably. This research also applied a feature selection approach to choose the best features.

Table 3. The accuracy after generating new features and using the random search optimizer.

Quartiles	Categories	Classifier	Feature Used	Accuracy
Before	Fail/Pass	DT	Engagement, Bhomepage	0.84
		RF	Num_Of_Prev_Attempts, Avereage, Total Number Of Clicks Befor, Engagement, Bdualpane, Bhomepage, Bpage, Bquiz	0.84
		SVM	Engagement, Boucontent, Bforumng, Avereage, Bhomepage, Bquestionnaire, Total Number Of Clicks After, Bquiz, Total Number Of Clicks Befor, Bsubpage,	0.83
		LR	Avereage, Total Number Of Clicks Befor, Engagement, Bfolder, Bforumng, Bglossary,	0.82
Q1		DT	Engagement, Bhomepage, Aq1:Oucontent	0.83
		RF	Imd_Band, Num_Of_Prev_Attempts, Studied_Credits, Disability, Avereage, Total Number Of Clicks Befor, Engagement, Bdataplus, Bdualpane, Gender, Region, Highest_Education, Boucollaborate, Boucontent, Bexternalquiz, Bfoler, Bhomeage, Bouwiki, Bpage, Bquiz, Bsharedsubpage, Aq1:Oucontent, Aq1:Page, Aq1:Quiz, Aq1:Rsource, Aq1:Sharedsubpage, Aq1:Subpage	0.84
		SVM	Performance, Engagement, Aq1:Quiz,Bdualpane,Boucollabrate,Bouctent,Bouelluminate,Aq1:Folder,Aq1:Forumng,Aq1:Glossary	0.81
		LR	Age_Band, Num_Of_Prev_Attempts, Studied_Credits, Disability, Engagement, Performance, Bhomeage, Bquiz, Aq1:Quiz	0.82
Q2		DT	Engagement, Aq2:Quiz, Bhomepage	0.84
		RF	Num_Of_Prev_Attempts, Avereage, Engagementtotal Number Of Clicks Befor, Bpage, Total Number Of Clicks After, Bdualpane, Boucontent, Bhomepage, Bquestionnaire, Bquiz, Brepeatactivity, Bresource, Aq2:Ouelluminate, Aq2:Quiz, Aq2:Repeatactivity, Studied_Credits, Disability,	0.84
		SVM	Engagement, Avereage, Total Number Of Clicks After, Aq2:Quiz, Bdualpane, Boucollaborate, Boucontent, Bouelluminate	0.84
		LR	Engagement, Bhomepage, Avereage, Total Number Of Clicks After, Region, Highest_Education, Aq2:Forumng, Aq2:Homepage	0.84

Due to the variation of the chosen features throughout the three quarters and their accuracy, all features that arise to be influencing features across the three quarters were chosen. Subsequently, these features were integrated into the model to validate their significance and the degree of precision that could be attained through their utilization. Particular features had an important influence impact on the accuracy according to a specific quarter (as shown in Table 4). The total influencing features across the module before the formal start are average, engagement, the entire number of activities before, the entire number of activities after, bforumng, bhomepage, bpage, and bquiz. Based on the features above, the accuracy was enhanced from 0.67% and 0.68% to 0.84%.

About Q1, the accuracy was enhanced from 0.69%, 0.71%, and 0.72% to 0.84% for the science and social science courses. The total influencing features are average, engagement, the entire number of activities before, the entire number of activities after, bhomepage, aq1:oucontent, aq1:oucontent, aq1:page, aq1:quiz, aq1:resource, aq1:sharedsubpage, aq1:subpage, age_band, num_of_prev_attempts, aq1:homepage.

In Q2, the accuracy was enhanced from 0.75%, 0.78%, 0.80%, and 0.82% to 84% for the science and social science courses where the total influencing features are average, total number of clicks before, total number of clicks after, engagement, bhomepage, aq2:forumng, aq2:homepage, aq2:quiz, aq2:subpage, aq2:oucontent, num_of_prev_attempts.

The overall accuracy in this study was 84%, as indicated in Table 3. The execution of several preprocessing stages before the building of the classification model could be the foundation for this. Furthermore, the usage of the random search optimizer and the produced features aided in improving the models' overall accuracy. Table 4 shows a comparison of the research's output findings with earlier works. It shows that the overall findings of this study outperformed previous work on the same dataset.

Table 4. A comparison between the findings of previous literature and this study.

Reference	Algorithm used	Accuracy %
[20]	Support Vector Machine (SVM)	The SVM 86.52%
[21]	Random Forest (RF), Naive Bayes (NB), K-Nearest Neighbor (KNN), and Linear Discriminant Analysis (LDA)	Random Forest (RF) 70%
[22]	decision trees, JRIP, J48, gradient-boosted trees, CART, and Nave Bayes	The J48 algorithm obtained the highest accuracy of 88.52%
[23]	Naïve Bayes, Decision Tree, and Support Vector Machine	Support Vector Machine (SVM) 83.33%
[3]	The ensemble method and the decision tree approaches were used	The accuracy 91.3829%
[24]	The Random Forest (RF)classifier was used	The Random Forest (RF) 84.2%
This study\without random search optimizer and feature selection	Random Forest (RF), logistic regression (LR), support vector machine (SVM), and Decision Tree (DT),	Logistic regression (LR) 0.82%
This study\with random search optimizer and feature selection	Random Forest (RF), logistic regression (LR), support vector machine (SVM), Decision Tree (DT), and Random Search optimizer (RS)	Random forest (RF), Logistic regression (LR), Support vector machine (SVM), Decision Tree (DT) 0.84%

5 Conclusions

This paper predicted students' performance over three quarters in a VLE course. An early identification of at-risk learners can help provide the required learning support and this, in turn, may lead to avoiding their academic failure. The present study proposed various predictive models based on machine learning algorithms namely, Random Forest (RF), Logistic Regression (LR), Support Vector Machine (SVM), and Decision Tree (DT). Three types of features were used in the prediction process which are demographics, performance scores, and learners' behavior in VLEs.

The overall findings suggested that the generation of new features helped improve the proposed methodology's performance. This is further enhanced by implementing the random search optimizer and feature selection techniques. The experimental results confirmed that the generated features were mostly selected by the feature selection technique in the three quarters. Hence, the highest prediction accuracy obtained was 0.84% for the three quarters.

Based on such outcomes, educational institutions must permit learners and encourage them to participate in the different activities that take place within the learning environment before the start of a course. As a result, educational institutions and policymakers must consider various activities and learning resources to provide successful online educational settings.

Regardless of the significant outcomes of this research, it is not without limitations that invite further research. First, generating other new features may enhance the prediction process and lead to obtaining more generalizable results. Second, the proposed model was implemented using traditional machine learning techniques, whereas applying other methods such as deep learning may help obtain better results. Finally, the model of the research was set on one dataset only, so using other datasets may further confirm the overall findings.

References

1. Waheed, H., Hassan, S.-U., Aljohani, N.R., Hardman, J., Alelyani, S., Nawaz, R.: Predicting academic performance of students from VLE big data using deep learning models. Comput. Human Behav. **104**, 106189 (2020)
2. Rivas, A., Gonzalez-Briones, A., Hernandez, G., Prieto, J., Chamoso, P.: Artificial neural network analysis of the academic performance of students in virtual learning environments. Neurocomputing **423**, 713–720 (2021)
3. Al-Azawei, A., Al-Masoudy, M.: Predicting learners' performance in virtual learning environment (VLE) based on demographic, behavioral and engagement antecedents. Int. J. Emerg. Technol. Learn. **15**(9), 60–75 (2020)
4. Muljana, P.S., Luo, T.: Factors contributing to student retention in online learning and recommended strategies for improvement: a systematic literature review. J. Inf. Technol. Educ. Res. **18**, 019–057 (2019)
5. Mogus, A.M., Djurdjevic, I., Suvak, N.: The impact of student activity in a virtual learning environment on their final mark. Act. Learn. High. Educ. **13**(3), 177–189 (2012)
6. Tan, P.-N., Steinbach, M., Kumar, V.: Introduction to Data Mining. Pearson Education Inc., New Delhi (2006)
7. Jović, A., Brkić, K., Bogunović, N.: A review of feature selection methods with applications. In: 2015 38th International Convention on Information and Communication Technology, Electronics and Microelectronics (MIPRO), pp. 1200–1205. IEEE (2015)
8. Liu, Y., Pan, Q., Zhou, Z.: Improved feature selection algorithm for prognosis prediction of primary liver cancer. In: Shi, Z., Pennartz, C., Huang, T. (eds.) ICIS 2018. IAICT, vol. 539, pp. 422–430. Springer, Cham (2018). https://doi.org/10.1007/978-3-030-01313-4_45
9. Miao, J., Niu, L.: A survey on feature selection. Procedia Comput. Sci. **91**, 919–926 (2016)
10. Darji, J., Nakrani, T., Sandhi, M.I.I., Prachi, M.: Machine learning based prediction technique for student's performance (2021)

11. Siregar, M.U., Setiawan, I., Akmal, N.Z., Wardani, D., Yunitasari, Y., Wijayanto, A.: Optimized random forest classifier based on genetic algorithm for heart failure prediction. In: 2022 Seventh International Conference on Informatics and Computing (ICIC), pp. 1–6. IEEE (2022)
12. Lee, C.S., Cheang, P.Y.S., Moslehpour, M.: Predictive analytics in business analytics: decision tree. Adv. Decis. Sci. **26**(1), 1–29 (2022)
13. Gheisari, M., et al.: Data mining techniques for web mining: a survey. In: Artificial Intelligence and Applications, pp. 3–10 (2023)
14. Kaul, A., Raina, S.: Support vector machine versus convolutional neural network for hyperspectral image classification: a systematic review. Concurr. Comput. Pract. Exp. **34**(15), e6945 (2022)
15. Roy, A., Chakraborty, S.: Support vector machine in structural reliability analysis: a review. Reliabil. Eng. Syst. Saf. **233**, 109126 (2023)
16. Zabor, E.C., Reddy, C.A., Tendulkar, R.D., Patil, S.: Logistic regression in clinical studies. Int. J. Radiat. Oncol. Biol. Phys. **112**(2), 271–277 (2022)
17. Bailly, A., et al.: Effects of dataset size and interactions on the prediction performance of logistic regression and deep learning models. Comput. Methods Programs Biomed. **213**, 106504 (2022)
18. Tarek, Z., et al.: Soil erosion status prediction using a novel random forest model optimized by random search method. Sustainability **15**(9), 7114 (2023)
19. Steinbach, M., Tan, P., Kumar, V.: Introduction to Data Mining. Pearson Education Inc., Boston (2006)
20. Daud, A., Aljohani, N.R., Abbasi, R.A., Lytras, M.D., Abbas, F., Alowibdi, J.S.: Predicting student performance using advanced learning analytics. In: Proceedings of the 26th International Conference on World Wide Web Companion, pp. 415–421 (2017)
21. Umer, R., Susnjak, T., Mathrani, A., Suriadi, S.: A learning analytics approach: Using online weekly student engagement data to make predictions on student performance. In: 2018 International Conference on Computing, Electronic and Electrical Engineering (ICE Cube), pp. 1–5. IEEE (2018)
22. Hussain, M., Zhu, W., Zhang, W., Abidi, S.M.R.: Student engagement predictions in an e-learning system and their impact on student course assessment scores. Comput. Intell. Neurosci. **2018** (2018)
23. Soni, A., Kumar, V., Kaur, R., Hemavathi, D.: Predicting student performance using data mining techniques. Int. J. Pure Appl. Math. **119**(12), 221–227 (2018)
24. Jawad, K., Shah, M.A., Tahir, M.: Students' academic performance and engagement prediction in a virtual learning environment using random forest with data balancing. Sustainability **14**(22), 14795 (2022)
25. Merchant, A., Shenoy, N., Bharali, A., Kumar, M.A.: Predicting students' academic performance in virtual learning environment using machine learning. In: ICPC2T 2022 - 2nd International Conference on Power, Control Computer Technology Processing (2022). https://doi.org/10.1109/ICPC2T53885.2022.9777008
26. Kuzilek, J., Hlosta, M., Zdrahal, Z.: Open university learning analytics dataset. Sci. data **4**(1), 1–8 (2017)
27. Aljohani, N.R., Fayoumi, A., Hassan, S.-U.: Predicting at-risk students using clickstream data in the virtual learning environment. Sustainability **11**(24), 7238 (2019)
28. Qasrawi, R., VicunaPolo, S., Al-Halawa, D.A., Hallaq, S., Abdeen, Z.: Predicting school children academic performance using machine learning techniques. Adv. Sci. Technol. Eng. Syst. J. **6**(5), 8–15 (2021). https://doi.org/10.25046/aj060502
29. Khalid, S., Khalil, T., Nasreen, S.: A survey of feature selection and feature extraction techniques in machine learning. In: Proceedings of 2014 Science and Information Conference, SAI 2014, pp. 372–378 (2014). https://doi.org/10.1109/SAI.2014.6918213

30. Ansari, G., Ahmad, T., Doja, M.N.: Hybrid filter–wrapper feature selection method for sentiment classification. Arab. J. Sci. Eng. **44**, 9191–9208 (2019)
31. Hossin, M., Sulaiman, M.N.: A review on evaluation metrics for data classification evaluations. Int. J. data Min. Knowl. Manag. Process **5**(2), 1 (2015)

YOLOv8-AS: Masked Face Detection and Tracking Based on YOLOv8 with Attention Mechanism Model

Shahad Fadhil Abbas[1]([✉]), Shaimaa Hameed Shaker[1], and Firas. A. Abdullatif[2]

[1] University of Technology-Iraq, Baghdad, Iraq
cs.20.30@grad.uotechnology.edu.iq

[2] College of Education for Pure Science/Ibn-Al-Haithem, Baghdad University, Baghdad, Iraq

Abstract. The development of intelligent surveillance systems relies significantly on the effectiveness of face detection. To identify suspicious individuals, this technology uses a face detection model that quickly and efficiently analyzes every frame generated by the system. This presents challenges for some public safety surveillance systems that rely on face detection and tracking. Hence, there is a growing need to develop effective algorithms for face detection and tracking, even when wearing masks. In this paper, we present a YOLOv8-AS model with an attention mechanism for real-time tracking designed for individuals with and without masks. This model was trained using publicly available face (ChokePoint and NRC-IIT) datasets that mostly feature individuals without masks. Although not directly trained on masked faces, our model demonstrated its strength by performing adeptly at accurately tracking individuals wearing masks.

Keywords: face tracking · mask face tracking · yolov8 · Attention Mechanism Model

1 Introduction

Surveillance, detection, and location systems have grown in many applications such as security, traffic surveillance, medical, and automated driving. Various methods have been employed to develop effective real-time systems. With more surveillance systems, control centers are struggling to monitor many cameras efficiently. Therefore, real-time detection and tracking systems are required. These systems can be used to detect foreign organisms and violent crimes, among other applications [1] and [2]. Recently, postprocessing methods have been used for face tracking. Sagonas et al. [3] used a re-fitting algorithm for the entire video, but this is limited to offline applications. Landmark localization may be inaccurate because face masks cover much of the face [4, 5]. Therefore, it is necessary to create algorithms for facial recognition and tracking of people wearing face masks. Online tracking examines current and previous frames to form trajectories. Offline tracking uses a stack of frames for the predictions. Classical methods focus on tracking-by-detection and assign detections across frames [1].

© The Author(s), under exclusive license to Springer Nature Switzerland AG 2024
A. M. Al-Bakry et al. (Eds.): NTICT 2023, CCIS 2096, pp. 267–275, 2024.
https://doi.org/10.1007/978-3-031-62814-6_19

This study proposes a face-tracking model that works with masks. The goal is to track a person in frames, even when they enter or exit rooms and their face size changes. This was performed using Yolov8-AS and attention mechanisms. The model is fast and works with both masked and unmasked individuals. Moreover, despite not being trained using specialized training images featuring face masks, this model performed satisfactorily in tracking the faces of individuals donning face masks. The contributions of this study can be concisely stated as follows:

The YOLOv8-AS detection model has an attention mechanism that collects contextual information from the underlying network through extra feature fusion and communication layers. This makes things much better as a new branch is added from the backbone to the detection head. The combination of channels and spatial attention mechanisms has created an attention model in each feature fusion layer, focusing on important information for face detection. This resulted in a 3.4% increase in precision. The organization of this paper is as follows: Sect. 2 examines previous research conducted in object tracking and detection. In Sect. 3, the datasets utilized are outlined. Section 4 describes the approach used in this paper. Section 5 presents the implementation and results of the study. Finally, conclusions are presented in Sect. 6.

2 Related Work

Tracking and surveillance systems typically include several stages, namely object tracking, detection, and recognition. Because moving objects form a region of interest (ROI), this study emphasizes the importance of the tracking phase. The objects are then observed after dividing them into predefined categories, such as cars and people, which is part of object recognition. Therefore, a thorough discussion of the current literature on detecting and tracking objects and people wearing masks is presented.

2.1 Face Tracking

In the realm of live tracking, our focus is on examining the present and preceding frames to reveal the relationships between objects, which are then utilized to form trajectories. Conversely, in offline tracking, attention is directed towards a collection of frames, enabling us to generate predictions. The concept of face tracking is closely associated with face detection, albeit with a reduced requirement for computational resources compared with running a face detector on each frame. In addition, it is linked to the broader field of visual object tracking (VOT), wherein the target of tracking is a specific face. The analysis and training of face-tracking models frequently employ RGB [2], HOG [3], and deep features [4].

The paper [5] presents a proposed real-time face-tracking algorithm designed for individuals wearing face masks. The algorithm comprises a face detector and a Kalman filter. Notably, this algorithm demonstrates an exceptional tracking speed and is suitable for individuals with and without face masks. Furthermore, despite not being trained in specialized images featuring face masks, this algorithm exhibited satisfactory performance in tracking individuals wearing face masks.

The research paper referred to in [10] presents a project that focuses on developing a face detection and recognition system on a Linux platform using Python. The system was built by training the face model with Eigen Face, Fish Face, and LBP algorithms, which are available in the OpenCV library. Experimental results revealed that the proposed approach displayed an impressive level of recognition accuracy. This study [11] developed a stable method for tracking a person's face in an unconstrained environment, which involves combining an offline detector, an online tracker, and an online recognizer. To achieve this, frame accuracy was improved using a well-trained face detector and Haar-like features for tracking. He also implemented a boosted classifier to reduce the impact of poor tracking output or when the target reappeared, thereby maintaining the overall accuracy.

2.2 Yolo

This study [12] focused on developing a low-cost parking time-violation tracking system using advanced technologies. By combining closed-circuit television (CCTV), deep learning models, and object-tracking algorithms, the researchers employed state-of-the-art detection techniques and methods that took temporal constraints into account. Vehicles in the parking lot were identified using YOLOv8 and tracked using DeepSORT and OC-SORT. The researchers [13] introduced a Yolov5-based multiple-object tracking system capable of identifying and tracking objects in real-time. It can recognize specific objects in crowded areas, track certain objects, identify object classes, or count items. Our system is efficient and works on both the CPU and GPU. This paper [14] proposes a face-tracking system that uses Gabor feature extraction to recognize faces, match faces based on correlation, and track faces using a Kalman filter. It is effective for real-life videos and resilient to changes in lighting, environment, scale, position, and orientation.

3 Dataset

This section focuses on the ChokePoint and IIT-NRC face datasets used in the experiments. These surveillance video datasets were used to assess the enhancement in verification performance from subset selection using the suggested quality method and other strategies. ChokePoint [15] is a video dataset of experiments related to person identification and verification in surveillance settings. Using three cameras above gates, natural choke points for pedestrian traffic capture images of people walking through each gate naturally. Figure 1 shows examples of dataset images.

Fig. 1. Mage example of chokepoint dataset

Facial images are captured sequentially as someone walks through a portal, creating a set of faces with variations owing to automatic detection, localization, and alignment. Using the three-camera setup, one camera is likely to capture a near-frontal subset of the faces.

The dataset contained (48) video sequences and (64,204) face images from 54 subjects across two portals. Each sequence was named based on recording conditions, with (P, S, and C) representing portals, sequences, and cameras, and E and L indicating entry or exit. The numbers correspond to the portal, sequence, and camera labels. P2L S1 C3 signifies a recording from Portal 2 by camera 3 of the people exiting during the first sequence. The IIT-NRC database [4, 16]: This database contains video clips of computer users' faces that have been encoded in low resolution using MPEG1 and captured with a webcam on a computer monitor, which show a variety of facial expressions and orientations. The video capture resolution was set at (160 × 120) pixels, and the face occupies (1/4 to 1/8) of the image width, reflecting a standard TV viewing experience in which an actor's face fills (1/8 to 1/16) of the screen. The purpose of this database was to evaluate a computer's proficiency in identifying faces in low-resolution environments, similar to the 12-pixel distance between human eyes. The images displayed in Fig. 2 serve as an illustration of the IIT-NRC dataset.

Fig. 2. Shows an example image of the IIT-NRC dataset.

Each video clip was 15 s, captured at 20 frames/s, and compressed using an AVI Intel codec at 481 kbps. The face video files of a person were reduced to a small size (less than 1 MB), resembling high-resolution face images commonly used in aviation. Single-image-based faces provide less information than video-based stored faces do. Thus, our database was more useful for testing. In addition, the small video file size makes our database easily downloadable, making testing more convenient.

4   YOLOv8-AS Model

The YOLOv8 backbone differs from that of YOLOv5, with a (3 × 3) convolution kernel in the first convolutional module and the C2f module replacing C3. This makes the model lightweight and enhances the gradient flow. The neck module has a dual-stream FPN, and the (1 × 1) convolution module is taken away before up-sampling. This makes the feature fusion better than in YOLOv5.To enhance the detection of people wearing masks, we incorporated an attention mechanism into the YOLOv8 detection approach. As shown in Fig. 3, three enhancements were made to the original YOLOv8 architecture. (1) The figure below shows a new branch added to the diagram, which is represented by a red rectangle. This branch is known as the feature fusion branch and has been introduced to improve the accuracy of the visual features of the face when a mask is present or absent.

(2) The feature fusion layers receive features from the backbone network (indicated by the red line from C2F) to minimize the loss of facial feature information that may occur because of their small size. (3) The attention model, shown in AT, was added to the blend layers to identify and emphasize the required information.

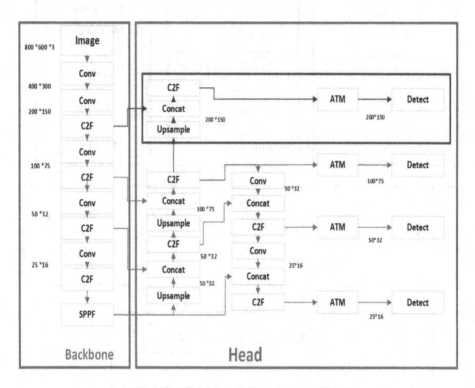

Fig. 3. The improved YOLOv8-AS method

To enhance YOLOv8's small face detection, a fusion layer was added to create a $152 \times 152 \times 255$ feature map. The new layer is shown in Fig. 3. Compared to the original YOLOv8, there were four fusion layers. We extracted fused feature maps from the original network, unsampled them, and combined them with a 152×152-pixel feature map from the backbone network. We used the CSP and CBL models for this process. An additional connection was added to the backbone network for pass feature information, which had dimensions of $152 \times 152, 76 \times 76, 38 \times 38$, and 19×19 pixels, to the sequence layers in the backbone of the head network. These connections are based on residual networks, which improve gradient backpropagation, prevent gradient fading, and preserve feature information for small faces. An attention model was added to the feature fusion layers to enhance the boulder information in the images. Specifically, we implemented an attention model using a combination of channel attention from the spatial transformer network (STN) [6] and spatial attention from the CBAM [7]. As shown in Fig. 3, the attention model was applied to the feature maps generated by the backbone (Fig. 4).

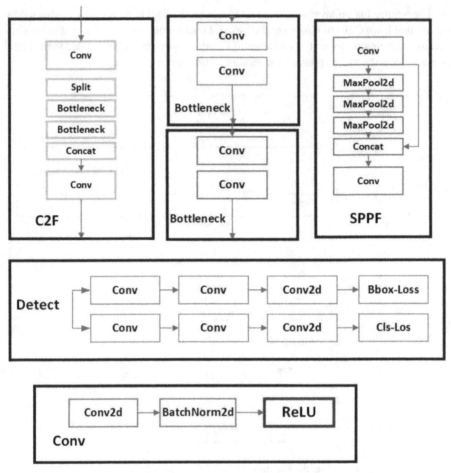

Fig. 4. Partial cell structure in network model: (**a**) structure of C2F, Bottleneck, SPPF model; (**b**) structure of the detection model, and (**c**) the structure of the Conv model.

5 The Implementations

In order to assess the efficacy of the proposed enhanced YOLOv8-AS, we compared its performance with those of other relevant object detection methods, including single-shot detector [8], Extend Single Shot Detector [9], Multi-scale Deconvolutional Single Shot Detector [10], You Only Look Once v3 [11], You Only Look Once v4 [12], You Only Look Once v5 [13], You Only Look Once v6 [14], You Only Look Once v7 [15], and You Only Look Once v8 [16]. Precision and Epochs were the evaluation metrics used. The ratio of correctly predicted positive samples to the total number of predicted positive samples was used to determine the accuracy of positive predictions.

$$Precision = \frac{TP}{TP + FN} \tag{1}$$

TP is the number of true positives correctly identified and FN is the number of false positives. If the predicted face was correct and the intersection over union (IoU) was above the designated threshold (0.6 in our trials), the detection was deemed precise. Epochs is a measure of real-time performance that reflects the number of images that object detection methods can process per second. A higher Epochs value indicated a higher detection speed (Table 1). Table 1 shows that YOLOv8-AS outperformed the other detection methods; for example, it was 3.4% better for ChokePoint and 1.5% better for NRC-IIT. YOLOv8-AS had excellent precision but lower Epochs than YOLOv8.

Table 1. Shows that YOLOv8-AS outperformed other detection methods

Dataset	Method	Accuracy	Epochs
ChokePoint	single-shot detector	43.1%	18
	Extend single shot detector	42.0%	22
	Multi-scale deconvolutional single shot detector	53.1%	22
	ResNet-18	69.1%	45
	You Only Look Once v3	60.0%	58
	You Only Look Once v4	72.20%	55
	YOLOv5	74.2%	69
	You Only Look Once v6	76.6%	66
	You Only Look Once v7	76.5%	70
	You Only Look Once v8	77.1	83
	You Only Look Once v8-AS	80.5%	102
NRC-IIT	single-shot detector	78.5%	18
	Extend single shot detector	78.4%	21
	Multi-scale Deconvolutional Single Shot Detector	77.6%	15
	ResNet-18	65%	24
	You Only Look Once v3	75.5%	43
	You Only Look Once v4	79.1%	40
	You Only Look Once v5	83.7%	54
	You Only Look Once v6	85.0%	57
	You Only Look Once v7	84.1%	60
	YOLOv8	84.5	60
	You Only Look Once v8-AS	87.8	75

6 Conclusions

This study improved the way to find people who are wearing masks in YOLOv8-AS by adding a new branch connection from the head module to use the module as a fusion layer to pull out the most basic facial features. This allowed for a reduction in feature information for small faces as a loss feature by incorporating the underlying network into the fusion layers. Additionally, the attention model highlights relevant information for detecting and tracking masked faces, implemented through a combination of channel and spatial attention mechanisms. The proposed approach was validated using the ChokePoint and NRC-IIT datasets, resulting in a 3.4% improvement in the YOLOv8-AS accuracy.

References

1. Kim, C., Li, F., Ciptadi, A., Rehg, J.M.: Multiple hypothesis tracking revisited. In: 2015 IEEE International Conference on Computer Vision (ICCV), pp. 4696–4704 (2015). https://doi.org/10.1109/ICCV.2015.533
2. Munaro, M., Basso, F., Menegatti, E.: Tracking people within groups with RGB-D data. In: Proceedings of the IEEE/RSJ International Conference on Intelligent Robots and Systems. IEEE/RSJ International Conference on Intelligent Robots and Systems, pp. 2101–2107 (2012). https://doi.org/10.1109/IROS.2012.6385772
3. Dalal, N., Triggs, B.: Histograms of oriented gradients for human detection. In: 2005 IEEE Computer Society Conference on Computer Vision and Pattern Recognition (CVPR 2005), vol. 1, pp. 886–893 (2005). https://doi.org/10.1109/CVPR.2005.177
4. Qi, Y., et al.: Hedging deep features for visual tracking. IEEE Trans. Pattern Anal. Mach. Intell. 41, 1116–1130 (2019). https://doi.org/10.1109/TPAMI.2018.2828817
5. Peng, X., Zhuang, H., Huang, G. Bin, Li, H., Lin, Z.: Robust real-time face tracking for people wearing face masks. In: 16th IEEE International Conference on Control, Automation, Robotics and Vision, ICARCV 2020, pp. 779–783. Institute of Electrical and Electronics Engineers Inc. (2020). https://doi.org/10.1109/ICARCV50220.2020.9305356
6. Jaderberg, M., Simonyan, K., Zisserman, A., Kavukcuoglu, K.: Spatial transformer networks. In: Advance in Neural Information Processing System, vol. 28 (2015)
7. Woo, S., Park, J., Lee, J.-Y., Kweon, I.S.: CBAM: Convolutional Block Attention Module. In: Ferrari, V., Hebert, M., Sminchisescu, C., Weiss, Y. (eds.) ECCV 2018. LNCS, vol. 11211, pp. 3–19. Springer, Cham (2018). https://doi.org/10.1007/978-3-030-01234-2_1
8. Liu, W., et al.: SSD: Single Shot MultiBox Detector. In: Leibe, B., Matas, J., Sebe, N., Welling, M. (eds.) ECCV 2016. LNCS, vol. 9905, pp. 21–37. Springer, Cham (2016). https://doi.org/10.1007/978-3-319-46448-0_2
9. Zheng, L., Fu, C., Zhao, Y.: Extend the shallow part of single shot multibox detector via convolutional neural network, vol. 10806, pp. 287–293 (2018). https://doi.org/10.1117/12.2503001
10. Cui, L., et al.: MDSSD: multi-scale deconvolutional single shot detector for small objects. Sci. China Inf. Sci. 63, 120113 (2020). https://doi.org/10.1007/s11432-019-2723-1
11. Redmon, J., ArXiv:1804.02767, A.F. ön baskı, 2018, U.: Yolov3: Artımlı bir gelişme. Arxiv.Org (2018)
12. Bochkovskiy, A., et al.: YOLOv4: optimal speed and accuracy of object detection. ArXiv. arXiv:2004.10934 (2020). https://doi.org/10.48550/ARXIV.2004.10934

13. Grekov, A.N., Shishkin, Y.E., Peliushenko, S.S., Mavrin, A.S.: Application of the YOLOv5 Model for the Detection of Microobjects in the Marine Environment. ArXiv. (2022). https://doi.org/10.33075/2220-5861-2024-4-XX

14. Li, C., et al.: YOLOv6: a single-stage object detection framework for industrial applications. ArXiv. (2022)

15. Wang, C.-Y., Bochkovskiy, A., Liao, H.-Y.M.: YOLOv7: trainable bag-of-freebies sets new state-of-the-art for real-time object detectors. In: 2023 IEEE/CVF Conference on Computer Vision and Pattern Recognition (CVPR), pp. 7464–7475 (2023). https://doi.org/10.1109/CVPR52729.2023.00721

16. Gan, N., et al.: YOLO-CID: improved YOLOv7 for X-ray contraband image detection. Electronics **12**, 3636 (2023). https://doi.org/10.3390/ELECTRONICS12173636

Symptom Principal Component Analysis (SPCA) for Dimensionality Reduction in Categorical Data: A Case Study on Breast Cancer

Fatema S. Al-Juboori[1] , Sinan A. Naji[1(✉)] , and Husam M. Sabri[2]

[1] University of Information Technology and Communications, Baghdad, Iraq
`dr.sinannaji@uoitc.edu.iq`
[2] College of Education Ibn Rushd for Human Sciences, University of Baghdad, Baghdad, Iraq

Abstract. The symptom analysis is a promising technique for the expansion of a set of random variables through linear combination over a finite field F_2. From the projective space, it is essential to choose the most useful subspaces for the reduction of the dimensionality of the categorical dataset. This paper presents a new method for obtaining the principal components based on symptom analysis, namely, super-symptoms generated from an iterative algorithm based on a set of rules for reducing the dimensionality in categorical data. The proposed method is called Symptom Principal Component Analysis (SPCA), which implies studying and analyzing linear data along with the projective subspaces to get the best results. The dataset includes 101 patients collected at Cancer Oncology Hospital, Baghdad Medical City. The proposed method has shown three major components that affect breast cancer, and it explains 92% of the total variations.

Keywords: Symptom analysis · Symptom Principal Component Analysis (SPCA) · Iterative algorithm · Breast cancer

1 Introduction

Many common statistical methods are not very efficient for high-dimensional data [1, 2]. Therefore, dimensionality reduction is one approach to solve this issue, which finds a few functions of factors to describe the risk group with the minimum loss of information [3]. There are numerous, potentially distinct models for such functions. The symptom-syndrome model [4] is a powerful technique to form a finite projective space [5]. Generally, studying a phenomenon requires defining the variables of that phenomenon as the first step to determining the most important variables and their count for the phenomenon under study [6]. Reducing too many input variables to a smaller number while preserving the principal amount of information about those variables without losing it can be achieved through the Principal Component Analysis (PCA) method [6]. In this study, the PCA is enhanced with symptom analysis to choose the weighty principal components obtained by the classical PCA analysis [7].

© The Author(s), under exclusive license to Springer Nature Switzerland AG 2024
A. M. Al-Bakry et al. (Eds.): NTICT 2023, CCIS 2096, pp. 276–287, 2024.
https://doi.org/10.1007/978-3-031-62814-6_20

The goal of this study is to develop a new method for finding the principal components based on symptom analysis, namely, super-symptoms generated from an iterative algorithm. The method had been applied to medical data collected from breast cancer patients at the Cancer Oncology Hospital, Baghdad Medical City. Unfortunately, breast cancer is one of the most common cancers that cause death among women [8].

The rest of this study is organized as follows: Sect. 2 gives a brief description of Symptom Principal Component Analysis (SPCA). Section 3 describes the application of the proposed method. Section 4 presents the experimental results using a medical dataset, and Sect. 5 concludes this paper with discussions and future work.

2 Symptom Principal Component Analysis (SPCA)

Super-symptom analysis is a modern and attractive technique due to its flexibility and ability to be used in many applications that are difficult to approach with the classical methods [4].

The main idea is similar to that of classical PCA, in which a large number of variables in the problem are transformed into a smaller number of variables. The resulting variables are the so-called main principal components (axes) that imply the largest amount of total variance in the dataset [9].

In other words, super-symptom analysis is a technique for transforming data from the original high-dimensional space into linear combinations of dichotomous variables across the field F_2. Projective space formation allows the selection of the most useful sub-space to reduce the dimensionality of categorical data.

The following sections present the Super-Symptom PCA and the iteration algorithm to construct the super-symptoms [10].

2.1 Symptom and Super-Symptom

Consider a random vector $X = (X_1, \ldots, X_m)^T$ with components taking values 0 and 1. Usually 0 and 1 mean absence and presence of factors, respectively. The new variable $X_i + X_j$(mod2) means the presence of any one in the absence of another factor. More than two variables are considered also. The definitions of the symptom and super-symptom and their justifications are described in Ref [4, 10]. Generally, the symptoms and the super-symptoms are expressed with a parameterized of dichotomous vectors $(\alpha_1, \ldots, \alpha_m)$ and $(\beta_1, \ldots, \beta_M)$, where $M = 2^m - 1$, $\alpha_k, \beta_i \in (0, 1)$, and all elements of the vector are mixture of 0's and 1's such that $\sum_{k=1}^{m} \alpha_k \neq 0$, $\sum_{i=1}^{M} \beta_i \neq 0$. Two other special parameters are introduced as follows: $a = \sum_{k=1}^{m} \alpha_k 2^{k-1}$ and $b = \sum_{i=1}^{M} \beta_i 2^{i-1}$ to represent the summation of the parameters, respectively. These are expressed, as follows [4, 10]:

$$G_a(X_1, \ldots, X_m) = \sum_{k=1}^{m} \alpha_k X_k (\text{mod } 2), a = 1, \ldots, M = 2^m - 1, \quad (1)$$

$$F_b(X_1, \ldots, X_m) = \sum_{a=1}^{M} \beta_a \prod_{k=1}^{m} X_k^{\alpha_k} (\text{mod2}), where\, b = 1, 2, \ldots, 2^M - 1 \quad (2)$$

where G_a represents the symptoms vector, F_b represents the super-symptoms vector.

3   Application of the Proposed Method

This section presents the application of the proposed method using breast cancer data collected at the Cancer Oncology Hospital, Baghdad Medical City.

3.1   Dataset Description

Breast cancer is considered one of the most commonly diagnosed malignant tumors in women throughout the world, as well as the main cause of death for malignant tumors [11]. In this study, the dataset includes 101 patients of breast cancer collected and documented in 2017 at Cancer Oncology Hospital, Baghdad Medical City. This hospital is the most registered and treated breast cancer hospital in the area [12]. In this study, the missing data were excluded and cleaned up for about 36 cases due to various situations, including the fact that the collected data contain some sections that the patient could not answer. For training phase, 80% of the dataset was used while the remaining 20% was used in the testing phase. More information about the dataset and the specific study's variable are shown in Table 1.

Table 1. Encoding variables of the dataset.

Code	Variables	Meaning
A	Age	1 - Age less than 59 (73%) 0 - Age greater than 59 (27%)
D	The Oncotype DX test	1 - ILC nodal severe type tumor (13%) 0 - IDC ducts type tumor in Pipe lactiferous (87%)
G	Grades of breast cancer	1 - Poorly differentiated tumor (67%) 0 - Moderately differentiated tumor (33%)
E	Estrogen receptor positive	1 - Yes (73%) 0 - No (37%)
P	Progesterone receptor positive	1 - Yes (75%) 0 - No (25%)
H	The human epidermal growth factor receptor	1 - HER2-positive breast cancer (73%) 0 - There is no antigen in tissue (27%)
K	Proliferative activity of cells	1 - Greater than 15 (60%) 0 - Ki-67 less than 15 (60%)
S	Surgery removal of the tumor or breast	1 - Mastectomy or Lumpectomy (56%) 0 - Excisional biopsy (44%)
T	Advanced type of tumor	1 - The size of the main tumor more than 3 cm (68%) 0 - Otherwise (32%)
L	Lymph nodes	1 - Tumor spreading to the lymph node (82%) 0 - No (18%)
M	Metastasis	1 - Distant metastasis(92%) 0 - No distant metastases (8%)

3.2 Iterative Algorithm for Constructing Super-Symptoms

In this study, the iterative algorithm for constructing super-symptoms was inspired by Ref. [10]. This algorithm was employed to suppress conducting a wide-ranging search with the least losses. In the first step, the algorithm accepts potential symptoms as input that can be created from $X = (X_1, X_2, \ldots, X_k)$ over F_2. These variables are the basic symptoms of linear syndrome. In general, any linearly independent symptoms can be considered the basic symptoms. So, the linear syndrome consists of $2^m - 1$ different components that represent all sorts of symptoms.

By applying the iterative algorithm for identifying the most three informative super-symptom variables R_1, R_2, R_3, the super-symptom $R_1 = 1 + E.P.S + P \pmod 2$, where E represents the Estrogen, P represents Progesterone, and S represents Surgery variables as indicated in Table 1, for which the UC with the variable $M.L$ is approximately 28.85%, where M represents the Metastasis and L represents the Lymph nodes, and UC represents the uncertainty coefficient [12]. According to the value of the coefficient of uncertainty with the dependent variable ML.

Thus, a number of all potential super-symptoms can be obtained from m of dichotomous variables $\mathbb{X}_p = (X_1, \ldots, X_k)$ is $2^{2^m - 1} - 1$. The algorithm was applied many times, selecting super-symptoms each time to obtain an approximate and accurate result.

3.3 PCA for Estimating the Matrix of Factor Loads

The initial solution had been obtained using the standard factor analysis. Generally, the aim of the initial solution is to estimate the value of the loads of factors composing the matrix. Then, estimate the other indicators according to the values of this matrix [13].

The steps for the initial solution can be summarized as follows:

1. Square Multiple Correlation (SMC) is calculated for each variable, with the rest of the variables as an initial estimate of the values of the communalities. Then we substitute the single elements on the main diagonal for the correlation matrix for commonality values. In another illustration, we put SMC_1 instead of r_{11}. Thus, we obtained the Reduced Correlation Matrix.

$$R^2 = \begin{bmatrix} \hat{h}_1^2 & r_{12} & . & r_{1p} \\ r_{21} & \hat{h}_2^2 & . & r_{2p} \\ . & . & . & . \\ r_{p1} & r_{p2} & . & \hat{h}_p^2 \end{bmatrix} \tag{3}$$

2. The eigenvalues are calculated from the reduced correlation matrix using the following equation:

$$|R_r - \lambda_j I| = 0 \tag{4}$$

Thus, we have obtained P of eigenvalues, and the number of these values is equal to the number of variables, where eigenvalues represent the amount of the factor's contribution to the total values of commonness (the importance of the factor).

3. We choose eigenvalues whose value exceeds 1, where their number represents the number of factors. In other words, we choose eigenvalues so that the ratio of joint variance explained by the extracted factors composes a large percentage.
4. The matrix of model A factors is estimated, where each of its factors is estimated in terms of (eigenvectors) a associated with each selected eigenvalues, and starting with the largest value according to the following:

Thus, the eigenvectors accompanying the largest eigenvalues of the values that are greater than one represent the estimated loads of the first factor, and the eigenvectors associated with the largest eigenvalues following the second factor represent the estimated loads of the second factor, and so on, and thus we have obtained the matrix of the first estimated factor loads A_1.
5. The values of commonality are extracted from the elements of the matrix A_1 as follows:

$$
\begin{aligned}
h_1^2 &= a_{11}^2 + a_{12}^2 + \ldots \ldots + a_{1m}^2 \\
h_2^2 &= a_{21}^2 + a_{22}^2 + \ldots \ldots + a_{2m}^2 \\
&\ldots \ldots \\
h_p^2 &= a_{p1}^2 + a_{p2}^2 + \ldots \ldots + a_{pm}^2
\end{aligned}
\tag{5}
$$

Then we replace the main diameter elements from Rr (reduced correlation matrix) with the new ones $h_1^2, h_2^2, \ldots, h_p^2$.
6. Find the A_2 matrix by iteratively repeating steps 2, 3, 4, and 5.
7. Continue to find the A_3 matrix by repeating the same previous steps, and so the process continues until the differences between h_j^2 and for two consecutive matrices are very small, and then we stop and consider the last matrix as the model matrix.

In this study, the Principal Components Method had been used depending on the iteration method, to build the highest super-symptoms $(R_1, \ldots, R_{10})$ because it is one of the most important methods that are used to study multivariate phenomena to reach an accurate identification of the most important basic variables that make up those phenomena to get to the important factors. That affects negatively or positively its behavior. The most important characteristic of this method is that it enables us to explain the interrelationships between the variables, which involves converting a set of correlated variables into a new set of uncorrelated (orthogonal) variables called principal compounds. Every principal compound is a linear combination of super symptom, and the variance of this component is part of the total variance, so the variance of the first principal compound is greater than that of any other principal compound, and the variance of the second principal compound is less than the variance of the first principal compound but greater than that of any other principal compound, and so on.

This can be summarized as the goal is to find linear factors or combinations called principal components that are few or less than the original variables to replace them based on the iteration algorithm, and these principal components are orthogonal, meaning there is no correlation between them. These components or factors are equal to the number of eigenvalues whose value exceeds one.

Therefore, the principal components model will be in the form of a linear mathematical relationship between the studied random variables, and it can be represented as follows:

$$PC_1 = a_{11}R_1 + a_{21}R_2 + \ldots\ldots + a_{p1}R_p$$
$$PC_2 = a_{12}R_1 + a_{22}R_2 + \ldots\ldots + a_{p2}R_p$$
$$\ldots\ldots$$
$$PC_j = a_{1j}R_1 + a_{2j}R_2 + \ldots\ldots + a_{pj}R_p \tag{6}$$
$$PC_j = \sum_{i=1}^{p} a_{ij}R_i, \ (i, j = 1, 2, \ldots\ldots p)$$

where PC_j: Principal Component$j$,

a_{ij}: Coefficient of variable j in component i, which represent the values of eigenvectors a_i related with the eigen roots λ_i of the matrix used.

3.4 Principal Component Analysis Based on Super-Symptom

Despite the fact that understanding the key points of principle component analysis requires a background of linear algebra, the basic issues can be clarified through simple geometric interpretations of the data. For explanation purpose, imagine that the symptom analysis in our example measured 10 super-symptoms $R_1, \ldots, R_{10}$ based on an iterative algorithm for constructing super-symptoms that has been applied repeatedly and trying to get the most accurate three super-symptoms each time.

In fact, the iterative algorithm selects super-symptoms based on a set of rules for reducing the dimensionality of categorical data, as stated in Sect. 3.2. The result is that the dimensionality can be reduced from the number of super-symptoms to reach three dimensions.

Prior to getting the PCA, eigenvalues and eigenvectors must be determined. Eigenvalues denote the variance for a given PC. Theoretically, they could be positive or negative values, but in practice, they act as variance, which is always positive. Furthermore, eigenvectors represent the direction of the principal components.

Figure 1 shows various red vectors that represent super-symptoms included in the analysis of the principal components. The cosine of the angle between two vectors represents the value of the correlation coefficient between them. This means that the super-symptom that separates their vectors by sharp angles are highly correlated (i.e., $\cos(0) = +1$).

Correspondingly, we can deduce the super-symptoms that are related to each other inversely (i.e., the higher the value of the first super-symptom, the lower the second, and vice versa). In other words, the angle separating the vectors representing the two super-symptoms compared to them must be an obtuse angle. As is the case for the super-symptoms $(R_1 = R)$ and $(R_2 = R_1)$ in Fig. 1, the correlation rises with the increasing of angle measure up to the straight angle, which measures 180 degrees (i.e., the two vectors are collinear but in the opposite direction), and then the correlation is a perfect inverse correlation where $\cos(180) = -1$.

Whereas, the super-symptoms that have vectors perpendicular or close to perpendicular to each other (i.e., the angle between them is close to $90°$), imply that the super-symptoms are not related to each other (i.e., they are independent symptoms).

As shown in Fig. 1, all vectors are drawn with equal lengths, where the standardization process was applied to the data before performing the principal component analysis PCA.

282   Fatema. S. Al-Juboori et al.

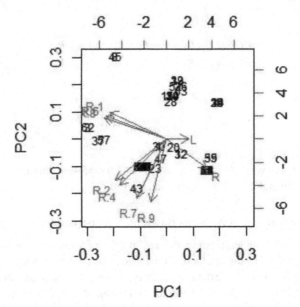

Fig. 1. PCA biplot of ten dimensions using their projections onto PC1 and PC2, for 10 super-symptoms.

3.5 Component Matrix of the 10-Component PCA

The Components Matrix can be explained as a loading matrix where each super-symptom has a loading in accordance for all of 10-component. For example, as shown in Table 2, the super-symptom R1 is correlated with 0.303 with the first component, −0.221 with the second component, −0.564 with the third, and so on.

The square value of each loading represents the portion of variance shown by a particular component. For the super-symptom R1, $(0.303)^2 = 0.091$ or 9% of its variance is revealed by the first component. Subsequently, $(-0.221)^2 = 0.049$ or 4% of the variance in super-symptom R1 is explained by the second component. The sum of the two square values of the both components is thus $9\% + 4\% = 13\%$. Therefore, the sum of all the squared loadings cumulatively down the components should result in 1 or 100%. This is also known as communality, and in a principal component analysis the communality for each super-symptom is equal to the total variance.

Table 2 shows the Component Matrix and by summing the squared component loadings across the components (i.e., each row), the communality is calculated for each super-symptom. Whereas, by summing the squared loading across the super-symptoms (i.e., each column), the eigenvalue for each component is calculated. For example, the first eigenvalue is calculated as follows:

$$(0.303)^2 + (-0.385)^2 + (-0.294)^2 + (0.312)^2 + (-0.250)^2$$
$$+(-0.398)^2 + (-0.398)^2 + (-0.149)^2 + (-0.411)^2 + (-0.060)^2 = 0.99$$

Table 2. Component Matrix Results

Super-Symptoms	Component Matrix									
	PC_1	PC_2	PC_3	PC_4	PC_5	PC_6	PC_7	PC_8	PC_9	PC_{10}
R1	0.303	−0.221	−0.564	0.201	−0.103	0.192	−0.270	0.579	−0.205	0.000
R2	−0.385	0.152	−0.158	−0.032	−0.874	−0.171	−0.070	0.003	0.055	−2.912
R3	−0.294	−0.384	0.214	0.368	0.026	0.142	0.007	0.293	0.690	−3.851
R4	0.312	−0.215	−0.558	0.083	−0.014	−0.245	0.117	−0.536	0.419	−5.089
R5	−0.250	−0.428	0.043	0.544	−0.064	0.163	0.078	−0.392	−0.514	2.956
R6	−0.398	0.111	−0.349	−0.040	0.229	−0.008	0.367	0.125	−0.042	7.071
R7	−0.398	0.111	−0.349	−0.040	0.229	−0.008	0.367	0.125	−0.042	−7.071
R8	−0.149	−0.507	0.071	−0.238	0.121	−0.749	−0.139	0.183	−0.166	−3.072
R9	−0.411	0.107	−0.215	−0.089	0.282	0.129	−0.77	−0.249	0.075	3.860
R10	−0.060	0.107	−0.034	−0.671	−0.137	0.497	0.114	−0.096	0.003	6.326

3.6 Total Variance Explained in the 3-Component PCA

Eigenvalues denote the variance for a given PC. Initially, by obtaining the PC_1; the successive component is found by partialing out the prior component. For example, Component 1 is 5.132, or 5.132 / 10 = 0.513 of the total variance (see Table 3).

Table 3. Results of Total Variance

Component	Eigenvalues		
	Total Variance	% of Variance	Cumulative %
PC_1	5.132	0.513	0.513
PC_2	3.025	0.302	0.815
PC_3	1.081	0.108	0.924
PC_4	0.371	0.037	0.961
PC_5	0.178	0.017	0.978
PC_6	0.133	0.013	0.992
PC_7	0.039	0.003	0.996
PC_8	0.030	0.003	0.999
PC_9	0.007	0.000	1.000
PC_{10}	0.000	0.000	1.000

Consequently, in this table, the PC_1 shows the highest variance, and the last component shows the least. Table 3 shows the total variance interpreted by each component.

In addition to that, the super-symptom variants most saturated in these components are PC_1, PC_2, PC_3:

First Component: This component is considered the most important component because it implies (51%) of the total variance, and consequently it is considered the most important component that influences breast cancer disease. This component implies super-symptoms R_4 and R_1 arranged according to their saturation values (i.e., the changes in the levels of female hormones progesterone P and estrogen E that can be the main cause of breast cancer), as shown in Table 4:

Table 4. The first component PC_1

Super-symptoms	Saturation
$R_4 = 1 + P + P.S(\text{mod } 2)$	0.312
$R_1 = 1 + E.P.S + P(\text{mod } 2)$	0.303

Second Component: This component represents the second rank in terms of importance as it implies (30%) of the total variance and this component implies each of the super-symptoms according to their saturation values which includes a number of symptoms, which are as follows: (grades of breast cancer G, proliferative activity of cells K, the oncotype DX test D, and the human epidermal growth factor receptor H) which increases the possibility of breast cancer as shown in Table 5.

Table 5. The second component PC_2

Super-symptoms	Saturation
$R_2 = GK + GKS + S(\text{mod } 2)$	0.152
$R_6 = DG + DGS + G + GS + S(\text{mod } 2)$	0.111
$R_7 = G + GHS + S(\text{mod } 2)$	0.111
$R_9 = 1 + DG + DGS + G + S(\text{mod } 2)$	0.107
$R_{10} = 1 + DPS + DS + P + PS(\text{mod } 2)$	0.107

Third Component: This component interprets (10%) of the total variance, and this component implies super-symptoms arranged according to their saturation values as shown in Table 6.

Table 6. The third component PC_3

Super-symptoms	Saturation
$R_3 = 1 + HPS + HS + P + PS \pmod 2$	0.214
$R_8 = 1 + E + EP + EPS + P \pmod 2$	0.071
$R_5 = 1 + HS + P \pmod 2$	0.043

3.7 Choosing the Number of Components

To fulfill the aim of principal component analysis, some kind of criterion should be used for choosing the ideal number of components (i.e., super-symptoms). One of the most used criteria is to select components that have a total variance greater than 1 as shown in Table 3. The scree plot had been plotted to visualize the proportion of variance explained by each individual PC, as shown in Fig. 2. This figure shows that the first component has the highest total variance, and the last component will always have the lowest.

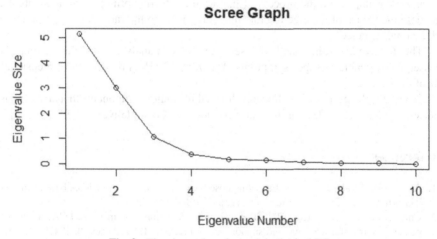

Fig. 2. The eigenvalues for each individual PC.

4 Results and Analysis

As shown in Table 4, PC_1 is the most important factor that affects breast cancer disease (i.e., in the super-symptoms R_4 and R_1). By analyzing the variants of the symptoms (estrogen receptor E, progesterone receptor P, and surgery S) of this component. In other words, the hormones progesterone receptor positive, estrogen receptor positive, and surgery (i.e., an excisional biopsy) can be the main causes of breast cancer.

Table 5 shows the second important factor PC_2 that affects breast cancer disease (i.e., the super-symptoms R_2, R_6, R_7, R_9 and R_{10}). By analyzing the variants of the symptoms (grades of breast cancer G, proliferative activity of cells K, the oncotype DX test D, and the human epidermal growth factor receptor H), of this component.

In addition, Table 6 shows the third important factor PC$_3$ that affects breast cancer disease (i.e., the super-symptoms R_3, R_8 and R_6). Interestingly, this factor includes the same symptoms mentioned previously (E, P, S, G, K, D and H).

In summary, it can be said that the variables affecting breast cancer disease were reduced by the SPCA method along with the iterative algorithm.

5 Conclusion

This study proposed a new method for analyzing the symptoms of breast cancer disease in the case of multivariate data sets for binary data. The proposed method is called Symptom Principal Component Analysis (SPCA), inspired by the classic PCA method, and aims to select super-symptoms generated from an iterative algorithm based on a set of rules for reducing the dimensionality of categorical data.

The results show high precision in selecting the super-symptoms that resemble the most informative symptoms and the appropriate way to demonstrate them. Furthermore, it has a clear impact on the process of finding the effective principal components that increase the values of the eigenvectors that correspond to the ratio in the interpretation of the total variance.

The iterative algorithm had been used effectively in analyzing the clinical data and consequently generates super-symptoms. Accordingly, SPCA reduced the dimensionality of the data.

Generally, the proposed method has shown three major components that affect breast cancer, and it explains 92% of the total variations (i.e., cumulative).

References

1. Preetha, R., Vinila, J.S., et al.: Early diagnose breast cancer with PCA-LDA based FER and neuro-fuzzy classification system, vol. 12, pp. 7195–7204 (2021)
2. Chiu, H.J., Li, T.H.S., Kuo, P.H.: Breast cancer–detection system using PCA, multilayer perceptron, transfer learning, and support vector machine. IEEE Access **8**, 204309–204324 (2020)
3. Bian, K., Zhou, M., Hu, F., Lai, W.: RF-PCA: a new solution for rapid identification of breast cancer categorical data based on attribute selection and feature extraction. Front. Genet. **11**, 566057 (2020)
4. N. J. R. E. S. P. o. t. S.-P. S. U. Alexeyeva, Saint-Petersburg, "Analysis of biomedical systems (2013)
5. Alexeyeva, N., Alexeyev, A., Gracheva, P., Podkhalyuzina, E., Usevich, K.: Symptom and syndrome analysis of categorial series, logical principles and forms of logic. I: 2010 3rd International Conference on Biomedical Engineering and Informatics, vol. 6, pp. 2603–2606. IEEE (2010)
6. B. M. S. Hasan, A. M. J. J. o. S. C. Abdulazeez, and D. Mining, "A review of principal component analysis algorithm for dimensionality reduction," vol. 2, no. 1, pp. 20–30, 2021
7. Alhussan, A.A., et al.: ForkJoinPcc algorithm for computing the PCC matrix in gene co-expression networks. Electronics **11**(8), 1174 (2022)

 8. Ibrahim, S., Nazir, S., Velastin, S.A.: Feature selection using correlation analysis and principal component analysis for accurate breast cancer diagnosis. J. Imaging **7**(11), 225 (2021)
 9. Kherif, F., Latypova, A.: Principal component analysis. Mach. Learn. 209–225 (2020)
10. Al-Juboori, F.S., Alexeyeva, N.P.: Application and comparison of different classification methods based on symptom analysis with traditional classification technique for breast cancer diagnosis. Periodicals Eng. Nat. Sci. **8**(4), 2146–2159 (2020)
11. Smolarz, B., Nowak, A.Z., Romanowicz, H.J.C.: Breast cancer—epidemiology, classification, pathogenesis and treatment (review of literature). Cancers **14**(10), 2569 (2022)
12. Alexeyeva, N., Al-Juboori, F., Skurat, E.: Symptom analysis of multidimensional categorical data with applications. Periodicals Eng. Nat. Sci. **8**(3), 1517–1524 (2020)
13. Mahimid, N.A., Saleh, A.H.: A statistical study of the factors affecting the mental health of the individual at the university of Mustansiriya. J. Admin. Econ. **26**(68) (2008)

An Improved Method for Retrieving Fashion Images Based on Parsing the Image Contents

Furqan Tahseen[1], Ahmed Talib[2(✉)], and Dalia A. Al-Ubaidi[3]

[1] Al-Ma'moon University College, 14th Ramadan Street, Baghdad, Iraq
[2] Middle Technical University (MTU), Technical College of Management, 10047 Bab Al-Muadham, Baghdad, Iraq
dr.ahmed.talib@mtu.edu.iq
[3] University of Information Technology and Communications, Baghdad, Iraq

Abstract. The field of fashion image retrieval is an active research area. In this paper, we rely on CBIR systems to retrieve fashion images. These systems depend on analyzing the contents of images in terms of features. We propose an improved fashion image retrieval method based on shape and local color features. In order to extract the shape feature, we have improved the VGG-16 model. To extract the local color feature, we use both the SRD and DCD from the MPEG-7 methods to remove the background and extract the dominant color. We also use the indexing method to quickly retrieve images. In order to retrieve the images that are most similar to the query image, we used the dissimilarity measure. The results showed that the proposed method determines the shape of a piece of clothing by 100/100 and that the percentage of time it takes (0.27%) to retrieve images is compared to a sequential search and is more accurate than a sequential search.

Keywords: MPEG-7 descriptors · Color indexing · salient object detection · neural network · distance measure

1 Introduction

The expansion of digital images and the expansion of huge storage spaces led to the emergence of huge collections of images called image libraries [1]. These digital image libraries have expanded on the Internet through the advancement of transmission technologies, but finding the required images has become a problem for users. The solutions to this problem lie in the management and organization of digital image libraries (databases), which are formulated in the form of two research concepts: "indexing" and "retrieval". The concept of indexing is how images are stored in the database (or in any other way) and how they are searched. While what is meant by the concept of retrieval is how to retrieve images related to the query image in the database [2].

There are two ways to retrieve images from digital libraries [2]. The first is text-based image retrieval; that is, the query is done by entering a description (text). The system analyzes the text entered by the user and then retrieves the images. There are two drawbacks to this method. First, it requires the annotation of all images in the database,

A. M. Al-Bakry et al. (Eds.): NTICT 2023, CCIS 2096, pp. 288–301, 2024.
https://doi.org/10.1007/978-3-031-62814-6_21

which is a tedious and time-consuming process. Second, the annotation process is usually inefficient because the user performs it in an unsystematic or irregular way, as different users use different words to describe the same image. The second method is content-based image retrieval (CBIR), that is, the user enters an image that represents the query process and retrieves images similar to the query image. The system analyzes the contents of the image, and these contents are represented in the form of extracted features such as color, shape, texture, and others. The process of feature extraction is done using descriptors. For example, a common descriptor for color feature extraction in image retrieval systems is dominant color descriptors (DCD). The image retrieval process is the process of identifying similar images stored in the database and displaying them to the user based on the similarity measure. The similarity measure has two types. The first is classification-based similarity measurement, which is the process of assigning a label (belonging to one of the pre-defined image categories) to the unlabeled query image. For example, this type is a convolutional neural network (CNN). The second type is based on the distance between images. Distance-based similarity measure. That is, the distance measure (for example, Euclidean distance) is relied upon to calculate the difference between the feature vector of the query image, Feature Vector (FV), and the feature vector of the database image (DB). It has become clear that the number of images has increased significantly, so the process of sequential searching the database affects the image retrieval time. That is, the larger the number of images, the higher time to retrieve images. This affects online applications, where image retrieval time is very important. Sequential search can be defined as "the process of comparing the query image with all images in the database." To avoid this problem, researchers resorted to indexing techniques. Indexing techniques are the process of reducing the whole search space, that is, instead of searching the all image in database, a specific set of images in the database is searched, or in other words, a new search space is produced that is smaller than the original search space. Therefore, in this paper, we use indexing technology to quickly retrieve images [2, 3].

CBIR systems first appeared in the 1990s. The "QBIC" system [4] is considered one of the first and most famous systems that retrieve images based on the content of the images. The simplicity of using CBIRs systems by the user and the satisfactory results are what led to the entry of CBIRs systems into many fields, including medical, commercial, security, and so on.

In this paper, we address CBIR systems in the fashion field, which is considered an active research area [5]. In addition to the Corona pandemic situation, e-purchasing has increased significantly around the world. Additionally, it has become possible to open an online store anywhere in the world. During the e-purchasing process, users face a problem choosing the appropriate items to purchase. The reasons for these problems include the size and variety of goods, culture, and language. CBIR systems can assist users in selecting clothes and help them make a better choice of clothes during the e-purchasing process [6]. Structure of this paper as follows: Sect. 2 explains related works, while Sect. 3 shows proposed method. Experimental results are detailed in Section 4 and concluding the paper is described in Sect. 5.

2 Related Work

Previous studies on CBIR systems in the field of fashion can be divided into two categories. The first category of studies is similarity based on distance. The second class of similarity studies is based on classification. The first category of studies uses general algorithms in order to extract features from fashion images, such as the dominant color descriptor or the color histogram. In order to retrieve fashion images that are most similar to the query image, distance measures are used, such as the Euclidean or Manhattan distance measure and other measures. The second category of studies uses special algorithms to extract features from fashion images, such as a convolutional neural network. In order to retrieve fashion images that are most similar to the query image, the category to which the query image belongs is determined.

This paragraph represents studies in the first category (distance-based similarity). Gupta and et al. [6] In this work, a clothing image retrieval system based on color and texture features was built. Pre-extract color and texture feature. The image is classified into two main categories. The first type is determining the shape of the piece (pants, shorts, etc.) through the length of the piece of clothing using the horizontal projection histogram. The second category is determining the sleeves of clothing images through the ratio of clothing color to skin color. After that, to extract the texture feature, the Gabor filter is used. To extract the color feature, the image is converted to the CIE lab color space, then the chroma difference similarity measure is used, and using the threshold, the color feature is extracted from the image. Nastia and et al. [7] proposed a method. The method proposed in this paper is based on the shape feature, where a new method called Pyramid Histogram of Oriented Gradient Cross Feature (PHOGXF) was hybridized to determine the shape of a piece of clothing in the image. The researcher hybridized both the Moore-Neighbor Tracing (MNT) method and the PHOG method. This method gave better results than the PHOG method. Mustaffa and et al. [8] Building a system based on color and shape features. First, the images are converted to HSV color space, and then the color and shape features are extracted. To extract the color feature, the color histogram of each color channel was used. To extract the shape feature, low and high thresholds were used for each color channel. The researcher used the Manhattan distance measure to calculate the similarity between images. Mutia [9]. Building a system based on color and shape features, to extract the dominant color from the image, DCD from MPEG-7 and PHOG were used to extract the shape feature.

This paragraph represents studies in the second category (classification-based similarity). Ziwei [10] Proposed method. This method is based on the shape feature using a ready-made model called VGG-16. Ding [11] Building a clothing image retrieval model based on color and shape features Data is labeled based on color and shape. To extract features from images through speed-up robust features (SURFs), a bilinear supervised hashing algorithm is used to classify images. Liu [12] to build a model for clothing image retrieval. The local, global, and color features are extracted using the Resnet50 network.

We also review some studies on reducing image retrieval time. Time is considered one of the factors affecting online applications. In order to reduce the time required, indexing methods are the domain of researchers in this field. Indexing methods are divided into two main parts. Multidimensional indexing and vector quantization techniques The first section is multi-dimensional indexing, which divides the data or space into small sections, such as the R-tree technique, R*-tree, B + -tree, and others. As for the second section, vector quantization techniques, this type of indexing does not divide the data or space but rather into clusters. The cluster is represented by a single value called the center. One of the techniques used in this field is hierarchical K-means clustering [2, 3].

3 Proposed Method

In this paper, we propose an improved method for retrieving fashion images based on shape and local color features. This proposed method is characterized by high accuracy and speed in retrieving fashion images. The proposed method consists of four main sequential operations:

1. Shape feature extraction
2. Local color feature extraction
3. Indexing method
4. Distance measure

3.1 Shape Feature Extraction

In this process, a category is determined from the dataset of images to which the query image belongs by specifying the shape of the piece of clothing. In order to determine the shape of a piece of clothing in this paper, we built a model based on convolutional neural networks. The model consists of the following:

- Data download.
- Data scaling.
- Data split.
- Improving and training the VGG-16 model.

Data Download
The process of downloading data is the first step in training the model, as it was obtained from the Kaggle website [13]. The dataset includes a total of 30,144 clothing images, divided into seven categories. Here are the categories of the dataset:

- Dresses (6,236) images
- Jackets Coats (1,895) images
- Pants (1,804) images
- Shorts (3,486) images
- Skirts (2,042) images
- Sweaters (3,036) images
- Shirt (11,642) images

The size of each image in the dataset is ($256 \times 256 \times 3$) pixels in JPEG format.

Data Sizing
The uploaded dataset is ready for direct use without any pre-processing. Only the size

of the images is changed from (256 × 256 × 3) to (224 × 224 × 3) in order to fit the size required by the VGG-16 model used.

Data Splitting

The researcher divided the original dataset into two parts:

- Training dataset: It is the data through which the model is trained, as the images and the category to which each image belongs are passed to the model. Data percentage amount (80%).
- Testing dataset: It is the data that is passed to the model without specifying the category to which it belongs in order for the model to recognize it and to know the extent of the model's performance in determining the images of the piece of clothing. Data percentage amount (20%).

Improving and Training the VGG-16 Model

The VGG(Visual Geometry Group in the University of Oxford)-16 model is considered one of the ready-made models that can be used [14], with the last layer changing according to the data set used. Table 1 represents the general structure of the VGG-16 model.

Table 1. Structure model VGG-16

Activation Function	Size Of Feature Map	Padding	Stride	Filter Size	Filters/Neurons	Layer
–	224 × 224 × 3	–	–	–	–	Input
RELU	224 × 224 × 64	1	1	3 × 3	64	Conv_1
RELU	224 × 224 × 64	1	1	3 × 3	64	Conv_2
–	112 × 112 × 64	–	2	2 × 2	–	MaxPool_1
RELU	112 × 112 × 128	1	1	3 × 3	128	Conv_3
RELU	112 × 112 × 128	1	1	3 × 3	128	Conv_4
–	56 × 56 × 128	–	2	2 × 2	–	MaxPool_2
RELU	56 × 56 × 256	1	1	3 × 3	256	Conv_5

(*continued*)

Table 1. (*continued*)

Activation Function	Size Of Feature Map	Padding	Stride	Filter Size	Filters/Neurons	Layer
RELU	56 × 56 × 256	1	1	3 × 3	256	Conv_6
RELU	56 × 56 × 256	1	1	3 × 3	256	Conv_7
–	28 × 28 × 128	–	2	2 × 2	–	MaxPool_3
RELU	56 × 56 × 512	1	1	3 × 3	512	Conv_8
RELU	56 × 56 × 512	1	1	3 × 3	512	Conv_9
RELU	56 × 56 × 512	1	1	3 × 3	512	Conv_10
–	14 × 14 × 512	–	2	2 × 2	–	MaxPool_4
RELU	14 × 14 × 512	1	1	3 × 3	512	Conv_11
RELU	14 × 14 × 512	1	1	3 × 3	512	Conv_12
RELU	14 × 14 × 512	1	1	3 × 3	512	Conv_13
–	7 × 7 × 512	–	2	2 × 2	–	MaxPool_5
RELU	4096	–	–	–	–	Fully Connected_1
RELU	4096	–	–	–	–	Fully Connected_2
Softmax	7	–	–	–	–	Fully Connected_3

The values that are set and given to the model in order to train the data entered into the model differ from one model to another. Here are the values that are set and given to the form:

Epoch: means the number of times to train and test the model; in this research, it was assigned a value of 64, meaning that the model will be trained on the training data set and tested on the test data set 64 times.

Patch size is set to 64. The larger the batch size, the more memory space is used for training.

The learning rate was set at 0.01% for the neural network.

When the value adjustment process was completed without any modifications to VGG-16, the model was trained on the training and test data sets.

The researcher faced a problem called over fitting. This problem is common in deep learning models and can be defined as "the large difference in both the value of the loss function and the accuracy of the model between the training data and the test data. That is, the value of the loss function in the model is small and the accuracy of the model is high on the training data, while the value of the loss function is large and the accuracy of the model is small on the test data. In order to overcome this problem, data augmentation and dropout technology were added to the VGG-16 model. The following is an explanation of these additions that the researcher included in order to improve the performance of the model:

1. Data augmentation is the process of artificially increasing the size of data using transformation methods on images of clothing. The purpose of data augmentation is to reduce the problem of model over fitting [15]. In this paper, the following methods were used:

   - Rotation: rotation of the image.
   - Height shift range: Shift the image horizontally.
   - shear: crop the image.
   - Zoom range: Zoom in on the image.
   - horizontal flip: flip the image horizontally.
   - Fill mode = nearest: Fill in the blank spaces in the image and make them equal to the nearest pixel.

2. Dropout: It is a technique used to reduce the problem of over fitting by randomly deleting the artificial cells (neurons) along with the cell links of the VGG-16 neural network in each iteration during the training process in each of the layers. The value of the deletion ratio is passed to the model (0.25) [16].

3.2 Local Color Feature Extraction

The process of extracting local color features from fashion images consists of two processes. The first process extracts the object from the images, and the second process extracts the color feature from the object. In this paper, the method proposed by Cheng [17], called salient region detection (SRD), was used. This method was applied to the set of images.

The next step in the object extraction process is to extract the color feature from the object. In this paper, the dominant color descriptor from MPEG-7 was used [18]. Provides an efficient and compact color representation of the color distribution in an image or object. This color descriptor is based on a clustering algorithm called generalized Lloyd algorithm (GLA) [2]. The general formula for the color descriptor is as follows:

$$\mathbf{DCD(I)} = \{(\mathbf{c_i}, \mathbf{p_i})\}, (\mathbf{i} = 1, 2, \ldots, \mathbf{N}) \tag{1}$$

N: number of dominant color

I: Represents images or objects

c_i: The extracted dominant color value is represented in the form of a 3D color vector

p_i: Represents the percentage of the dominant color, the percentage value ranges within a range $(1 >= p_i > 0)$

The sum of the percentages is equal to one. The maximum number of dominant colors extracted from images (or objects) is eight.

3.3 Indexing Method

In order to avoid the problem of sequential searching in a large database that causes slow display of results, an indexing method known as dominant color-based indexing methods (DCBIM) was used for the color space (RGB) that was proposed by the researcher Abdul Amir [2]. This method works by narrowing the search space or creating a new search space that includes images similar to the query image, meaning that instead of searching the entire database, a group of images is searched. The smaller the resulting area, the faster the clothing images can be retrieved. This method works on all major color descriptors. AS show in Fig. 1. In this method, the last five bits of the dominant color are relied upon. The last five bits of the three channels (R, G, B) are extracted from each dominant color and converted into three dimensions using the "Extract_Index_Dimensions_algorithm", The first dimension represents the last three bits, whose values range from (0–512) according to the value of the dominant color. The second dimension represents the fourth bit, whose value ranges from (0–8). The third level represents the third bit, which ranges from (0–8). The structure of this method consists of a fixed-size matrix of four levels: the first level represents the first dimension, the second level represents the second dimension, and the third level represents the third dimension. The fourth level aims to exclude images that are not similar based on percentages, as images in which the difference between the percentages of the indexed dominant colors is large are excluded. The fourth level is a B + -tree that includes four nodes; each node has a limit; the limits of the first node are (0–0.25) and the limits of the second node are (0.25–0.5). The limits of the third node (0.5–0.75) and the limits of the fourth node (0.75–1.00) In addition, this method includes a process of recognizing and even tolerating the indexed colors of the second and third levels, where indexed colors are tolerated within a limited range (0–8-24). The goal of this process is to include images similar to the query image in the new search space.

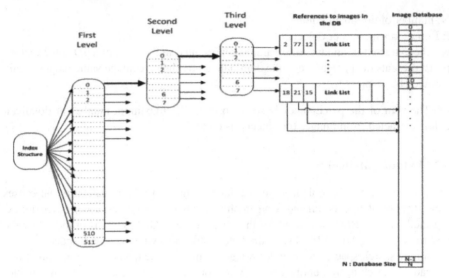

Fig. 1. Structure of Proposed RGB Indexing Method

3.4 Distance Measure

After the search space is narrowed using the indexing method, From the rest of the images, the ones that are most similar to the query image are selected. In order to determine the images that are most similar to the query image, we use the dissimilarity measure [19]. Its mathematical formula is:

$$D(F1, F2) = \sum_{i=1}^{N1} P1_i^2 + \sum_{j=1}^{N2} P2_j^2 - \sum_{i=1}^{N1} \sum_{i=2}^{N2} 2 * a_{ij} * P1_i P2_j \qquad (2)$$

where F1 is object extract from query image, F2 object extract from image DB. Where P1, P2 represent percentage of DC in the two objects of F1 and F2.

On the other hand, a_{ij} represents color similarity between two color Ci and Cj

$$a_{ij} = \{ \begin{matrix} 1 - \frac{d_{i,j}}{d_{max}} & d_{i,j} \leq Th \\ 0 & d_{i,j} > Th \end{matrix} \qquad (3)$$

The threshold *Th* represents the maximum distance whereby the two colors are considered similar. $d_{max} = \alpha * Th$, $\alpha = 1$ or 1.2 Where $d_{i,j}$ represents Euclidean distance between Ci and Cj. Abbreviation of the C represents 3-D color values in (R,G,B color space). The mathematical formula for the Euclidean distance is as follows:

$$d_{i,j} = \sqrt{\left(c_i^R - c_j^R\right)^2 + \left(c_i^G - c_j^G\right)^2 + \left(c_i^B - c_j^B\right)^2} \qquad (4)$$

4 Experiment Results

In this section, we review the performance results of the proposed system. The results of each of the following will be reviewed:

- performance of the VGG-16.
- Performance of the indexing method.
- Performance of dissimilarity measure.

Each of the above processes has its own performance evaluation metric. Also, ten samples will be taken randomly from each category of the dataset. In addition, each process has random samples that differ from the others.

The dataset used to evaluate the performance of the VGG-16 model is mentioned in Sect. 3.1. To evaluate the performance of the indexing method and the dissimilarity measure, three categories from the same dataset were used: skirts, shorts, Jackets coats.

4.1 Performance Results of the VGG-16

In order to measure the performance of the VGG-16 model, we will use the accuracy and loss function (Categorical Cross Entropy) metric [12]. Can be obtained according to following equation.

$$\textbf{Accuracy} = \frac{\textbf{True}_{positive} + \textbf{True}_{negative}}{\textbf{True}_{positive} + \textbf{True}_{negative} + \textbf{False}_{positive} + \textbf{False}_{negative}} \quad (5)$$

$$loss = -\sum_{i=1}^{N} y_i \log(y_i^{\wedge}) \quad (6)$$

We trained VGG-16 directly on the dataset. The results showed that the model suffers from an over fitting problem. This problem is explained in Sect. 3.1. Here are the results of the model:

- Model results on training data

Accuracy	0.8530
Loss function	0.4073

- Model results on testing data

Accuracy	0.3870
Loss function	2.4563

After adding dropout and data augmentation technology to VGG-16, the results showed a significant improvement as follows:

- Model results on training data

Accuracy	0.9012
Loss function	0.4073

- Model results on testing data

Accuracy	0.8555
Loss function	0.5566

4.2 Performance Results of the Indexing Method

To measure the performance of the indexing method, we use two types of metrics: accuracy metrics and efficiency metrics.

Efficiency measures: The goal of the indexing method is to reduce the retrieval time of fashion images. The method used in this paper reduces retrieval time by reducing the search space. In other words, instead of searching the entire database, a group of images is searched in the database. The smaller the search area, the less time is required to retrieve images. In addition, we calculate the time it takes for the indexing method to reduce the search space. Accordingly, we will use two measures. The first measure is a measure called the search space ratio (SSR) [2] can be obtained according to following equation.

$$\textbf{SSR = Reduced Search Space (RSS)/Whole Search Space (WSS)} \quad (7)$$

The second measure is time, using the unit millisecond.

Accuracy metrics: We use two metrics to measure the accuracy of the new search space: recall and precision. Recall is a measure of the system's ability to present all similar images as shown in Eq. (9). The precision measure indicates the system's ability to provide only similar images, can be obtained according to following equation [2].

$$\textbf{Precision} = \frac{number\ of\ relevant\ image\ retrieved}{total\ number\ of\ images\ retrieved\ from\ the\ database} \quad (8)$$

$$\textbf{Recall} = \frac{number\ of\ relevant\ image\ retrieved}{total\ number\ of\ relevant\ images\ in\ the\ database} \quad (9)$$

Table 2 includes the results of the indexing method, which represents the results of the categories of shorts, skirts, and Jackets Coats. Based on standards of accuracy and efficiency The results show that the indexing method reduces the time required compared to a sequential search by 31%. Based on accuracy measures (precision and recall), about half of the resulting area is similar to the query image.

Table 2. Result of indexing method.

Time_DC	Time	SSR	Recall	Precision	Classes
151.4	4	45%	0.63	0.47	Shorts
79.6	3	30%	0.33	0.62	Skirts
64	4	20%	10.6	0.33	Jackets Coats
98.3	3.67	31.6%	0.52	0.47	Average

4.3 Performance Results Dissimilarity Measure

The image retrieval process (displaying the images most similar to the query images to the user), represented by the distance measure, is considered the most important process. In this section, we evaluate the performance of the dissimilarity measure in retrieving similar images using accuracy measures, precision, and recall. The results are shown in the table.

The system has also been fully tested. In addition to evaluating the difference measure, the accuracy of a model-based VGG-16 in determining the shape of a piece of clothing is calculated, and the effect of the indexing method on the speed of retrieving fashion images compared to a sequential search is calculated.

Table 3 includes the results of the proposed method with a comprehensive search on the following categories: shorts, skirt, coat. The results showed that the () model achieved average results in all categories (100/100). The indexing method achieved good results in reducing the time taken to retrieve images by an average percentage of (27%). The results were also more accurate. The proposed method is more accurate and less time-consuming.

Table 3. Result of proposed method.

Time	Recall	Precision	Classification_accuracy	Method
Shorts				
50	0.5	0.75	100/100	Proposed Method
176	0.5	0.74	100/100	DCD
Skirts				
30	0.48	0.7	100/100	Proposed Method
86.2	0.45	0.68	100/100	DCD
Jackets Coats				
25	0.66	0.8	100/100	Proposed Method
71	0.66	0.8	100/100	DCD

5 Conclusion

In this paper, we propose an improved method for retrieving fashion images in CBIR systems based on shape features and local color features. Adding the dropout and data augmentation techniques to the model improved the results and reduced the problem of over fitting. Also, relying on indexing technology reduced the time required to retrieve fashion images and increased accuracy compared to a comprehensive search. In addition, the dissimilarity measure played its role effectively in this paper to retrieve images of colored clothes.

References

1. Attig, J., Ann, C., Michael, P.: Context and meaning: the challenges of metadata for a digital image library within the university. Coll. Res. Libr. **65**(3), 251–261 (2004)
2. Ahmed, T.A.: Colour-based image retrieval algorithms based on compact colour descriptors and dominant colour based indexing methods. Universiti Utara Malaysia (2014)
3. Talib, A., Mahmuddin, M., Husni, H., George, L.E.: Dominant color-based indexing method for fast content-based image retrieval. In: Boonkrong, S., Unger, H., Meesad, P. (eds.) Recent Advances in Information and Communication Technology. AISC, vol. 265, pp. 135–144. Springer, Cham (2014). https://doi.org/10.1007/978-3-319-06538-0_14
4. Niblack, C.W., et al.: QBIC project: querying images by content, using color, texture, and shape, Storage and retrieval for image and video databases (1993)
5. Latha, D., Raj, T.J.V.: Different types of CBIR applications. Int. J. Res. Eng. Appl. Manag. (IJREAM) (2019)
6. Gupta, M., Bhatnagar, C., Jalal, A.S.: Clothing image retrieval based on multiple features for smarter shopping. Procedia Comput. Sci. **125**, 143–148 (2018)
7. Nasita, I., Muchtar, K., Saddami, K., Arnia, F.: Cross-domain clothing image retrieval based on hybrid shape feature. In: 2019 IEEE International Conference on Cybernetics and Computational Intelligence (CyberneticsCom), pp. 98–102. IEEE (2019)
8. Mustaffa, M.R., Wai, G.S., Abdullah, L.N., Nasharuddin, N. A.: Dress me up! content-based clothing image retrieval. In: Proceedings of the 3rd International Conference on Cryptography, Security and Privacy, pp. 206–210 (2019)
9. Mutia, C., Akmal, M.: content based image retrieval Busana Muslimah Menggunakan Fitur Kombinasi PHOG dan DCD. (JurTI) Jurnal Teknologi Informasi 5(1), 70–76 (2021)
10. Liu, Z., Luo, P., Qiu, S., Wang, X., Tang, X.: Deepfashion: powering robust clothes recognition and retrieval with rich annotations. In Proceedings of the IEEE Conference on Computer vision And Pattern Recognition, pp. 1096–1104 (2016)
11. Ding, Y., Wong, W.K.: Fashion outfit style retrieval based on hashing method. In: Wong, W.K. (ed.) AITA 2018. AISC, vol. 849, pp. 187–195. Springer, Cham (2018). https://doi.org/10.1007/978-3-319-99695-0_23
12. Liu, A.A., Zhang, T., Song, D., Li, W., Zhou, M.: FRSFN: a semantic fusion network for practical fashion retrieval. Multimedia Tools Appl. **80**, 17169–17181 (2021)
13. Xu, X.: Kaggle. https://www.kaggle.com/datasets/xxc025/fashiondataset
14. Simonyan, K., Zisserman, A.: Very deep convolutional networks for large-scale image recognition. arXiv preprint arXiv:1409.1556. (2014)
15. Taylor, L., Nitschke, G.: Improving deep learning with generic data augmentation. In: 2018 IEEE Symposium Series on Computational Intelligence (SSCI). pp. 1542–1547. IEEE (2018)

16. Srivastava, N., Hinton, G., Krizhevsky, A., Sutskever, I., Salakhutdinov, R.: Dropout: a simple way to prevent neural networks from overfitting. J. Mach. Learn. Res. **15**(1), 1929–1958 (2014)
17. Cheng, M.M., Mitra, N.J., Huang, X., Torr, P.H., Hu, S.M.: Global contrast based salient region detection. IEEE Trans. Pattern Anal. Mach. Intell. **37**(3), 569–582 (2014)
18. Sikora, T.: The MPEG-7 visual standard for content description-an overview. IEEE Trans. Circuits Syst. Video Technol. **11**(6), 696–702 (2001)
19. Deng, Y., Manjunath, B.S., Kenney, C., Moore, M.S., Shin, H.: An efficient color representation for image retrieval. IEEE Trans. Image Process. **10**(1), 140–147 (2001)

Computer Networks

A Robust Image Cipher System Based on Cramer-Shoup Algorithm and 5-D Hyper Chaotic System

Zainab Khalid Ibrahim$^{(\boxtimes)}$ and Ekhlas Abbas Albahrani

Mustansiriyah University, Baghdad, Iraq
Zainab86khalid48@gmail.com

Abstract. The presented study suggests a novel image encryption method depending on a modified Cramer-Shoup encryption algorithm and a 5D hyper chaotic system. The first public key encryption algorithm to be shown secure towards adaptive selected cipher-text attack is the Cramer-Shoup system. The encryption algorithm encrypts and decrypts images with different sizes using the Cramer-Shoup substitution technique and the 5D hyper chaotic permutation technique. A modified Cramer-Shoup algorithm is used to substitute the input image after it has first been permuted with the use of hyper chaotic system. The suggested approach will quadruple the size of the encrypted image in comparison with the original image The proposed system had been tested using a variety of methods, and results show that it has a substantially uniform histogram for the dataset. The smallest correlation horizontally is -0.03894, vertically is -0.00898, and diagonally is -0.004478 for all tested images. The greatest entropy is 7.691532, the greatest maximum deviation is 54959, the greatest irregular deviation is 92039, and the greatest peak signal-to-noise ratio (PSNR) is 7.703765, while the greatest mean square error (MSE) is 40.427. Furthermore, the unified average changing intensity (UACI) is close to 0.339956, and the number of pixels change rate (NPCR) is 0.996238.

Keywords: Image encryption · Cramer-Shoup encryption algorithm · 5D hyper chaotic system

1 Introduction

ElGamal cryptosystem is expanded upon by the Cramer-Shoup algorithm, which was developed in the year 1998 by Victor Shoup and Ronald Cramer. Dissimilar to ElGamal, Cramer-Shoup adds additional components to assure non-malleability even against a resourceful attacker. A ciphertext that is double the size of ElGamal [1] is produced by using a universal one-way hash function plus additional computations to accomplish non-malleability. The official name for the definition of Shoup's security is "indistinguishability under adaptive chosen cipher-text attack." The attacker is assumed to have access to decryption oracle that can decrypt any cipher-text using the technique's secret decryption key, making this the strongest security definition about a public key cryptosystem. The attacker can access decryption oracle before and following viewing a certain target ciphertext to attack due to the "adaptive" element of security definition. The less reliable defense against attacks using non-adaptive selected ciphertext just permits the attacker to obtain the decryption oracle before looking at the target ciphertext. Even though it has been commonly known that several widely used cryptosystems were susceptible to the attacker, system designers long dismissed the attack as theoretically interesting but impracticable. This started to change in the late 1990s, especially after Daniel Bleichenbacher showed how to successfully attack SSL servers with an adaptive selected ciphertext attack employing RSA encryption [2]. The chaotic systems-based cryptosystem was used and approved as a result of the primary characteristics of chaotic systems, like periodicity, dynamics estimation property in cryptography systems, and sensitivity to initial conditions. The performance of chaotic base image encryption techniques was heavily researched in an effort to tackle the problem of real-time secure image transmission [3]. An adequate image encryption technique must operate more effectively and quickly in real-time applications in modern security systems.

2 Related Works

Depending on the concepts of cryptosystems, encryption algorithms could be classified as either asymmetric (private and public key) or symmetric (single key). About the latter, communication parties use the same key for decryption and encryption. Numerous image encryption studies depending on public key methods and chaos theory have been conducted recently. A two-key cryptosystem is an asymmetric encryption scheme where the transmitter encrypts secret plain message with the use of the open public key and recipient decrypts it by using a separate secret private key [4]. Both public and private keys are given by the receiver, who secretly holds onto the private keys. Due to the fact that asymmetric crypto-systems are mainly dependent upon the non-deterministic polynomial (NP) problem, an attacker can't deduce the correct private key. Symmetric crypto-systems face problems with key transmission and distribution, even though they are often effective. A cryptosystem's security only depends on safeguarding secret keys. Any additional transmission regarding the secret key in a method of symmetric encryption results in an incorrect symmetric cipher. To tackle the key management as well as distribution in symmetric cipher, various studies used asymmetric structure [5–7] for encrypting the secret images. In [8], Shakiba put forth a Chebyshev map-based asymmetric chaotic encryption method. Even though testing shows it to have excellent durability and security, it can take up to 13.9 s on average to encrypt a 512 by 512 image. Reference [9] developed an asymmetric cryptography algorithm depending on quantum chaotic systems that generate private and public key pairs using the RSA method. The pixels in the odd column and odd row are first diffused, after that the directions in the plain image's column and row are scrambled. Second, the even row and even row pixels are subjected to the same confusion operation, which is applied to both column and row pixels. The Hill cipher's security for image encryption is increased and enhanced by switching it from symmetric to asymmetric encryption, according to reference [10], which indicated an image encryption method ECCHC integrating Hill cipher and elliptic curve cryptography. Reference [11] provides a satisfactory evaluation of numerous attacks against Reference [10]. The length of the encryption key that has been utilized in Reference [10] was found to be not enough to fend off aggressive attacks. Elliptic curve integrated encryption system (ECIES), which combines an elliptic curve with a linear multiplication matrix to create a secret key matrix, was suggested by Reference [11] as a solution to such problems. The key length problem was thus overcome. An asymmetric picture encryption method depending on SHA-3, information hiding technology (IHT), and compressive sensing (CS) is presented by the authors in Reference [12]. The preprocessed image is assessed with the use of the SHA-3 algorithm to obtain the hash value throughout encryption, and a randomly generated matrix is created for modular operation with the plain image. RSA algorithm after that calculates cipher parameters with the use of such hash values as plain parameters. The initial chaotic map values are after created for both of them using a new mathematical model. A new asymmetric chaotic image encryption method depending on RSA as well as IWT with confusion-diffusion structure is also proposed in Reference [13]. The relevant plain image components are incorporated into key generating procedure throughout encryption. Thus, the plain secret image that was created along with the keystream will be able to successfully fend off both known-plaintext attacks (KPA) and chosen-plaintext attacks (CPA). Double operation station encryption is also made possible by the introduction of IWT. The 5D hyper chaotic system and Cramer-Shoup algorithm have been combined in order to create a new algorithm of image encryption that is suggested in the present research for image decryption and encryption. Image encryption, key generation, and image decryption are the three main methods that make up the suggested system. A modified Cramer-Shoup algorithm is used to substitute the input image after it has first been permuted with the use of hyper chaotic system. The suggested approach will quadruple the size of encrypted image in comparison with the original image since Cramer-Shoup algorithm produces very huge numbers. Because every statistical analysis of the encrypted image will be completely different from the original image as a result of this procedure, the algorithm will gain strength.

Contributions: Our primary goal is to show the effectiveness of the Cramer-Shoup public key scheme in image encryption:

1. The Cramer-Shoup algorithm is a public key algorithm that encrypts a single value using a prime number and its primitive root. The Cramer-Shoup algorithm has been modified in this paper so that it can encrypt an image by converting the resulting large numbers into a binary string and then into values ranging from 0 to 255.
2. A novel encryption algorithm is suggested, which consists of two methods: the hyper chaotic permutation method and the Cramer-Shoup substitution method.
3. The proposed image algorithm is distinguished by the fact that the encrypted image will be four times larger than the original image. This means that the original image's computational properties will vanish, making it extremely difficult for the algorithm to attack.
4. All of the keys required for the hyper chaotic permutation method and the Cramer-Shoup substitution method are generated using the 5D hyper chaotic system.

5. In addition to that, the suggested color image encryption approach has acceptable encryption overall performance and can effectively deal with a variety of commonplace place attacks.

3 Paper Organization

The basic hypothesis regarding the Cramer-Shoup encryption algorithm and the hyperchaotic system are shown in Sect. 4. Section 5 gives the suggested encryption system, and Sect. 6 illustrates how the suggested algorithm is put into practice. Before the conclusion, Sect. 7 shows the statistical and security analysis of the suggested encryption system.

4 The Related Theory

In the presented work, the proposed algorithm has been based upon a new five-dimensional Lorenz model and Cramer-Shoup Encryption Algorithm.

4.1 Hyper Chaotic System

Traditional Lorenz system is provided by [14]:

$$
\begin{aligned}
\dot{x} &= \sigma y - \sigma x, \\
\dot{y} &= rx - y - xz, \\
\dot{z} &= xy - bz,
\end{aligned}
\tag{1}
$$

A fluid enclosed between two plates that are positioned perpendicular to Earth's gravitational attraction is cooled from above and heated from below in a 2D Rayleigh-Bénard thermal convection process. Reference [15] investigated a new Lorenz model in five dimensions, which is depicted by the equation below [16]:

$$
\begin{aligned}
X_1 &= \sigma (y_0 - x_0), \\
Y_1 &= rx_0 - y_0 - xz_0, \\
Z_1 &= x_0y_0 - bz_0 - x_0v_0, \\
V_1 &= x_0z_0 - 2x_0u_0 - (1 + 2b)v_0, \\
U_1 &= 2x_0v_0 - 4bu_0
\end{aligned}
\tag{2}
$$

In which $(x_0, y_0, z_0, v_0, u_0)$ represent initial conditions and the system is hyper chaotic when the constant parameters are $\sigma = 10, r = 100, b = 8/3$. Equations (1) and (2) differ in that 2 new 1st-order ordinary differential equations as well as two extra dynamical variables, v and u, are added. Additionally, there are two new nonlinearities: xu in the v equation and xv in the z and u equations. It is predicted that system solutions (2) will become more chaotic compared to the solutions for system (1) because nonlinearities are the origin of chaos.

4.2 Cramer-Shoup Encryption Algorithm

Victor Shoup and Ronald Cramer first presented the Cramer-Shoup encryption method in 1998. It is a public key encryption system. It is based upon two concepts frequently employed in modern cryptography: decisional Diffie-Hellman assumption and the discrete logarithm problem. It is a unique combination of security, efficiency, and flexibility. Unlike some other encryption schemes, Cramer-Shoup provides provable security guarantees against chosen ciphertext attacks, which makes it a popular choice in applications where security is critical. An interactive chosen-ciphertext attack in which the attacker provides several ciphertexts to be decrypted and then utilizes the outcomes of these decryptions to choose new ciphertexts. Cramer-Shoup is also quite effective and could be utilized to swiftly encrypt huge volumes of data. The Cramer-Shoup algorithm is also adaptable enough to support a broad variety of cryptographic primitives, including secure hash functions and digital signatures [17]. Following are the steps of Cramer-Shoup encryption/decryption algorithm:

A. *Key Generation*: let G be a group of prime order q, where q represents huge and cleartext messages that are parts of G. In addition, a universal 1-way hash function family which maps lengthy bit strings to Zq elements is employed.

- Public key generation: first selects two random numbers, $g1, g2 \in G$, five numbers $x1, y1, x2, y2, z \in Zq$ and H a one-way hash function. The public key consists of $g1, g2, x1, x2, z, H, y1$, and $y2$.
- Private key generation: computes additional values, c, d, and h, using the following equations [18]:

$$c = g_1^{x_1} g_2^{x_2} \tag{3}$$

$$d = g_1^{y_1} g_2^{y_2} \tag{4}$$

$$h = g_1^z \tag{5}$$

The private key consists of c, d, and h.

B. Encryption: when the plain text is $m \in G$ and $r \in Zq$ where r is a random number, then compute [18]:

$$u_1 = g_1^r \tag{6}$$

$$u_2 = g_2^r \tag{7}$$

$$e = h^r m \tag{8}$$

$$\alpha = H(u_1, u_2, e) \tag{9}$$

$$v = c^r d^{r\alpha} \tag{10}$$

The resulting cipher text is (u_1, u_2, e, v)

C. Decryption: let the cipher-text (u_1, u_2, e, v), then the plain text m can be retrieved by first computing α using Eq. (9) and check if the condition $u_1^{x_1+y_1\alpha} u_2^{x_2+y_2\alpha} = v$ is not true, then the decryption algorithm returns "reject"; otherwise, it returns the original message m as shown [18]:

$$m = \frac{e}{u_1^z} \qquad (11)$$

5 The Proposed Image Encryption/Decryption System

The encryption phase, key generation phase, and decryption phase are the three crucial phases that make up the suggested system.

5.1 Key Generation Phase

Permutation operations and the Cramer-Shoup method require keys in the suggested algorithm. The steps listed below will be used to produce these keys depending on hyper chaotic:

1- **Generating the keys for Cramer-Shoup algorithm:-** the initial values $(x_0, y_0, z_0, v_0, u_0)$ and control parameters $(\sigma, r, \text{and } b)$ are inputs to the hyper- chaotic system Eq. (2). The system will show chaotic behavior when $(\sigma = 10, r = 100, b = 8/3)$. The initial values are floating point numbers with a (10^{16}) precision with a range [0, 1]. The Cramer-Shoup keys are created by iterating the 5D hyper chaotic system several times until generating all Cramer-Shoup keys which are: prime number q, two integer numbers (g1, g2) < q, and five numbers x1, x2, y1, y2, z $\in$ Zq. In each iteration, the resulting five floating points are converted into five integer numbers using the Eq. (12).

$$N_i = \text{floor}\left(C_i \times 10^2\right) \qquad (12)$$

where N_i is the resulting integer number, C_i is one of the resulting five floating point numbers, and i = 1,2...5.

2- **Generating permutation keys for the permutation method:-** the hyper chaotic system is iterated for generating the two permutation keys which are the permutation row array and the permutation column array. Each array contains integer numbers without repeating with the size of original image dimensions.

5.2 The Suggested Encryption Algorithm Phase

The steps in the suggested image encryption technique are as follows:

1- Begin by dividing the original color image into three color channels (which are, red, green, and blue). Each channel is arranged in a two-dimensional array Pnxm.
2- Using the key generation algorithm mentioned in Sect. 5.1, generate the Carmer-Shoup keys and permutation keys required for the encryption algorithm.

3- Every color channel of the image is subjected to the permutation transformation. The permutation transformation enables the transformation of each channel's column and row into new column and row positions, respectively, depending on the permuted column array and permuted row array.

4- After mapping the pixels in the color image, we perform the substitution transformation for each channel of the color image using the Cramer-Shoup algorithm.

5- The Cramer-Shoup algorithm produces huge integers that are outside the range [0–255] that the image's pixel values should be. Each resulting value will then be converted into a binary string as a result. Eight-bit strings have been created from this string. Every eight-bit string is converted into an integer number between 0 and 255. As a result, the encrypted image will be twice as large as the original image, making it hard for an attacker to tell the difference between the two.

6- Concatenate the three color channels to produce the encrypted image.

  The Fig. 1 shows the flowchart of encryption algorithm.

5.3 The Suggested Decryption Algorithm Phase

The decryption algorithm can be described as inverse of the encryption method in that each step is reversed, as stated by the steps below:

1- The encrypted image is separated into three color channels.

2- The keys for the permutation and substitution transformations are generated in the same way as in the encryption process.

3- Using rows and columns permutation arrays, the inverse permutation transformation is applied to each image color channel to get the original pixel locations.

4- The decryption algorithm of Cramer-Shoup algorithm is performed to decrypt the resulting pixels for each image channel.

5- The three-color channels that result are combined to produce the original image.

6 Simulation Results

This section reports the simulation outcomes for the suggested image decryption and encryption. The simulation setting for the algorithm proposed in this study includes Windows 11, a Core (TM) i5-10500 H CPU operating at 2.50 GHz, and MATLAB (version R2021B) software. Let the red channel image Rc $[3 \times 3] = [144\ 145\ 146;\ 162\ 163\ 161;\ 155\ 158\ 159]$, the chaotic initial values be: $x0 = 0.8769584987437682$, $y0 = 0.91295867463291275$, $z0 = 0.7768956438674328$, $v0 = 0.9674388976348647$, $u0 = 0.9867538556785432$, and the constant parameters are $\sigma = 10$, $r = 100$, and $b = 8/3$. The first step is generating the keys for the permutation and substitution process. The permutation keys are three-row arrays ([2, 3, 1], [2, 3, 1], [1, 3, 2]) and three column arrays = ([3, 1, 2], [1, 2, 3], [1, 3, 2]. So, the pixel [1, 1] = 144 in the Red array will be permuted to location [2, 3] in the Permuted Red array and the pixel [1, 2] = 145 will be in location [3, 1], and so on. The result is Permuted Red array = [155 146 161; 162 159 144; 145 163 158]. The Carmer-Shoup keys are also generated which are $q = 17$, $g1 = 2$, $g2 = 3$, $x1 = 5$, $x2 = 8$, $r = 5$, and $z = 7$. The second step is the substituted process in which each pixel in the Permuted Red array will be encrypted using the Carmer-Shoup algorithm. The first pixel Rc [1] = 155 will be substituted as shown: c = g1^x1*g2^x2

= 209952, h = g1^z = 128, ee [1] = (h^r1) * Rc [1] (128^5) * 155 = 5.325759447e + 12. The ee[i] value is then converted into a 64-bit binary string [10011010 11111111 11111111 11111111 11111011 000]. As indicated by [154, 255, 255, 255, 251, 0], each 8-bit will be transformed into an integer value. This implies that 6 integer integers will be used to replace each pixel. The image's pixels will all undergo these procedures. Three color images are encrypted and decrypted in Fig. 2. All plain images, as seen in the figure, are encrypted without any useful information, leading to noise-like images, however, decrypted images could be efficiently recovered.

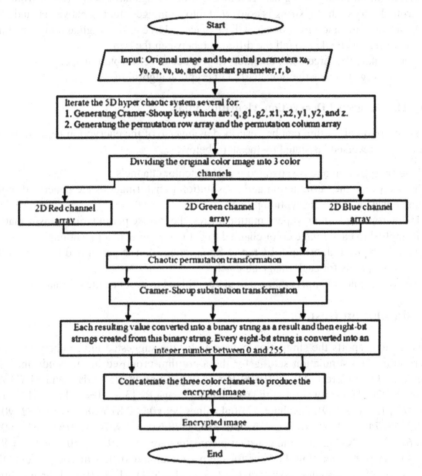

Fig. 1. The flowchart of proposed encryption algorithm

7   Experiment Result

The suggested cipher system could be verified using various tests, including histogram analyses, key space analyses, mean square errors, correlation coefficients, maximum deviations, information entropies, NPCR, irregular deviations, PSNR, and UACI.

Fig. 2. Example of the proposed encryption algorithm: (a) original parrot image, (b) the encryption of (a).

7.1   Key Space Analyses

The size of a key space has a direct impact on the encryption algorithm's strength. The encryption is more resistant to attack the larger the key space is. The allowable key space is 2^{128}, which is computationally safe against brute-force attacks, to prevent them [18]. The suggested method's total key space is $(10^{16})^5 \approx 2^{266}$. The encryption method could withstand brute-force attacks since the key space is sufficiently vast.

7.2   Histogram Analysis

Histogram analysis has been utilized in order to show the distribution related to pixel values between plain and cipher images. We could conclude that the cipher image doesn't contain any information for carrying out a crypto analysis attack on the encryption technique [19, 20] in the case when the histograms regarding the cipher and plain images are significant. Indicating that the images "Eye" and "Lena" are original images, Figure 3 shows the original image histogram and related ciphered images. In this Figure, the histograms of output encryption images are spread rather consistently on a scale, and the original image analysis histogram contains no useful information.

7.3 Correlation Coefficient Analysis

The association between two variables A and B is calculated using the correlation coefficient, which yields a result range of $[-1, 1]$, in which a value of 1 represents a perfect positive correlation, a -1 value represents a perfect negative correlation, and the value of 0 represents no correlation [21]. The following connection could be utilized for the assessment of correlation [24]:

$$C_{x,y} = \frac{\sum_{i=0}^{q-1}(x_i - x^-).(y_i - y^-)}{[\sum_{i=1}^{q-1}(x_i - x^-)^2]^{\frac{1}{2}}.[\sum_{i=1}^{q-1}(y_i - y^-)^2]^{\frac{1}{2}}} \tag{13}$$

where the mean values of x and y are:

$$x' = \sum_{i=0}^{q-1}\frac{xi}{q} \text{ and } y' = \sum_{i=0}^{q-1}\frac{yi}{q}$$

$C_{x,y}$ denotes the correlation coefficient, y, and x are 2 adjacent pixels' gray values, and q represents the total number of adjacent pixels that have been taken from the image.

The correlation between consecutive pixels in a diagonal, horizontal, and vertical direction is shown in Table 1 for the encrypted images "Lena," "eye," and "Parrot." Plain image's correlation coefficient is close to 1, while encrypted image's correlation coefficient is closer to 0, highlighting the fact that the algorithm of encryption satisfies 0-co-correlation, which means that an attacker is not capable of obtaining any valuable information through using a statistical attack.

7.4 Information Entropy Analyses

The level of suspicion in a random variable is evaluated by entropy. The Shannon entropy, which measures the predictable value regarding color image information, is what the phrase most commonly refers to [22]. Information entropy H (m) of a message could be computed as follows [24]:

$$H(m) = \sum_{i=1}^{2^{n-1}} p(m_i)log2\frac{1}{p(m_i)} \tag{14}$$

where N is the message's bit number, 2N denotes every character that might be used, p (mi) shows the possibility that each character will be used, and the entropy is represented as bits. In a case when the likelihood of each one of the gray values in a 256-grayscale image is the same, the entropy information is equal to 8, indicating that the image is entirely random. Table 2 displays entropy for 3 cipher images (eye, Lena, and parrot). The suggested solution is secure against the entropy attack since all of the entropy values are near 8, indicating that an encrypted image is approaching some random source.

7.5 Image Quality

The encryption quality can be determined by measuring Maximum Deviation and Irregular Deviation.

- **Maximum Deviation:** the largest deviation between the associated encrypted image and plain image. In the case when the deviation estimate is higher, the encrypted image is more off than the original image [23]. Table 3 shows the highest deviation for the three cipher pictures (Lena, Eye, and Parrot). These results reveal that all of the photos have higher maximum deviation values, with the bigger maximum deviation providing the highest encryption quality [24]. The steps in this measure are:

  1. It creates a histogram for both the encrypted and original versions of the images before figuring out the difference between the two histograms and leaving it be.
  2. The area under absolute difference curve is numbered, and summation of the deviations (MD) denotes encryption quality where the MD equation is [24]:

$$\text{Md} = \frac{Ad_0 + Ad_{255}}{2} + \sum_{i=1}^{254} Ad_i \qquad (15)$$

- **Irregular Deviation:** A reliable encryption procedure should randomly create the values of pixel information. The statistical deviation distribution tends to be uniformly distributed in the case where encryption algorithm treats pixel values indiscriminately. To evaluate statistical distribution regarding histogram deviation in the nearly uniform distribution, we use an irregular deviation measure. The irregular deviation for the 3 cipher images (eye, Lena, and parrot) has been listed in Table 3. The encryption quality should be higher the lower the irregular number is. The findings show that the proposed algorithm contains few out-of-the-ordinary values, indicating strong encryption quality. The approach used to calculate irregular deviation is [25]:

  - Step1: Calculating the absolute difference between cipher image C and plain image P as

$$D = |P - C| \qquad (16)$$

  - Step2: Calculated is the D's histogram (H)
  - Step3: Calculating Ah which is the average value of the histogram amplitude hi at index i using [25]:

$$\text{Ah} = \frac{1}{256} \sum_{i=0}^{255} h_i \qquad (17)$$

  - Step4: the absolute deviation from the mean value calculated for the histogram as [25]:

$$|AD = |h_i - Ah| \qquad (18)$$

  - Step5: irregular deviation ID is calculated by:

$$\text{ID} = \sum_{0}^{255} AD_i \qquad (19)$$

While ID estimation is low, the encryption quality is higher.

7.6 Peak Signal-to-Noise Ratio (PSNR)

The difference in pixel properties between cipher image and the plain image is measured by the PSNR [26]. The following equation [16] demonstrates this mathematically [16]:

$$PSNR = 10\log_{1\alpha}\left[\frac{R * C * 255^2}{\sum_{i=0}^{R-1}\sum_{J=0}^{C-1}(PI(i,j) - CI(i,j))^2}\right] \tag{20}$$

PI (i, j) stands for original image pixel at matrix PI (i, j), CI (i, j) for the encrypted image pixel, and R and C represents image width and height, respectively. PSNR for the 3 cipher images (eye, Lena, and parrot) has been listed in Table 3. All of the numbers in Table 3 are lower values, proving that the proposed cipher is sufficient. A good encryption algorithm should have a lower PSNR estimation.

7.7 Avalanche Effect

The avalanche effect could be verified with the use of NPCR, MSE, and UACI.

- MSE: calculates the squared error between cipher and plain images. The MSE between a pair of images is determined by this equation [23]:

$$MSE = \frac{1}{H * W}\sum_{j=0}^{H-1}\sum_{j=0}^{W-1}\left[E_1(i,j) - E_2(i,j)\right] \tag{21}$$

where H, W stands for the digital image's width and height. Difference in the quality between the two images is obvious if the MSE > 30 dB [27]. As you can see from Table 4, the suggested algorithm's encryption is quite good because the resultant MSE is larger than 30 dB.

- NPCR: the percentage of pixels found by NPCR varies between 2 encrypted images. NPCR value must be near 99.6. The NPCR has been estimated with the use of the equation [28]:

$$NPCR = \frac{1}{W \times H}\sum_{i=1}^{W}\sum_{j=1}^{H}f(i,j) \times 100\% \tag{22}$$

- UACI finds ratio of the changes amongst the pair of the encrypted images. UACI value must be near 36.7. The UACI is calculated using the equation [28]:

$$UACI = \frac{1}{W \times H}\sum_{i=1}^{W}\sum_{j=1}^{H}\frac{|U1(i,j) - U2(i,j)|}{255} \times 100\% \tag{23}$$

As you can see from Table 4, the proposed algorithm's encryption is quite good because the results of the three images for NPCR and UACI are good.

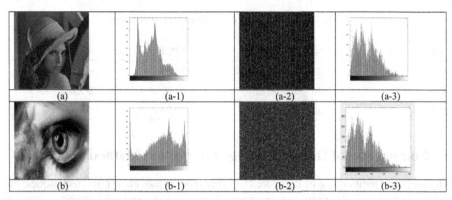

Fig. 3. Histogram analysis of (a) original Lena image, (a-1) histogram of (a), (a-2) encrypted of (a), (a-3) histogram of (a-2). (b) Original Eye image, (b-1) histogram of (b), (b-2) encrypted of (b), (b-3) histogram of (b-2).

Table 1. Results of correlation analysis of the suggested image crypto-system

Direction	Lena plain image	Lena Cipher Image	Eye plain image	Eye Cipher image	Parrot Plain Image	Parrot Cipher image
Horizontal	0.9729	−0.0389	0.9761	0.0411	0.9634	0.0425
Diagonal	0.9703	−0.0044	0.9715	0.0098	0.9673	0.0222
Vertical	0.9845	−0.0089	0.9749	0.0022	0.9629	0.0165

Table 2. Results of entropy analysis of the suggested image crypto-system

Image name	Entropy
Lena	7.6500
Eye	7.6915
Parrot	7.6627

Table 3. Results of the Measures of the Encryption Quality of the suggested Image Cryptosystem

Image	Max. Deviation	Irregular deviation	PSNR
Lena	26716	60077	9.2008
Eye	26681	38058	7.7911
Parrot	54959	92039	9.0013

Table 4. The avalanche effect results of three images

Name image	MSE	NPCR	UACI
Lena	38900 = 45 db	99.66	37.7
Eye	40339 = 46 db	99.60	35.1
Parrot	39129 = 45 db	99.50	36.3

8   Comparison of Different Image Encryption Methods

The proposed method can be compared to many other image encryption techniques, but only contemporary algorithms relying upon chaotic systems and asymmetric encryption algorithms. Table 5 contrasts the proposed method with five different image cryptosystems [9, 10, 13, 14, 24]. We selected a few critical criteria allowing us to compare algorithms from various perspectives, such as key space size, correlation analysis, entropy analyses, MSE, NPCR, PSNR, and UACI. As seen from the comparison in Table 5, all of the algorithms have passed every test with success. However, the evaluations differ based on the kind of test; for instance, our suggested algorithm has a lower PSNR than other algorithms and has been capable of approaching the ideal value for some tests, like UACI and NPCR. This suggests that both in terms of efficiency and security, the suggested image cipher is similar to its rivals.

Table 5. Results of compression obtained using five different cipher systems and the suggested image encryption algorithm

	Key space size	Lower Correlation analysis			Largest Entropy analysis	Average of MSE	Lower PSNR	Average of NPCR	Average of UACI
		Horizontal	vertically	diagonally					
Proposed system	2^{266}	−0.03894	−0.00898	−0.0044	7.69	45 db	7.70	99.58	36.36
Ref. [9]	2^{ε} where ε is a random bit length	0.0051	0.0050	−0:0027	7.99	47 db	36.05	99.59	50.01
Ref. [10]	–	−0.0027	0.0015	0.0012	7.99	–	–	99.61	33.47
Ref. [13]	2^{256}	0.0008	−0.0031	0.0003	7.59	36 db	43.11	–	–
Ref. [14]	2^{149}	−0.0014	0.0006	0.0016	7.99	–	–	99.61	33.48
Ref. [24]	2^{346}	0.0102	0.0097	−0.00009	7.69	40 db	7.92	–	–

9   Future Work

To be fair in this field of study, one can apply the proposed algorithm to a 3D gray image with three different keys one for each dimension. In addition, multimedia applications in fields like satellite imaging and medicine are developing at a rapid pace. The development of the image encryption techniques for these applications will require a significant amount of computational power because they demand high-resolution images. This problem can therefore be solved by applying parallel image encryption techniques.

10   Conclusion

In today's world, where data breaches and cyber-attacks are becoming more common, it's crucial to have strong encryption methods to protect sensitive information. That's why the Cramer-Shoup cryptosystem is so important. It provides a reliable and secure way to encrypt data, making it virtually impossible for anyone to intercept or decipher the information without the proper decryption key. The Cramer-Shoup algorithm and a 5-D hyper chaotic system are combined to form the basis of a new image encryption approach presented in this research. The idea is to use chaotic permutation as well as the Carmer-Shoup substitution technique for encrypting and decrypting images. The chaotic system and the Cramer-Shoup method were used to improve the cryptography, making it more complicated for an attacker to encrypt the cipher image because the encrypted image will be twice as big as the original. Based on statistical and security investigation, the suggested method for encrypting and decrypting color images is a very good algorithm. These tests' findings demonstrate that the suggested encryption is more complex and secure than those now in use, as well as having a large key space and limited correlation (vertical, horizontal, and diagonal) with the original image.

References

1. ElGamal, T.: A public key cryptosystem and a signature scheme based on discrete logarithms. In: Blakley, G.R., Chaum, D. (eds.) CRYPTO 1984. LNCS, vol. 196, pp. 10–18. Springer, Heidelberg (1985). https://doi.org/10.1007/3-540-39568-7_2
2. Alvarez, G., Li, S.: Some basic cryptographic requirements for chaos-based cryptosystems 16(08), 2129–2151 (2006). https://doi.org/10.1142/S0218127406015970
3. Amigó, J.M., Kocarev, L., Szczepanski, J.: Theory and practice of chaotic cryptography. Phys. Lett. A 366(3), 211–216 (2007)
4. Yuan, H.-M., Liu, Y., Lin, T., Hu, T., Gong, L.-H.: A new parallel image cryptosystem based on 5D hyper-chaotic system. Sig. Process. Image Commun. 52, 87–96 (2017). https://doi.org/10.1016/j.image.2017.01.002
5. Liu, M., Ye, G.: A new DNA coding and hyperchaotic system based asymmetric image encryption algorithm. Math. Biosci. Eng. 18(4), 3887–3906 (2021). https://doi.org/10.3934/mbe.2021194
6. Yadav, A.K., Singh, P., Saini, I., Singh, K.: Asymmetric encryption algorithm for colour images based on fractional Hartley transform. J. Mod. Opt. 66(6), 629–642 (2019)
7. Wu, C., Hu, K.-Y., Wang, Y., Wang, J., Wang, Q.-H.: Scalable asymmetric image encryption based on phase-truncation in cylindrical diffraction domain. Opt. Commun. 448, 26–32 (2019). https://doi.org/10.1016/j.optcom.2019.05.009
8. Rakheja, P., Singh, P., Vig, R.: An asymmetric image encryption mechanism using QR decomposition in hybrid multi-resolution wavelet domain. Opt. Lasers Eng. 134, 106177 (2020). https://doi.org/10.1016/j.optlaseng.2020.106177

9. Shakiba, A.: A randomized CPA-secure asymmetric-key chaotic color image encryption scheme based on the Chebyshev mappings and one-time pad. J. King Saud Univ. Comput. Inf. Sci. **33**(5), 562–571 (2021). https://doi.org/10.1016/j.jksuci.2019.03.003

10. Ye, G., Jiao, K., Huang, X.: Quantum logistic image encryption algorithm based on SHA-3 and RSA. Nonlinear Dyn. **104**(3), 2807–2827 (2021). https://doi.org/10.1007/s11071-021-06422-2

11. Dawahdeh, Z.E., Yaakob, S.N., Razif bin Othman, R.: A new image encryption technique combining Elliptic Curve Cryptosystem with Hill Cipher. J. King Saud Univ. Comput. Inf. Sci. **30**(3), 349–355 (2018). https://doi.org/10.1016/j.jksuci.2017.06.004

12. Benssalah, M., Rhaskali, Y., Drouiche, K.: An efficient image encryption scheme for TMIS based on elliptic curve integrated encryption and linear cryptography. Multimedia Tools Appl. **80**(2), 2081–2107 (2021). https://doi.org/10.1007/s11042-020-09775-9

13. Huang, X., Dong, Y., Zhu, H., Ye, G.: Visually asymmetric image encryption algorithm based on SHA-3 and compressive sensing by embedding encrypted image. Alex. Eng. J. **61**(10), 7637–7647 (2022). https://doi.org/10.1016/j.aej.2022.01.015

14. Du, S., Ye, G.: IWT and RSA based asymmetric image encryption algorithm. Alex. Eng. J. **66**, 979–991 (2023). https://doi.org/10.1016/j.aej.2022.10.066

15. Lorenz, E.N.: Deterministic nonperiodic flow. J. Atmos. Sci. **20**(2), 130–141 (1963). https://doi.org/10.1175/1520-0469(1963)020%3c0130:DNF%3e2.0.CO;2

16. Felicio, C.C., Rech, P.C.: On the dynamics of five- and six-dimensional Lorenz models. J. Phys. Commun. **2**(2), 025028 (2018). https://doi.org/10.1088/2399-6528/aaa955

17. Shen, B.-W.: Nonlinear feedback in a five-dimensional Lorenz model. J. Atmos. Sci. **71**(5), 1701–1723 (2014). https://doi.org/10.1175/JAS-D-13-0223.1

18. Cramer, R., Shoup, V.: A practical public key cryptosystem provably secure against adaptive chosen ciphertext attack. In: Krawczyk, H. (ed.) Advances in Cryptology—CRYPTO 1998, vol. 1462, pp. 13–25. Springer, Cham (1998). https://doi.org/10.1007/BFb0055717

19. Jalil, L.F., Saleh, H.H., Albhrany, E.A.: New pseudo-random number generator system based on Jacobian elliptic maps and standard map. Iraqi J. Comput. Commun. Control Syst. Eng. **15**(3), 77–89 (2015). https://www.uotechnology.edu.iq/ijccce/issues/2015/vol15/no.03/full-text/07.pdf

20. Yadava, R.K., Singh, B.K., Sinha, S.K., Pandey, K.K.: A new approach of color image encryption based on Henon like chaotic map. J. Inf. Eng. Appl. **3** (2013)

21. Albahrani, A., Alshekly, T.K.: New chaotic substation and permutation method for image encryption. Int. J. Appl. Inf. Syst. **12**(4), 33–39 (2017)

22. Bora, K., et al.: Analysis of crypto components of a chaotic function to study its random behavior. Int. J. Comput. Sci. Inf. Technol. **5**(2), 2566–2568 (2014)

23. Albahrani, A., Alshekly, T.K., Lafta, S.H.: A review on audio encryption algorithms using chaos maps-based techniques. J. Cyber Secur. Mobility **11**(1), 53–82 (2022). https://doi.org/10.13052/jcsm2245-1439.1113

24. Abbas, E.A., Karam, T.A., Abbas, A.K.: Image cipher system based on RSA and chaotic maps. Eurasian J. Math. Comput. Appl. **7**(4), 4–17 (2019). https://doi.org/10.32523/2306-6172-2019-7-4-4-17

25. Albahrani, E.A., Maryoosh, A.A., Lafta, S.H.: Block image encryption based on modified playfair and chaotic system. J. Inf. Secur. Appl. **51**, 102445 (2020). https://doi.org/10.1016/j.jisa.2019.102445

26. Ragab, H., Alla, O.S., Noaman, A.Y.: Encryption quality analysis of the RCBC block cipher compared with RC6 and RC5 algorithms. Cryptology ePrint Archive (2014), https://ia.cr/2014/169

27. Alshekly, T.K., Albahrani, E.A., Ayed, L.B.: A Novel parallel gray image cryptosystem based on Hyperchaotic System and Learning with Error Post Quantum Cryptography. In: 2022 3rd

Information Technology to Enhance e-Learning and Other Application (IT-ELA) (2022). https://doi.org/10.1109/IT-ELA57378.2022.10107955
28. Al-Mhadawi, M.M., Albahrani, E.A., Lafta, S.H.: Efficient and secure chaotic PRNG for color image encryption. Microprocess. Microsyst. **101**, 104911 (2023)

Variable-Length Encryption Key Algorithm Based on DNA Tile Self-assembly

Basim Sahar Yaseen$^{(\boxtimes)}$ ⓘ

Shatt Al-Arab University College, Basra, Iraq
`basimsahar@sa-uc.edu.iq`

Abstract. In this paper, a new approach to creating secure encryption and decryption keys using DNA is proposed. This method is unique as it utilizes tiling theory and the DNA self-assembly model. To create the key, a numeric value is entered, which represents the initial value of the key. This value is then used to produce a digital sequence of congruent equations. The resulting sequence is then converted into a double-stranded DNA sequence using dual, triple, or other patterns. The strength and complexity of the key can be adjusted by producing a stream of binary blocks that flow based on the matching between the block and the pattern of the produced strand. This method has several advantages, including its dependence on the strength of the selected patterns, the simulation of biological storage of DNA information, and the principle of timing and synchronization, which ensures that the double-strand segments match a specific pattern. Overall, this method provides powerful security specifications beyond what physical components such as shift registers and memories, as well as mathematical equations, can offer. It can produce the same number and complexity of keys as these devices but with additional security benefits. For instance, 2–12 chosen patterns can be equivalent in complexity to 2^{154}.

Keywords: DNA Self-assembly · Tilles · Variable length · Encryption Keys · Patterns

1 Introduction

DNA computing [1] is widely used to construct self-assembled structures due to its biocompatibility and programmability. Self-assembled DNA nanostructures are used to explore the power and programmability of kinetically controlled structural evolution in molecular models. The concept of tile self-assembly is derived from the DNA self-assembly model and is related to Wang's tiling theory. Wang's theory [2] showed that assembling tiles with colored edges according to their matching colors and certain patterns can simulate the Turing machine. This rule implies that the edges of the same color face each other, which can be understood intuitively by considering a particular row of tiles to represent the state of a Turing machine [3]. The use of color coding plays the role of matching rules and indicates that computing using Wang tiles, if combined with a powerful computing tool such as the DNA self-assembly computation model,

© The Author(s), under exclusive license to Springer Nature Switzerland AG 2024
A. M. Al-Bakry et al. (Eds.): NTICT 2023, CCIS 2096, pp. 322–333, 2024.
https://doi.org/10.1007/978-3-031-62814-6_23

will be a universal complex problem-solving technique. DNA cryptography is rapidly developing in the field of sensor network security [4]. DNA nanotechnology modeling is an interdisciplinary research field with great development potential in the field of information security technology [5] such as information encryption, hiding, authentication, and compression [6]. It is a new approach to the evolution of modern cryptography [7]. The self-assembly of DNA has become the focus of scientific research to solve security and limited storage problems in sensor networks. The tiling self-assembly model distinguishes itself from the rest of the DNA computing models in that it self-assembles structures and patterns according to predefined formulas and shapes, analogous to the principle of tiling theory. In addition to the main goal of protecting stored and transmitted information, the use of new technology to reduce storage requirements and optimize data transfer and data compression can lead to cost savings. The additional goal is to develop sensor networks and cloud computing [8]. The novelty and contributions of the work are embodied by producing a binary block sequence of unknown length for attackers from a specific initial numerical value and a set of specified tiles. The entire key sequence produced has no known period, as do shift registers, and is therefore difficult to trace. Compared with the requirements for inputs, it is highly complex.

2   Literature Review

The use of the biological environment has inspired the development of powerful features in cryptography. However, not all works carried out during this era fully utilized the advantages of this environment [9]. The benefits of biological technology were limited to encoding using genetic bases of DNA, which only added coding complexity. Table 1 presents a powerful work that is considered a real and tangible achievement in the recent development of DNA computing models in the field of cryptography.

Table 1. Among the powerful works that are considered a real, tangible achievement in exploiting DNA computing models recently within the field of cryptography

Reference No.	Author(s)	Work Type	Summarize	Year
[10]	Caixia Li	Design	Subjecting the image pixels to molecular procedures of DNA Nano self-assembly to reduce the correlation between them and thus achieve randomness of a kind	2021

(*continued*)

Table 1. (*continued*)

Reference No.	Author(s)	Work Type	Summarize	Year
[11]	P.Surendra Varma, K.Govinda Raju	Design	The inputs to the algorithm are 128 bits of plaintext and 128 bits of the key. The algorithm performs 4 * 4 matrix operations on the inputs for the purpose of combining them. They are later converted into DNA codes through a digital DNA coding where each digit has a value of 0, 1, 2, 3 is the DNA symbol A, C, G, T	2014
[12]	E. Vidhya, R. Rathipriya	Design	By using genetic algorithm, generate a population of binary-valued chromosomes and performing algorithmic operations to obtain chromosomes with a high fitness are used as keys and inputs for the Diffie-Hellman algorithm. After encrypting it, the plain text is converted into binary numbers, then into a string of ASCII, and finally it is converted into a sequence of Genetic bases with the help of a digital DNA scheme	2020
[13]	A. S. Polenov	Cryptanalysis	The paper deals with the components of a sequence of cipher bits, including KEYSTREAM. The basic of the processing work is to convert operations from propositional logic to DNA logic and execute it in polynomial time	2014

(*continued*)

Table 1. (*continued*)

Reference No.	Author(s)	Work Type	Summarize	Year
[14]	S. B. Sadkhan, Basim S.Yaseen	Cryptanalysis	The search constructs a composite database of keystream block files, connected through DNA sticker model operations, in order to access the correct key initials by associating the components of the FBSRS output. The search is a parallel ATTACK	2018
[15]	S. B. Sadkhan, Basim S.Yaseen	Cryptanalysis	The suggested method adopts a technique to combine between the genetic algorithm and the DNA sticker computer, to execute the parallel search attack of a cryptosystem	2019
[16]	S. B. Sadkhan, Basim S.Yaseen	Cryptanalysis	The research proposes a modified digital simulation of the DNA sticker model and use it to attack FBSRs generators	2019
[17]	Basim S.Yaseen	Cryptanalysis	Based on DNA splicing MODEL, implement a method to attack unknown stream cipher. The method deals with final output of the stream key generator	2021

3 Methodology

3.1 Basics

In the self-assembly mechanism of DNA strands, a numerical value is transformed into a sequence of genetic bases based on a mathematically congruent system. The double strand of DNA consists of a main strand and a complementary strand. The main strand has a 5' phosphate bond end, while the complementary strand has a 3' bond end (as shown in Fig. 1-a). DNA consists of four genetic bases: A, T, C, and G. The genetic bases in the main strand are linked to their complements in the complementary strand. For example, A is linked to T, and C is linked to G, and vice versa. Table 2 shows the conversion from genetic bases to binary. There are two types of bonds within the strand of genetic bases: hydrogen bonds (double or triple) that connect the main bases with their complements, and triple carbon (3' end) or five-phosphate (5' end) that connect the main bases and their complements with the other pairs within the double strand.

3.2 The Simulation

In order to create an algorithm, it is important to first translate the biological structure of genetic bases and their various bonds into a digital visualization. To achieve this, we use a color-coding system and shapes to represent the different types of bases and their corresponding bonds. For instance, we use two yellow-colored squares linked by two lines to represent the first type of bases and their complements (A–T), which are bonded by a double hydrogen bond. On the other hand, the second type of base and its complement (C–G) are represented by two green-colored squares, connected by three lines representing the triple hydrogen bond.In addition to the color-coding system, we also position the phosphate bonds to the left of the interconnected squares. Figure 1 shows a visual representation of this structure. To further refine the digital structure of genetic makeup, we have established a set of discriminating features for images.

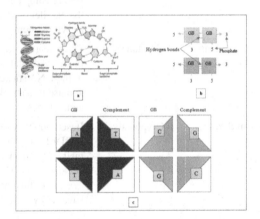

Fig. 1. a - Simulation of the biological representation. b - A pictorial diagram that is closer to the digital models. c - Representation of the basic genetic bases and their complements as tiles (Color figure online)

For citations of references, we prefer the use of square brackets and consecutive numbers. Citations using labels or the author/year convention are also acceptable. The following bibliography provides a sample reference list with entries for journal articles [1], an LNCS chapter [2], a book [3], proceedings without editors [4], as well as a URL [5].

For example, if the genetic base is main, the upper (or lower) and left edges belong to an uncolored triangle. Conversely, if the genetic base is complemented, the upper (or lower) and right edges belong to an uncolored triangle. Figure 2 provides an illustration of this feature. Overall, this approach allows for the creation of a basic work structure that can store and represent genetic information in a digital format.

Upper

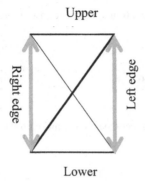

Lower

Fig. 2. If the genetic base is main, the upper (or lower) and left edges belong to an uncolored triangle. If the genetic base is complemented, the upper (or lower) and right edges belong to an uncolored triangle. (Color figure online)

3.3 Tiles and Patterns

In light of this, the tiles formed will be colored squares of the same colors opposite the main genetic bases, with squares representing the complementary bases, which are four types and correspond to the four genetic bases. Figure 1c depicts the four tiles of the genetic bases. Patterns begin to form when dealing with groupings of colored squares that represent the genetic bases in double-stranded DNA. When considering two tiles, the space of possibilities consisted of $2^4 = 16$ patterns, all of which were possible sequences of two genetic bases (tiles) within the double strand. The probability spaces of the two tiles (16 patterns) are shown in Fig. 3. The proposed algorithm considers patterns with a diverse number of tiles. Therefore, increasing one base at a time, starting from the two tiles, will lead to the formation of a space of possibilities of size 2^i, where i = 6, 8, 10, 12, ... Therefore, adding one tile, in a new stage, to the set of tiles in the previous stage leads to an increase in the probabilities space by (the size of the previous probabilities space multiplied by 2^2). In Table 3, we note that an increase in the number of patterns by one corresponds to an increase in the number of diverse types by 2^2.

3.4 Key Blocks Generation

When producing key blocks of variable length, it's important to consider that matching between the patterns in random access memory and the DS-DNA parts is not limited to consecutive genetic bases only. It also includes matching with genetic bases separated by spaces (various GBs). The process of finding the totality of matches between the patterns of the random store and the DS-DNA segments with all possible distances between the GBs is shown in Fig. 4. When a match is found, the output is a sequence of bits representing the DS-DNA part with the GBs, whether they belong to the pattern or not. Figure 5 provides a more detailed description of the overall processing flow and control flow paths for the production of key blocks.

Table 2. The conversion from the genetic bases representation to the binary

	0	1
0	A	C
1	G	T

Table 3. Shows the number of genetic bases that make up the tile, the number of patterns, and the complexity of each tile.

No. of genetic bases (tile)	No. of patterns	Equivalent complexity
2	16	2^4
3	64	2^6
4	256	2^8
5	1024	2^{10}
6	4096	2^{12}
7	16384	2^{14}
8	65536	2^{16}
9	262144	2^{18}
10	1048576	2^{20}
11	4194304	2^{22}
12	16777216	2^{24}
Total patterns space for tiles (2–12)		2^{154}

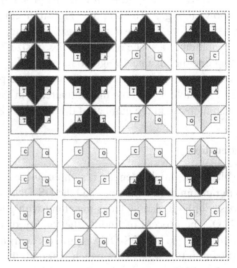

Fig. 3. Probabilities space for 2-tiles patterns

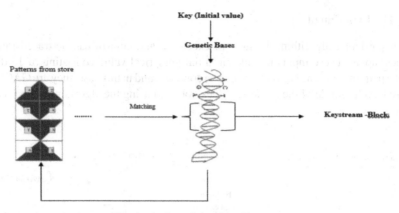

Fig. 4. A plan of algorithm Processes

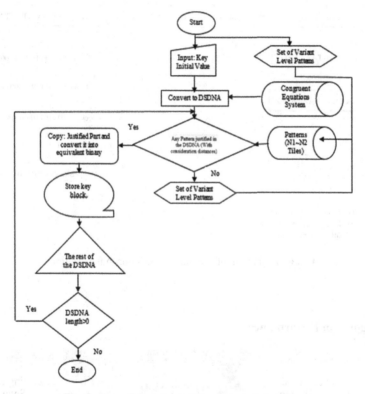

Fig. 5. Overall flowchart of the proposed algorithm

3.5 The Experiment

In this particular algorithm, the flow of execution and control can be traced through various inputs. These inputs include an initial numerical value consisting of 12 digits, and the patterns chosen. The patterns can be complex, and in this case, they are GAGGAA and AG. An example of the numeric values obtained using the algorithm can be seen in Fig. 6.

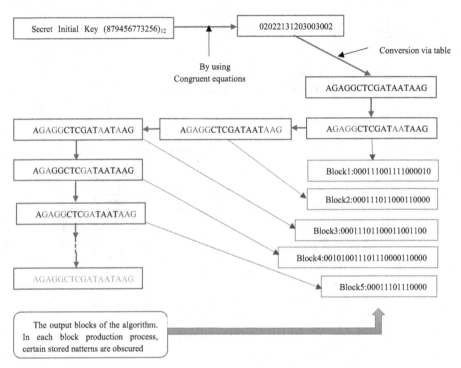

Fig. 6. A diagram of example of the algorithm execution

3.6 Algorithm Performance

Fig. 7. The main three stages of the proposed algorithm

The algorithm itself can be viewed as a three-stage process, as shown in Fig. 7. This division helps to identify the strengths and weaknesses of the algorithm and how it compares to established standards for evaluating cryptographic algorithms. The first stage involves determining the initial value of the key and choosing patterns at different levels. This stage is crucial as the key consists of two parts: the initial value and the chosen patterns. The initial value is directly proportional to the security of the algorithm and forms a double-stranded DNA structure. Choosing patterns is also important as it determines the randomness of the output. This part can be invariant to multiple primary key values. The second stage is the construction of the double-stranded DNA, which primarily depends on the initial value of the key. This stage reflects the strength and comprehensiveness of the encryption key. The third and final stage involves producing a sequence of final key blocks for encryption. This sequence depends entirely on the patterns chosen in the first stage. The randomness, comprehensiveness, and diversity in the selection of patterns are reflected in the randomness of the block sequence. This randomness affects the overall security and performance of the algorithm, as shown in Fig. 8.

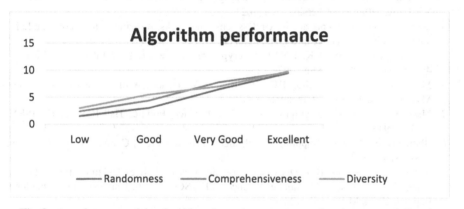

Fig. 8. A performance of the algorithm via randomness, comprehensiveness, and diversity

Performance was evaluated based on three selection features: randomness, comprehensiveness, and diversity. It's clear from the evaluation that when the initial value of the key and the patterns are excellent, the algorithm's performance is high and excellent, and vice versa. Poorly chosen features result in poor algorithm performance.

4 Discussion and Conclusions

In this paper, a new algorithm for producing variable-length block keys has been designed, tested, simulated, analyzed, and compared with other designs. The results of the simulation show that the proposed algorithm, which is based on tilling DNA self-assembly, performs well and meets the requirements of real-world cryptography [18]. It is feasible and demonstrates the great potential of DNA molecules in information security. Encryption tools and techniques can be classified into traditional and non-traditional techniques. Traditional techniques, such as AES, RC5, and A5x, are built from physical components and software may be built for them. They have constant secrecy and complexity, which cannot be increased during work according to their requirements. On the other hand, non-traditional techniques, such as the proposed algorithm, can be produced biologically and simulated programmatically. Their complexity varies during operation, and the key in them is non-periodic, making it easier to adapt to the confidentiality requirements of any system. The algorithm's complexity is evaluated as $O(N1)$ * $O(2^{N2})$, where N1 is the value of the initial key and N2 is the equivalent value of the number of patterns.

References

1. Binti Abu Bakar, R., Watada, J.: DNA computing and its applications: survey. ICIC Express Lett. **2**(1), 101–108 (2008)
2. Kari, J.: Recent results on aperiodic Wang tilings. In: Carbone, A., Gromov, M., Prusinkiewicz, P. (eds.) Pattern Formation in Biology, Vision and Dynamics. World Scientific (2000)
3. Hefferon, J.: Theory of Computation: Making Connections. University of Vermont, Jericho, VT, USA, hefferon.net Version 1.10 (2023)
4. Ukil, A.: Security and privacy in wireless sensor networks. ResearchGate, Innovation Labs, Tata Consultancy, Services, Kolkata, India (2010)
5. Kaundal, A.K., Verma, A.K.: DNA based cryptography: a review. Int. J. Inf. Comput. Technol. **4**(7), 693–698 (2014). ISSN: 0974-2239
6. Menezes, A.J., van Oorschot, P.C., Vanstone, S.A.: Handbook of Applied Cryptography. Massachusetts Institute of Technology (1996)
7. Mao, W.: Modern Cryptography: Theory and Practice. Prentice Hall PTR (2003). ISBN: 0-13-066943-1
8. Bhowmik, S.: Cloud Computing. Cambridge University Press, Cambridge (2017). ISBN: 978-1-316-63810-1
9. Sadkhan, S.B., Yaseen, B.S.: DNA-based cryptanalysis: challenges, and future trends. In: 2nd Scientific Conference of Computer Science, IEEE-SCCS 2019. University of Technology, Bagdad, 27–28 March 2019 (2019)
10. Li, C.: Software implementation and design of information encryption algorithm based on DNA nano self-assembled sensor array. J. Sens. **2021**, 5800620 (2021)
11. Varma, P.S., Raju, K.G.: Cryptography based on DNA using random key generation scheme. Int. J. Sci. Eng. Adv. Technol. (IJSEAT) **2**(7), 168–175 (2014)
12. Vidhya, E., Rathipriya, R.: Key generation for DNA cryptography using genetic operators and Diffie-Hellman key exchange algorithm. Int. J. Math. Comput. Sci. **15**(4), 1109–1115 (2020)
13. Polenov, S.: The computing of NP-complete problems in polynomial time using DNA – logic. World Appl. Sci. J. **30**(9), 1188–1192 (2014)
14. Sadkhan, S.B., Yaseen, B.S.: A DNA-sticker algorithm for cryptanalysis LFSRs and NLFSRs based stream cipher. In: International Conference on Advanced Science and Engineering, IEEE-ICOASE 2018, University of Zakho - Duhok Polytechnic 12University, and Submitted to the IEEE Xplore Digital Library, 9–11 October 2018 (2018)
15. Sadkhan, S.B., Yaseen, B.S.: DB based DNA computer to attack stream cipher. In: International Conference, IEEE- ICECCPCE 2019, Mosul and Erbil,13–14 February 2019 (2019)

16. Sadkhan, S.B., Yaseen, B.S.: Hybrid method to implement a parallel search of the cryptosystem keys. In: International Conference on Advanced Science and Engineering, Springer-ICOASE 2019, 2–3 April 2019. University of Zakho - Duhok Polytechnic University, and Submitted to the IEEE Xplore Digital Library (2019)
17. Yaseen, B.S.: Splicing DNA model for unknown stream cipher cryptanalysis. In: Information Technology to Enhance E-Learning and Other Application (IT-ELA), 28–29 December 2021. IEEE-Explorer (2021)
18. Abboud, U.W., Shuwandy, M.L.: SDA Plus: improving the performance of the system determine algorithm (SDA) of the switching between AES-128 and AES-256 (MOLAZ method). In: 2023 IEEE 13th International Conference on System Engineering and Technology (ICSET), pp. 61–65 (2023)

Performance Evaluation Comparison Between Central SDN Network and DSDN

Wed Kadhim Oleiwi[1] and Alharith A. Abdullah[2(✉)]

[1] Department of Computer, College of Science for Women, University of Babylon, Babylon, Iraq

[2] Department of Networking, College of Information Technology, University of Babylon, Babylon, Iraq

alharith@uobabylon.edu.iq

Abstract. Software-Defined Networking (SDN) is a cutting-edge technology for increasing network programmability. Because of its centralized nature, SDN is particularly susceptible to single point of failure (SPoF) and scalability issues. Distributed SDN (DSDN) offers to eliminate the single point of failure; security issue, and scalability that present in the centralize SDN controller. The main concept is to have numerous controllers that can share the burden on the network, and one controller can take over another controller when it breaks.in this paper we work on Performance Evaluation comparison between central SDN network and DSDN, The Result Of Performance Evaluation Comparison Between Central SDN network and DSDN show that the dropped packet of (SPoF) in central SDN was 100% while the percentage of Dropped Packet decrease to 12%, and 22% as one controller, two controllers respectively fail in DSDN. The performance evaluation of Topology Discover shows there are a Progressive relationship between the time of Topology Discover and number of nodes in data plane. The Throughput of distributed Opendaylight controllers is 16.8% points higher than that of the centralized controller, while the centralized Opendaylight controller has a 22.4% greater Latency than distributed SDN controllers; where each of Throughput and Latency are SDN network Scalability metrics.

Keywords: SDN · DSDN · Opendaylight · GNS3

1 Introduction

To make the design, monitoring, and management of next-generation networks easier, software-defined networking separates a conventional network into a centralized control plane and a remotely programmable data plane [1]. An intelligent, centralized SDN control plane guides the actions of forwarding devices during packet processing and provides a bird's-eye view of the entire network. As a result of centralized management, programmable networks are possible, as are the implementations of adaptive and autonomous network control. Traditional networks are hard to manage because they are not centralized and have a lot of moving parts. SDNs are trying to solve this problem [1, 2].

© The Author(s), under exclusive license to Springer Nature Switzerland AG 2024
A. M. Al-Bakry et al. (Eds.): NTICT 2023, CCIS 2096, pp. 334–345, 2024.
https://doi.org/10.1007/978-3-031-62814-6_24

A major vulnerability of the centralized SDN strategy is that the whole network is managed by a single point of control. In order to scale effectively, a centralized entity has to have enough computing capacity and effective data management methods to process flow requests from forwarding switches [3]. Flow setup requests have the potential to overwhelm the centralized controller and slow down response time if they arrive at a high rate and there are frequent changes to the network. The central SDN controller may become a bottleneck as the network expands in terms of the number of switches and end hosts. In order to address these problems, it has been suggested that many controllers be used to distribute the control plane, with one controller being replaced by another in the event of a failure and the controllers working together to distribute network load [3]. Use horizontal and hierarchical topological models to give tasks to controllers on the control plane [4].

The issues of centralized SDN may be mitigated when several controllers are used in SDN architecture to manage scalability, reliability, performance, and single-point failure. The distributed SDN control plane is more responsive, scalable, and reliable than conventional SDN control planes, allowing it to quickly respond to a variety of networking events such as connection failures, requests for new flow configurations, intrusions, etc. Distributed control plane architecture can handle the millions of networking events that happen in dynamic places like multi-tenant data centers because its solutions can be scaled up and changed [3–6].

There are some works that work on evaluate the performance of central SDN's network as in [7], It analyzes and compares the throughput and latency performance of numerous well-known open-source controllers, including ONOS, Ryu, Floodlight, and OpenDayLight, using the OpenFlow benchmarking tool Cbench. Best practices for quantitative controller evaluation are outlined, and we go into the possibilities of benchmarking tools for SDN controllers in [8] proposed Best practices for quantitative controller evaluation are outlined, went into the possibilities of benchmarking tools for SDN controllers, the most important result of this study was The literature is full with suggested controllers that never see the light of day because their authors fail to disclose enough details for a third party to implement them, This means that any evaluation of them must be based exclusively on theoretical concerns. In this study [9], OpendayLight and open networking operating system (ONOS), two of the most potent and well-known SDN controllers, are experimentally compared. Mininet is used as an experimenting platform. Wireshark is used to record and examine live packet traffic. In [10] they take a close look at three different OpenFlow-based SDN controllers, The latency, throughput, and scalability of the controllers FloodLight, OpenDayLight (ODL), and Ryu are measured. To do this, Mininet is utilized as a platform to run the Cbench tool in a virtual setting. A comparison study between two (Onos, ODL) SDN's controller performance proposed, the well-known ping and Iperf tools employs in his research [11] to determine the network's latency and bandwidth.

In [12] and [13] the researchers work on topology discovery performance evaluation, they conduct hands-on experiments to compare the topology discovery and topology updating capabilities of the ONOS and OpenDaylight controllers.

All the work above works with central SDN's network except [14], which works with SDN's network with two SDN 's controllers, but this does not represent a distributed

SDN network where the minimum number of SDN's controllers in a DSDN network is three[15], In [16] the researcher work on one experiment, a centralized controller was utilized to compare the proactive mode to the reactive method, while in another experiment, a separate controller was employed to compare the two modes and their respective circumstances.

The rest of this paper shows comparison performance evaluation result between the central SDN's network and DSDN's network.

2   Testing Environment and Methodology

2.1   Virtual Environment

Lists the software that was used for these experiments in Table 1. This software are run on a HP DESK-TOP-F5259A5, with a Processor Intel (R) Core (TM) i7-10750H CPU @2.60 GHz, RAM 16.0 GB. The GNS3 is used as the environment on which all other software is set up. The Ubuntu-servers type 18.04.3LTS are used, an OpenDayLight controller is installed through, the configuration of a custom network topology using mininet. The design of the central SDN in GNS3, using Beryllium SR4 OpenDayLight controllers and the Python script to build a custom topology in mininet and the applications of the OpenDaylight controller. This topology remotely controlled by the ODL controller. All data networks make use of Open Flow switches (OVS switch 1.3).

Table 1. The software used in the experiments of this work.

Software	Function
GNS3	Graphical network simulator
Mininet	Custom Network topologies
Open Flow Switch	Virtual SDN Switch
OpenDayLight	SDN controller Platform
VMware Workstation	Virtualization Software
Ubuntu-server 18.04.3LTS	Host Operation System
Python	Programming language
Wireshark	Packet Capture and analyzer
ping all Tool	Connect test among each hosts in data plane

2.2 Methodology

The techniques/methods that used to build the Distributed SDN Network (DSDN), as it shown in the below steps:

Step 1: Traditional Software Define Network (single controller) is built according to the specified environment as Fig. 1 shows.

Step 2: An examination of a specific use case demonstrates the security issues associated with traditional software-defined networks that represent single point of failure (SPOF) and fault tolerance.

Step 3: Study of a specific use case illustrating the scalability issues; performance limitations introduced by the increasing number of devices on the data plane in a Traditional software defined network.

Step 4: Distributed Software Define Network is built according to the specified environment as Fig. 2 shows.

Step 5: Study of a specific use case illustrating the security issues that represent single point of failure (SPOF) and fault tolerance with distributed software-defined networks compared to traditional software-defined networks.

Step 6: Study of a specific use case illustrating the scalability issues and other performance metrics of a distributed software-defined network compared to a traditional software-defined network.

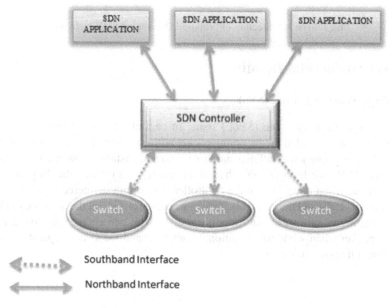

Fig. 1. The Architected of Traditional SDN Network Technology.

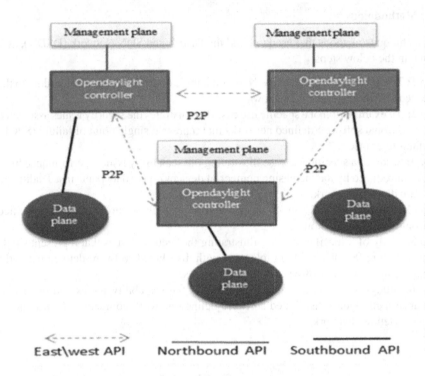

Fig. 2. The Architecture of proposed Distributed SDN Networks.

3 Experiments and Results

3.1 Single Point of Failure (SPoF)

The first experiment examine (SPoF) issue, in central SDN's network, where OVS switches, fail in forwarded the incoming packet without connect with remote controller where all the flow entry was deleted and there is no new installed flow entry the dropped packet was 100%,while in DSDN when all controller was operate the dropped packet was 0%,while when one of this three controller was logout the percentage of dropped packet raise to 12%, and this as result to the system not fulfillment the required of DSDN with the minimum number of controller is three. And this percentage raise to reach 22% dropped packet when we logout additional controller and the system operate with one controller.as it shows in Fig. 3.

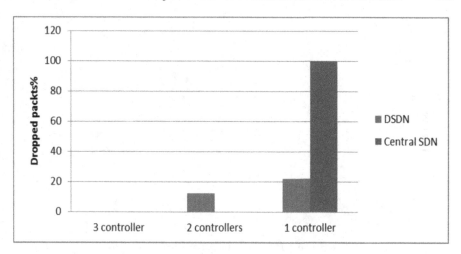

Fig. 3. The Single Point of Failure in Central SDN Controller VS. Distributed SDN Controller.

3.2 Topology Discover

On VM of the mininet, start the tree topology with custom no. of switch and we use the Wireshark to capture real time traffic between mininet and ODL controller, we compute the average of link Layer Discovery protocol (LLDP)cycle recording the sending time of the first LLDP, the Link Layer Discovery Protocol where each switch had sent a PacketIN encapsulated LLDP message back to the controller. These PacketINs are answering the LLDP messages the controller has send to discover the network, for each cycle, we calculate the time of the LLDP cycle which is the difference between the sending time of the first LLDP message sent by the controller and the reception time of the last LLDP message where this cycle happened every 5 s in Opendaylight controller.

Figure 4 shows that the performance of the central controller is better than the distributed until the 158 switches in data plane this can be because of consistency among the controller in the distributed, but when the infrastructure increase the performance of distributed ODL become better than central controller, where its performance become more stable.

Figure 5 shows there are a Progressive relationship between the time of Topology Discover and no. of nodes in data plane. The performance of central controller is better than distributed until the no. of node near to 160 nodes, while the Topology Discover of distributed ODL after this number be more stable. In general, the time required for discover the linear topology is less than time of tree topology with 5%.

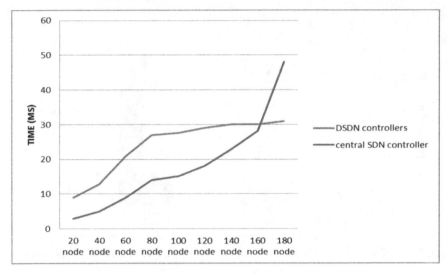

Fig. 4. The Tree Topology Discover

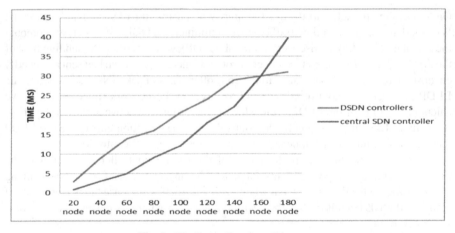

Fig. 5. The linear Topology Discover.

3.3 Scalability

To analyze the network performance when the number of OVS Switches is increased where we can define the scalability as the ability of the controller to control and manage large number of the nodes without an obvious change in performance, for this experiment we use each of the throughput and latency as metrics to scalability of Opendaylight controllers.

(1) Throughput

As show in Fig. 6 and Fig. 7, the rate at which the controller, the single controller, and the distributed controllers handle flow requests is proportional to the rate at which PACKET

IN events with an ARP payload are generated. As can be seen in Fig. 5 and Fig. 6, the throughput of decentralized OpenDaylight controllers is 16.8% points higher than that of the centralized controller. In addition, when the number of devices in the data plane grows, performance remains more consistent with distributed SDN, while throughput drops down as the number of devices grows in the central controller. Finally, we observe that the throughput of SDN networks with a tree topology is more consistent where there than that of networks with a linear topology because there is a decrease in overall bandwidth utilization and an increase in the propagation time between nodes.

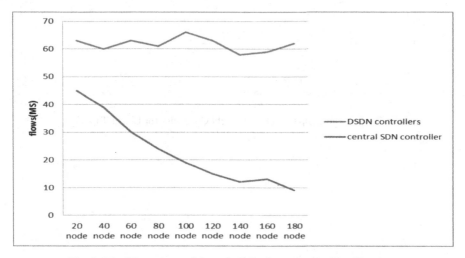

Fig. 6. The Throughput of Opendaylight Controller for Tree Topology

(1) Latency
The time between the packets is sent to the controller and the response on it, is a Latency. To compute the latency, we use the ping Toole between the first host in the network and the farthest one after the end of the first cycle of the topology discover. According to the data, the centralized opendaylight controller has a 22.4% greater latency than distributed SDN controllers Fig. 8 and Fig. 9. Distributed SDN maintains a more consistent latency as the number of devices in the data plane grows, in contrast to the centralized controller, where latency grows as the number of devices grows. The latency with the linear topology is also clearly seen to be greater than the latency with the tree topology by a margin of 12.46%,where a packet transmission times are much higher in a linear topology than in a tree topology. This is because the time it takes for intermediary nodes to propagate a packet to its destination grows proportionally with the number of hops that exist between the source and the destination.

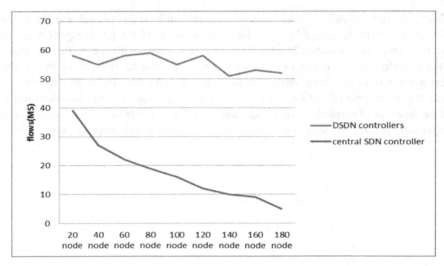

Fig. 7. The Throughput of Opendaylight Controller for Linear Topology

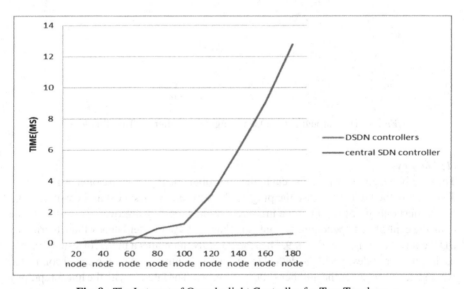

Fig. 8. The Latency of Opendaylight Controller for Tree Topology

3.4 Server Source Consumer (CPU and Memory)

Figure 10 and Fig. 11 show that as the number of controllers in the network grew, so did the number of users who could make use of the system's memory and its storage space. Experiment results show that after a few minutes of network activities as it explain in GNS3, CPU consumption may drop by half, and the system becomes more stable, but the main problems lie with memory.

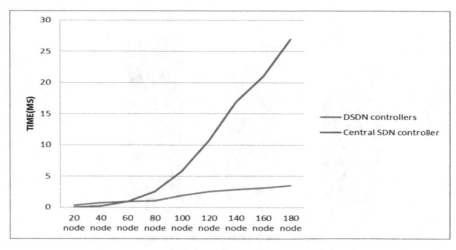

Fig. 9. The Latency of Opendaylight Controller for Linear Topology

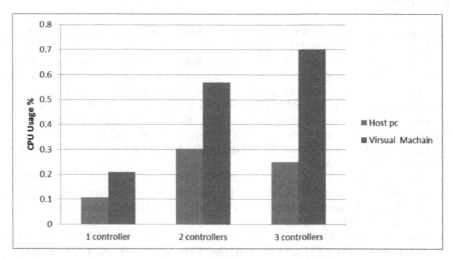

Fig. 10. The servers CPU consumer

4 Conclusions

The suggested testbed; virtual environment, is an excellent choice for a DSDN network. Concerns with central SDN network security, scalability, and dependability are highlighted by the proposed architecture. The suggested solution would improve DSDN's safety and uniformity, as the suggested solution lessens the load on the central processing unit (CPU) used by the network's leader controller.

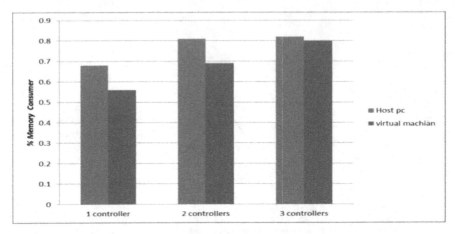

Fig. 11. The Servers Memory Consume

References

1. Herrera, J.G., Botero, J.F.: Resource allocation in NFV: a comprehensive survey. IEEE Trans. Netw. Serv. Manag. **13**(3), 518–532 (2016)
2. Di Martino, C., Giordano, U., Mohanasamy, N., Russo, S., Thottan, M.: In production performance testing of SDN control plane for telecomoperators. In: Proceedings IEEE/IFIP International Conference Dependable Systems and Network, pp. 642–653 (2018)
3. Sadkhan, S.B., Abbas, M.S., Mahdi, S.S., Hussein, S.A.: Software-defined network security-status, challenges, and future trends. In: 2022 Muthanna International Conference on Engineering Science and Technology (MICEST), pp. 10–15. IEEE (2022)
4. Tank, G.P., Dixit, A., Vellanki, A., Annapurna, D.: Software-defined networking-the new norm for networks (2012)
5. Mahdi, S.S., Abdullah, A.A.: Survey on Enabling Network Slicing Based on SDN/NFV. In: Al-Emran, M., Al-Sharafi, M.A., Shaalan, K. (eds.) International Conference on Information Systems and Intelligent Applications: ICISIA 2022, pp. 733–758. Springer, Cham (2022). https://doi.org/10.1007/978-3-031-16865-9_59
6. Chiosi, M., et al.: Network functions virtualisation: an introduction, benefits, enablers, challenges and call for action. In: SDN and OpenFlow World Congress, pp. 22–24 (2012)
7. Dixit, A., et al.: Towards an elastic distributed SDN controller. In: ACM SIGCOMM Computer Communication Review, vol. 43, no. 4, pp. 7–12. ACM (2013)
8. Mamushiane, L., Lysko, A., Dlamini, S.: A comparative evaluation of the performance of popular SDN controllers: 2018 Wireless Days (WD), pp. 54–59 (2018). https://doi.org/10.1109/WD.2018.8361694
9. Zhu, L., Karim, Md.M., Sharif, K., Li, F., Du, X., Guizani, M.: SDN controllers: Benchmarking & performance evaluation. arXiv preprint arXiv:1902.04491 (2019)
10. Badotra, S., Panda, S.N.: Evaluation and comparison of OpenDayLight and open networking operating system in software-defined networking. Cluster Comput. **23**(2), 1281–1291 (2020)
11. Mendoza, D.H., Oquendo, L.T., Marrone, L.A.: A comparative evaluation of the performance of open-source SDN controllers. Latin-American J. Comput. **7**(2), 64–77 (2020)
12. Dissanayake, M.B., Kumari, A.L.V., Udunuwara, U.K.A.: Performance comparison of ONOS and ODL controllers in software defined networks under different network typologies. J. Res. Technol. Eng. **2**(3), 94–105 (2021)

13. Azzouni, A., Nguyen, M.T., Pujolle, G.: Topology discovery performance evaluation of Open-Daylight and ONOS controllers. In: 2019 22nd Conference on Innovation in Clouds, Internet and Networks and Workshops (ICIN) (2019). https://doi.org/10.1109/icin.2019.8685915
14. Wazirali, R., Ahmad, R., Alhiyari, S.: SDN-openflow topology discovery: an overview of performance issues. Appl. Sci. **11**(15), 6999 (2021)
15. Ahmad, S., Mir, A.H.: Scalability, consistency, reliability and security in SDN controllers: a survey of Diverse SDN controllers. J. Netw. Syst. Manage. **29**, 9 (2021). https://doi.org/10.1007/s10922-020-09575-4
16. El-Geder, S.: Performance evaluation using multiple controllers with different flow setup modes in the software defined network architecture, Doctoral dissertation, Brunel University London (2017)

Enhancing the Performance of Wireless Body Area Network Routing Protocols Based on Collaboratively Evaluated Values

Sabreen Waheed Kadhum[(✉)] and Mohammed Ali Tawfeeq

Computer Engineering Department, Mustansiriyah University, Baghdad, Iraq
{sabreenwhd,drmatawfeeq}@uomustansiriyah.edu.iq

Abstract. Wireless Body Area Sensor Network (WBASN) is gaining significant attention due to its applications in smart health offering cost-effective, efficient, ubiquitous, and unobtrusive telemedicine. WBASNs face challenges including interference, Quality of Service, transmit power, and resource constraints. Recognizing these challenges, this paper presents an energy and Quality of Service-aware routing algorithm. The proposed algorithm is based on each node's Collaboratively Evaluated Value (CEV) to select the most suitable cluster head (CH). The Collaborative Value (CV) is derived from three factors, the node's residual energy, the distance vector between nodes and personal device, and the sensor's density in each CH. The CEV algorithm operates in the following manner: CHs are dynamically selected in each transmission round based on the nodes' CVs. The algorithm considered the patient's condition classification to guarantee safety and attain a response speed appropriate for their current state. So, data is categorized into Very-Critical, Critical, and Normal data classes using the supervised learning vector quantization (LVQ) classifier. Very Critical data is sent to the emergency center to dispatch an ambulance, Critical data is transmitted to a doctor, and Normal data is sent to a data center. This methodology promotes efficient and reliable intra-network communication, ensuring prompt and precise data transmission, and reducing frequent recharging. Comparative analyses reveal that the proposed algorithm outperforms ERRS (Energy-Efficient and Reliable Routing Scheme) and LEACH (low energy adaptive clustering hierarchy) regarding network longevity by 27% and 33%, augmenting network stability by 12% and 45% over the aforementioned protocols, respectively. The performance was conducted in OMNeT++ simulator.

Keywords: Cluster head · QoS · Routing · Residual Energy · Wireless Body Area Networks

1 Introduction

Wireless body area sensor networks (WBASNs) refer to a specific type of sensor networks that employ wireless bio-sensors positioned on or within the human body for the purpose of measuring vital physiological parameters, including but not limited to blood

A. M. Al-Bakry et al. (Eds.): NTICT 2023, CCIS 2096, pp. 346–361, 2024.
https://doi.org/10.1007/978-3-031-62814-6_25

pressure, heart rate, body temperature, and glucose level. This technology facilitates the remote monitoring of a patient's health status [1]. WBANs comprise several biological sensors. These sensors can be placed at various locations on the body, either worn externally or implanted beneath the skin. Each has special requirements and is utilized for distinct missions. These gadgets are used to detect changes in vital signs as well as emotions or emotional states such as fear, tension, happiness, etc. The connection is established using a dedicated coordinator node that has greater processing capability and is not limited by power limitations. It is in charge of transmitting biological signals from the patient to the medical doctor to deliver real-time medical diagnostics and allow him to decide on the correct choices. [2]. A simple WBAN architecture is shown in Fig. 1 where the architecture is divided into several sections.

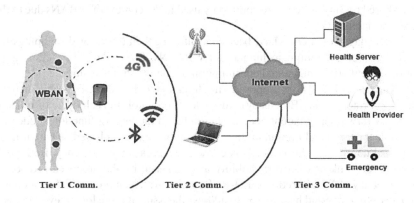

Fig. 1. Wireless Body Area Network architecture [3]

From Fig. 1, the WBASN architecture can be segmented into three distinct tiers.

- Intra (tirel): refers to the communication between the sensors and the personal device as well as between the sensors themselves.
- Inter (tire2): Include the communication between the personal device and the access point using wireless technologies.
- Beyond (tire3): refers to the communication between the access point and the remote health medical center.

WBASNs can also greatly lower patient treatment expenses by continuously monitoring the patient's vital sign-related data. Body sensor networks employ a star topology architecture, which sets them apart from wireless sensor networks. In this configuration, a proximal point (G) is situated on the body to collect data from nodes. The Bio-nodes in the network function using limited power resources. In physical sensor networks, sensors (like motion sensors) are installed on the body of a patient to observe the vital signs or detect motion. By monitoring a patient's vital signs, the wireless body sensor network provides an instant response to the user through which the user can follow the progression of a patient's disease and take the necessary precautions [4]. The energy consumption of sensors plays a crucial role in this particular network, as the depletion of the energy source directly impacts the overall lifespan of the network, leading to a reduction in its duration [5].

In WBAN-related applications, routing protocols based on a clustering technique are appropriate. The network is partitioned into separate logical sub-networks termed clusters in the clustering approach. Each cluster consists of a single header node known as the cluster head (CH) and multiple cluster member nodes. The nodes within the cluster broadcast their sensed data to the cluster head (CH) in a single-hop manner. The cluster head (CH) is responsible for aggregating the received data from each cluster into a single packet referred to as a datum [6]. CH then relays this data to the G (Hub). In this way, the clustering strategy is designed to minimize energy consumption while maximizing data delivery within a suitable end-to-end delay. Cooperative control constitutes the essential operational aspect of clustering and routing inside a Wireless Body Area Network. Cluster-based routing systems are commonly used in the context of WBANs due to their inherent scalability [7].

In recent days, researchers have been concentrating their efforts on developing effective CH selection and routing techniques for WBASNs. The goal is to boost throughput, duration of network stability, and consumption of energy. In this realm:

A. Khanna et al., 2018 [8] proposed a modification to the SIMPLE (Stable Increased Multi-Hop Protocol Link Efficiency) cluster-level protocol. In each cycle of transmission, a new forwarder or parent node is chosen based on the cost function. Nodes with high residual energy and minimum distance from the sink are elected as parent nodes. An optimal number of cluster heads is chosen to increase the efficiency. Both packet delivery to the sink and network stability are enhanced by the proposed protocol. Z. ULLAH, et al.2019 [9] introduced EH-RCB, an innovative approach aimed at enhancing energy efficiency and harvesting. To address the issue of route loss, they employed a clustering strategy that involves the creation of two clusters. These clusters are formed with the assistance of sink nodes, which act as pre-defined cluster heads. The proposed technique calculates a cost function for each SN in the network to determine the most optimal retransmission node for data transfer. About stability, throughput, and packet delivery ratio, the research findings indicate that EH-RCB outperforms protocols of a similar nature. F. Ullah, 2020 [10] proposed an Energy-Efficient and Reliable Routing Scheme (ERRS) to mitigate reliability and the period stability for the WBAN-constrained resource. ERRS is made up of two main solutions, Forwarder Node Rotation and the Forwarder Node Selection methods. The ERRS proposed uses the clustering routing technique that has adaptive/static advantages and gains a longer lifetime for a network and an enhanced stability period, as well as maximized reliability. N. Bilandi et al., 2020 [11] compared the effectiveness of the latest three meta-heuristic algorithms that take their cues from the natural world in selecting relay nodes. Grey wolf optimization was found to reduce total energy usage by 23% and 16% as compared to particle swarm and ant lion optimization, respectively. S. E. Pradha et al., 2021 [12] introduced a novel low-power media access control (MAC) protocol called Scheduled Access MAC (SAMAC) to keep networks running longer without compromising on quality of service. This study considers two scenarios. Keeping the same volume of traffic while adjusting the number of sensors constitutes the first scenario. The second scenario involves maintaining a consistent number of sensors while altering the traffic load. The simulation results show that the proposed SAMAC technique significantly improves the quality of service and uses less energy than the Baseline MAC protocol and the ZigBee protocol. S. Firdous et al., 2022 [13] presented a new algorithm titled Power-Efficient Cluster-based Routing (PECR). This algorithm incorporates several key components, including K-Means clustering, optimal route selection, energy-based communication, and the rotation of cluster heads and primary cluster heads. These elements are integrated to enhance energy efficiency within the routing process. PECR reduces energy consumption, minimizes traffic congestion, and ultimately increases network lifetime. Sensors detect data and transfer it to a Base Station (BS) via a valid channel. The results show that it reduces traffic overhead as well as effectively leverages energy assets. K. Guo and S. Syed, 2022 [14] focus on the effective Choosing the cluster head node and fairly rotating the cluster head, taking into consideration four key parameters: the remaining power of the node, the average Euclidean distance, the priority level determined by the sensitivity of data, and the physical capabilities of the device. A Bayesian clustering approach with energy efficiency considerations was presented to facilitate the optimal selection of the cluster head node. The performance of the algorithm is evaluated in comparison to the LEACH protocol. This technique enhances the network's total lifespan by nearly 1.5 times while concurrently diminishing the average energy usage of the network. B. Vahedian and M. Nasr, 2022 [15] utilized two approaches, namely the Earliest Deadline First (EDF) real-time scheduling technique and its integration with the Least Laxity First (LLF) scheduling technique, to prioritize nodes for the transmission of data packets. By combining the scheduling methods EDF and LLF, collisions can be decreased along with packet retransmission and the energy required to operate sensor nodes. Three scenarios were used (random, EDF, and the proposed technique). When compared to the other two alternatives, the proposed technique has lower average power consumption. S. S. Oleiwi et al., 2022 [16] employed the cooperative routing protocol (CRP) to reduce end-to-end latency and hence improve package delivery and network longevity. CRP allows a variety of sources to communicate with several wireless networks; The architecture incorporates several relay channels that offer distinct pathways for message transmission, hence mitigating the risk of a singular point of failure. The sink merged the original message to produce the original data. When the CRP scheduling process is completed, Data analysis is the initial step carried out during the transaction. In WBAN, the critical data is processed first, followed by the non-critical data group, and finally by the group of normal data. It makes its own decisions that are immediately connected to the network data center and provides an efficient solution via a personal device for non-critical data, and it has a direct connection to the physician. The results show that the proposed protocol significantly reduces the end-to-end delay, as well as calculation time and error tolerance in WBASN users. B. Ghosh et al., 2023 [17] presented the effects of various absorption rates on the body and offer recommendations for deployment of linear and non-linear relays, as well as the best payload size in different scenarios. Additionally, the energy efficiency of a relay-assisted WBASN has been evaluated under several conditions, including varying distances among the source and destination, different modulation techniques, and varied payloads. The results of the study suggest that the center of the source and the destination nodes are not the best places to put the relay. If the relay is positioned properly, the overall lifespan of WBAN will be much longer. R. Dass, and M. Narayanan,2023 [18] introduced a routing system, known as the safe Optimal Path-Routing (SOPR) protocol, which utilizes clustering techniques to identify and discover secure routes in WBANs that are resilient to avoid black-hole threats. Furthermore, the protocol has been seamlessly incorporated into the Balanced Energy-Efficient and Reliable (BEER) strategy to facilitate the dependable and secure transfer of data. When compared to M-ATTEMPT and ATTEMPT, the results demonstrate that the suggested protocol exhibited enhancements in the overall performance of the network through the augmentation of the packet success ratio, the overhead reduction of attack detection, latency, detection time, and power consumption.

Existing research in wireless body area sensor networks (WBASNs) has made significant contributions to this field, such as designing new routing methods, cross-layer optimization methods, and QoS provisioning mechanisms in WBANs, although they have some limitations. Much existing research focuses on individual features of WBAN routing, such as energy saving or delay optimization, without taking into account the overall system performance, communication continuity, or severity of the patient's condition.

The primary innovation in this study involves the introduction of a stable routing protocol and a classification procedure that prevents false alarms while ensuring efficient communication within the network. This guarantees the rapid and accurate transmission of critical data, leading to high throughput and the preservation of residual energy, thereby extending the network's lifespan. Additionally, considering the dynamic nature of WBANs operating in ever-changing environments, the system can swiftly adapt to network changes to meet application requirements. The dynamical selection through cluster heads facilitates alternative path selection, creating an energy-efficient system by distributing the load among sensors. These combined contributions significantly enhance the performance of WBANs, making the network more efficient, reliable, and capable of supporting a diverse range of healthcare applications.

This work proposes a WBAN with 10 Bio-sensors that can cover most human body analysis strategies including electroencephalography (EEG), electrocardiogram (ECG), oxygen saturation (SO2), respiration (BRTH), blood pressure (BP), glaucous (Gloc), temperature (Temp), pulse rate (PR), electrooculography (EOG), and heart failure (HF). The proposed model can categorize human cases into three classes based on health status that includes critical, very critical (requiring an ambulance dispatch to the patient), and normal (data sent to the data center). These cases are determined based on sensor reading with the Leaning Vector Quantization (LVQ) classifier employed for this purpose. LVQ is a supervised machine learning algorithm used for classification tasks, learning from labeled examples to classify new, unseen examples [20, 21].

In the proposed algorithm, robust collaborative evaluated values are computed using three-parameter vectors; node residual energy, distances from the nodes to the personal device, and sensors density in each CH. The nodes with the minimum collaborative evaluated values are elected as cluster heads. Since the remaining energy of the Bio-nodes (including CHs and non-CHs) has changed, the collaborative values are recalculated after each round.

In the subsequent sections of this work, Sect. 2 introduces the adopted network models, demonstrates the proposed technique, and the details on how to extract the collaborative values. The findings of the simulation are reported and analyzed in Sect. 3, along with a comparison to specific classical techniques. Ultimately, the findings of this study are expounded upon in Sect. 4, wherein the conclusions are presented.

2 System Model

The proposed WBASN system comprises ten homogeneous Bio-sensors having restricted resources for hardware due to their compact size. These Bio-nodes are deployed stationary on the human body along with one personal device (G), and one sink that collects and classifies (CC) the aggregated data. The CC node is placed near of human body whereas (G) is placed on either the left or right side of the human waist as shown in Fig. 2. The communication between the sensors in a WBAN is typically wirelessly, using short-range wireless technology ZigBee [1, 22].

The Bio-sensors detect physiological information from the human body and transmit it to (G) via the chosen CHs. This data is then aggregated by G, which subsequently feeds it to the sink node CC. Upon receiving the data from G, the sink CC classifies it into one of three categories: **critical data (CD), very critical data (VCD), and normal data (ND).** The data is then forwarded to the medical server. Table 1 presents a comprehensive and detailed explanation of the nodes that have been utilized in this study.

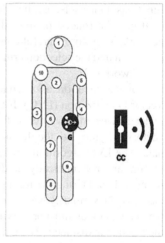

Fig. 2. Nodes deployment location

Table 1. Detailed description of employed nodes

No	Abbreviations	Description	Normal	Critical	Very Critical
1	EEG	Electroencephalography	Higher than 8 Hz	7–8 Hz	Lower than 7 Hz
2	ECG	Electrocardiogram	60–100 bpm	Less than 60 bpm	greater than 100 bpm
3	SO2	Oxygen saturation	94–99%	60–80%	Below 60%
4	BRTH	Respiration	12–20 breaths/minute	15–25 breaths/minute	under 5 breaths/minute
5	BP	Blood(circularity) Pressure	less than 120/80 mmHg	120–139 mmHg	140 mmHg and greater
6	Gloc	Diabetes sensor	less than 100 mg/dl	Between the range of 101–126 mg/dl	greater than 126 mg/dl
7	TEMP	Temperature	37 C	38 C	39 C or more
8	PR	Pulse Rate	40–100 bpm	101–109 bpm	130 bpm or more
9	EOG	Electrooculography	1.80 or greater	1.65 to 1.80	lower than 1.65
10	HF	Heart Failure (B-type natriuretic peptide)	(BNP) below 100 pg/mL	BNP ranges from 101–400 pg/mL	(BNP) greater than 400 pg/mL

A cluster head is a designated node that acts as a coordinator for a group of other nodes, known as cluster members. The cluster head is responsible for managing the communication within its cluster, including data aggregation and forwarding, as well as routing data to the sink node CC or other cluster heads. The process of selecting a cluster head is typically based on certain criteria, such as its remaining energy level, its proximity to the other nodes in the cluster, or its ability to perform data aggregation functions. Once selected, the cluster head may remain in that role for a certain period, or it may be rotated periodically to distribute the workload among the nodes in the network. Therefore, careful consideration must be given to the design of the clustering algorithm and the selection criteria for the cluster heads, to optimize the performance of the network [23].

For classification tasks, a machine learning algorithm was proposed named Learning Vector Quantization (LVQ). It is a supervised learning algorithm that learns from a set of labeled examples to classify new, unseen examples. LVQ uses a set of prototype vectors to symbolize the distinct classes (VCD, CD, and ND) present within the input data. The LVQ algorithm operates through an iterative process wherein the prototype vectors are systematically modified to reduce the classification error observed in the training data. During the training process, the algorithm updates the prototype vectors based on the similarity between the input data and the current prototypes. The similarity between an input and a prototype is typically measured using a distance metric, such as the Euclidean distance [21]. One of the advantages of LVQ is its ability to handle noisy data and overlapping classes. Additionally, LVQ is a relatively simple algorithm that requires only a small number of parameters to be tuned, making it easy to implement and interpret.

The system model consists of one stationary WBAN as shown in Fig. 3. The star topology consists of one getaway node (G) and ten terminal Bio-nodes (TNs) the variable d is defined to represent the fixed distance between the terminal nodes (TNs), and the target domain G.

In the proposed model, cluster head selection is based on a continuous relationship that integrates the residual energy, the distances among the Bio-nodes and the hub (G), and the node density within each CH. According to this relationship, collaborative evaluated values (CEVs) are derived from the nodes exhibiting the lowest CEVs are subsequently selected to function as Cluster Heads (CHs). To calculate CEVs, a WBAN with N nodes and one hub (G) is considered. Every individual node denoted as "i," possesses an initial energy level, represented as E_i, i = 1, 2,..., N. The formula for the Euclidean distance d_i between vertices i and G is computed as:

$$d_i = \|P_i - P_G\| \tag{1}$$

where P_i and P_G are the coordinates of nodes i and G respectively. In contrast, the residual energy E_i of the node is typically smaller than d. Therefore, by normalizing the distance measurements, a consistent CEV can be obtained based on d, E, and the cluster node density N_c. A suitable normalization technique for the variable d can be achieved by subtracting the mean of d from each of its values and dividing the resulting values by the standard deviation. [24].

$$\overline{d} = \frac{1}{n} \sum_{i=1}^{n} d_i \tag{2}$$

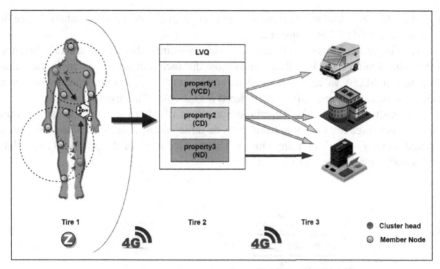

Fig. 3. Collaborative Evaluated Value architecture

$$dn_i = \frac{\left(d_i - \overline{d}\right)}{d_{std}} \tag{3}$$

$$d_{STD} = \sqrt{\frac{1}{(n-1)} \sum_{i=1}^{n} (d_i - \overline{d})^2} \tag{4}$$

The normalized distance vector, denoted as d_n, is a mathematical representation used to quantify the distances between objects or points. The mean distance, denoted as $\overline{d}$, represents the average value of these distances, while the distance's standard deviation denoted as d_{std}, quantifies the dispersion or variability of the distances from the Mean. The formula of CEVs used in [25] is modified to meet the requirements of this study. The extracted collaborative evaluated value CEV_i for i = 1, 2, ..., N, is equivalent to the product of Nc by the square of the product of (E_i and dn_i) normalized by the product of the sum of the squared residual energy vector E_j and the sum of the vector of squared distances dn_j, that is;

$$CEV_i = \frac{(E_i * dn_i)^2}{\sum_{j=1}^{N} E_j^2 * \sum_{j=1}^{N} dn_j^2} Nc \tag{5}$$

where Nc is the total number of nodes in every single cluster and N is the total number of nodes in the entire network.

To determine if a sensor node qualifies as a Cluster Head, the following conditions must be met:

Remaining Energy > threshold1: This criterion ensures that the sensor's residual energy surpasses the assumption that 'threshold1' is equivalent to half of the average energy of all nodes This threshold would represent an energy level that the sensor should have or more, to be considered as a cluster head.

Distance To Personal Device < threshold2: This measures whether the sensor's distance to the central hub (G) is less than the average distance that is equivalent to the value of the threshold2. A value less than 'threshold2' sets the upper limit for the distance a sensor can be from G and still be considered for the role of a cluster head.

Sensor Density > threshold3: Threshold 3 assesses whether the sensor density within a specific coverage area is equal to or exceeds three nodes, indicative of the minimum required number of nodes for any cluster. The detailed steps of the proposed algorithm are illustrated by the flowchart in Fig. 4:

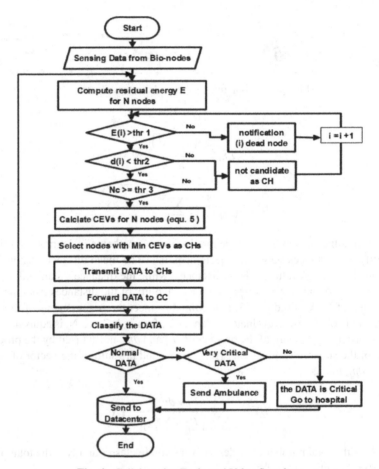

Fig. 4. Collaborative Evaluated Value flowchart

The algorithm under consideration employs iterative processes, wherein each iteration involves the identification and selection of nodes with the lowest CEVs to serve as CHs. Subsequently, a loop is implemented to assess each elected cluster head (CH) to execute the following tasks disseminates information regarding their respective time allocations for transmitting data, receiving data, and transmitting the aggregated information to the central hub (G). Following the completion of each iteration, the values of thresholds 1 and 3 are investigated, however, the value of threshold 2 (which indicates the distances between the Bio-nodes and the hub (G)) has not been examined because it is considered to be static in this study. Then The CEVs vector is subsequently updated for all nodes, taking into account their updated residual energy. At the conclusion of each cycle, critical data as well as very critical data are also delivered to the database for storage.

3   Performance Metrics and Simulation Results

A comprehensive series of experiments utilizing the Objective Modular Network (OMNET++) simulation tool was conducted to assess the functionality of the proposed CEV algorithm. The WBAN topology used for the CEV algorithm comprises ten SNs positioned at distinct static locations on a human body, along with one hub and one sink node as shown in Fig. 2. Detailed information can be found in Table 1. Physiological measurements (BP, Gloc, ECG, EOG, HR, etc.) are all sensed by these Bio-sensors. The procedure under consideration is evaluated against existing protocols found in the literature, including LEACH [26] and ERRS [10]. Table 2 displays comprehensive details regarding the simulation parameters employed, including the initial energy allocated to all sensor nodes, the dimensions of the simulated area, the quantity and spatial distribution of deployed nodes, and other relevant information.

Table 2. Simulation parameters

Parameters	Values
Simulation Tool	OMNET++
Simulated area	$2 \times 2 \, m^2$
Environment	Indoor
WBAN standard	IEEE 802.15.6
Topology	Multi-hop Star topology
Number of humans	01
Number of Bio-nodes	10
Number of hubs	01
Number of sinks	01
Total number of rounds	1402
Initial energy	1 $J/node$
$E_{Tx-electronics}$	16.7 nJ/bit
$E_{Rx-electronics}$	36.1 nJ/bit
$E_{Amplifier}$	1.98 J/bit
Payload	4000 bits

Following the completion of each round, all nodes remaining energy and density are evaluated, which represent thresholds 1 and 3 in the given order. However, because it is considered static in this study, the value of threshold 2, which indicates the distance, has not been examined. The CEV vector is then updated for all nodes. Through the iterative evaluation of several experimental tests, performance was assessed based on the following essential metrics: Network lifetime, stability, throughput, and residual energy. Using these parameters as a guide, simulation-based tests were conducted to evaluate the effectiveness of the CEV algorithm compared to two established benchmark routing protocols, LEACH and ERR.

3.1 Network Lifetime

The lifetime of a network is its entire amount of time in operation. The clock starts ticking when the first node is deployed and stops when the last node dies. Figure 5 shows the lifetime of the proposed model in the three protocols.

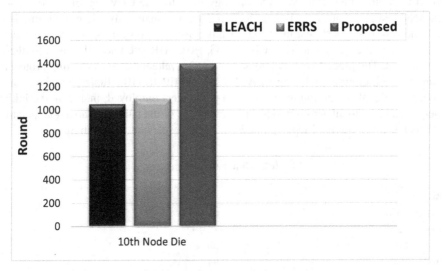

Fig. 5. Network Lifetime

In this figure, a correlation is observed between the number of active nodes and the duration of rounds. The performance of the LEACH protocol appears to be unsatisfactory, with a network operation duration of 1052 rounds. In contrast, the ERRS protocol exhibits a better network lifetime, recorded at 1100 rounds. Meanwhile, the proposed protocol extends the network lifetime to a noteworthy 1402 rounds. Compared to ERRS and LEACH protocols, the CEV technique increases network lifespan by 27% and 33% respectively.

3.2 Network Stability

The period preceding the demise of the initial node within the network is commonly denoted as stability. Dead nodes as a function of rounds are shown in Fig. 6.

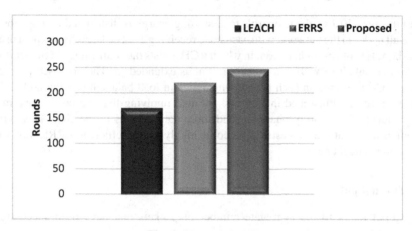

Fig. 6. Network Stability

The comparative analysis reveals that the LEACH procedure exhibits the lowest level of performance. The stability period in this procedure is limited to a mere 171 rounds, resulting in heightened levels of instability. The duration of the stability period in the ERRS protocol spans a total of 221 rounds. In the proposed protocol, the first node dies after 248 rounds. The stability period is improved by 12% and 45% over the ERRS and LEACH protocols, respectively.

3.3   Residual Energy

Residual energy is the network's average total remaining energy after each round. This is illustrated in Fig. 7.

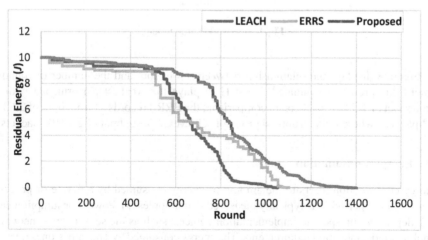

Fig. 7. Analysis of Residual Energy

In this test, 10 J of the network's total starting energy is distributed homogenously across all nodes in the network, with each node receiving 1J. The LEACH protocol uses a multi-hop style of communication, in which a CH assists the source node in transmitting data to the sink. The WBAN's execution time is extended. In The proposed protocol, dynamic CHs selected in each round, which aids in load balancing on the nodes. The simulation results illustrated in Fig. 7 show that, implying that the nodes have more energy during the stability phase to send more data packets to the sink. This serves to demonstrate that the suggested protocol is highly energy efficient. ERRS shows the minimum residual energy.

3.4 Throughput

Throughput refers to the aggregate number of packets that have been successfully received at the destination.

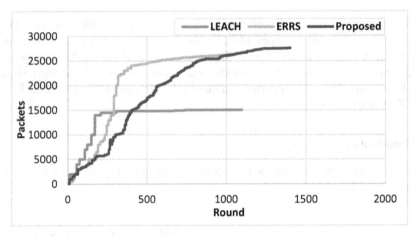

Fig. 8. Network Throughput

Figure 8 displays the relationship between throughput and the number of rounds. Among the protocols evaluated, LEACH displays the weakest performance with a throughput of $1.5*10^4$ packets. In comparison, the ERRS records a throughput of $2.63 * 10^4$ packets, whereas the proposed protocol achieves a throughput of $2.7*10^4$ packets.

3.5 Energy Consumption

Energy consumption refers to the amount of energy consumed by a device or process over a specific period. It is typically measured in unit joules (J). Power consumption can vary depending on specific implementation choices, such as the sensor types, network topology, and application requirements. The energy consumed by Transmission is related to the difference in distance between terminal nodes and the destination.

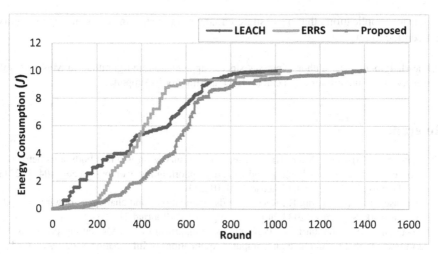

Fig. 9. Analysis of Energy Consumption

In Fig. 9, a comparison is conducted between the energy usage of the proposed CEV protocol and both the LEACH and ERRS protocols. It's evident that CEV significantly outperforms ERRS and LEACH schemes due to the algorithms employed, specifically in the LEACH scheme, CH is selected randomly, leading to a rise in the consumption of energy. As for ERRS, it is important to note that duplicate data was not taken into account, and the utilization of a coordinator introduced additional complexities, delays, and heightened energy usage. So, the power consumption by the proposed protocol outperforms ERRS by 15% and LEACH protocol by 22%.

4   Conclusion

This research introduces a Collaborative Evaluated Value Protocol for Wireless Body Area Networks. This protocol primarily presents two key strategies aimed at enhancing the performance of Wireless Body Area Networks (WBANs). The first mechanism involves periodically selecting the optimal number of Cluster Heads (CHs) from among the members of each cluster in every round. The second mechanism focuses on classes the data. The utilization of this methodology has facilitated effective and dependable communication within the network, hence guaranteeing the prompt and precise transmission of vital information to the emergency department or ambulance. The primary objective of this research endeavor is to augment the longevity, robustness, and remaining energy of the network. The protocol under consideration is being compared to two prominent protocols in the field, namely LEACH and ERRS. The findings indicate that the CEV protocol demonstrates superior performance compared to other protocols concerning network lifetime, network stability, and residual energy. In light of the diverse array of potential applications for Wireless Body Area Networks (WBANs), our forthcoming research will address the significant concern of mobility. The evaluation of the security component of the suggested method may be conducted in further studies, particularly for life-critical applications that necessitate enhanced precision in terms of dependability and confidentiality.

Acknowledgment. The author expresses sincere appreciation and gratitude to Mustansiriyah University. (www.uomustansiriyah.edu.iq) Baghdad-Iraq for its support.

References

1. Hajar, M.S., Al-Kadri, M.O., Kalutarage, H.K.: A survey on wireless body area networks: architecture, security challenges, and research opportunities. Comput. Secur. **104**, 102211 (2021). https://doi.org/10.1016/j.cose.2021.102211
2. Dhanasekar, M., Rangaraj, R.: Recent trends, applications, and challenges in wireless body area network, September 2023. https://doi.org/10.1729/Journal.31316
3. Alzahrani, A.S., Almotairi, K.: Performance comparison of WBAN routing protocols. In: 2nd International Conference on Computer Applications & Information Security (ICCAIS), ICCAIS 2019, pp. 1–5 (2019). https://doi.org/10.1109/CAIS.2019.8769594
4. Taleb, H., Nasser, A., Andrieux, G., Charara, N., Cruz, E.M.: Wireless technologies, medical applications and future challenges in WBAN: a survey. Wirel. Netw. **27**(8), 5271–5295 (2021). https://doi.org/10.1007/s11276-021-02780-2
5. Boikanyo, K., Zungeru, A.M., Sigweni, B., Yahya, A., Lebekwe, C.: Remote patient monitoring systems: Applications, architecture, and challenges. Sci. African **20**, e01638 (2023). https://doi.org/10.1016/j.sciaf.2023.e01638
6. Culpepper, J., Dung, L., Moh, M.: Design and analysis of hybrid indirect transmissions (HIT) for data gathering in wireless micro sensor networks. In: 2003 IEEE 18th Annual Workshop on Computer Communications, CCW 2003 - Proceedings, vol. 3 (2003). https://doi.org/10.1109/CCW.2003.1240800
7. Arafat, M.Y., Pan, S., Bak, E.: Distributed energy-efficient clustering and routing for wearable IoT enabled wireless body area networks. IEEE Access **11**, 5047–5061 (2023). https://doi.org/10.1109/ACCESS.2023.3236403
8. Khanna, A., Chaudhary, V., Gupta, S.H.: Design and analysis of energy efficient wireless body area network (WBAN) for health monitoring. In: Gavrilova, M.L., Tan, C.J.K. (eds.) Transactions on Computational Science XXXIII. LNCS, vol. 10990, pp. 25–39. Springer, Heidelberg (2018). https://doi.org/10.1007/978-3-662-58039-4_2
9. Ullah, Z., Ahmed, I., Ali, T., Ahmad, N., Niaz, F., Cao, Y.: Robust and efficient energy harvested-aware routing protocol with clustering approach in body area networks. IEEE Access **7**, 33906–33921 (2019). https://doi.org/10.1109/ACCESS.2019.2904322
10. Ullah, M.F., Khan, Z., Faisal, M., Rehman, H.U., Abbas, S., Mubarek, F.S.: An energy efficient and reliable routing scheme to enhance the stability period in wireless body area networks. Comput. Commun. **165**, 20–32 (2021). https://doi.org/10.1016/j.comcom.2020.10.017
11. Bilandi, N., Verma, H.K., Dhir, R.: Performance and evaluation of energy optimization techniques for wireless body area networks. Beni-Suef Univ. J. Basic Appl. Sci. **9**(1), 1–11 (2020). https://doi.org/10.1186/s43088-020-00064-w
12. Pradha, S.E., Moshika, A., Natarajan, B., Andal, K., Sambasivam, G., Shanmugam, M.: Scheduled access strategy for improving sensor node battery life time and delay analysis of wireless body area network. IEEE Access **10**, 3459–3468 (2022). https://doi.org/10.1109/ACCESS.2021.3139663
13. Firdous, S., Bibi, N., Wahid, M., Alhazmi, S.: Efficient clustering based routing for energy management in wireless sensor network-assisted Internet of Things. Electronics **11**(23), 3922 (2022). https://doi.org/10.3390/electronics11233922

14. Guo, K., Syed, S.A.S.: Energy efficiency based lifetime improvement for wireless body area network. IET Commun. **16**(7), 795–802 (2022). https://doi.org/10.1049/cmu2.12381

15. Vahedian, B., Mahmoudi-Nasr, P.: Toward energy-efficient communication protocol in wireless body area network: a dynamic scheduling policy approach. Int. J. Eng. Trans. A Basics **35**(1), 191–200 (2022). https://doi.org/10.5829/IJE.2022.35.01A.18

16. Oleiwi, S.S., Mohammed, G.N., Al-Barazanchi, I.: Mitigation of packet loss with end-to-end delay in wireless body area network applications. Int. J. Electr. Comput. Eng. **12**(1), 460–470 (2022). https://doi.org/10.11591/ijece.v12i1.pp460-470

17. Ghosh, B., Adhikary, S., Chattopadhyay, S., Choudhury, S.: Achieving energy efficiency and impact of SAR in a WBAN through optimal placement of the relay node. Wirel. Pers. Commun. **130**, 1861–1884 (2023). https://doi.org/10.1007/s11277-023-10361-z

18. Dass, R., et al.: A cluster-based energy-efficient secure optimal path-routing protocol for wireless body-area sensor networks. Sensors **23**(14), 6274 (2023). https://doi.org/10.3390/s23146274

19. Nilashi, M., et al.: Electroencephalography (EEG) eye state classification using learning vector quantization and bagged trees. Heliyon **9** (2023). https://doi.org/10.1016/j.e15258

20. Semadi, P.N.A., Pulungan, R.: Improving learning vector quantization using data reduction. Int. J. Adv. Intell. Informatics **5**(3), 218–229 (2019). https://doi.org/10.26555/ijain.v5i3.330

21. Nova, D., Estévez, P.A.: A review of learning vector quantization classifiers. Neural Comput. Appl. **25**(3–4), 511–524 (2014). https://doi.org/10.1007/s00521-013-1535-3

22. Al_Barazanchi, I., et al.: A survey on short-range WBAN communication; technical overview of several standard wireless technologies. Period. Eng. Nat. Sci. **9**(4), 877–885 (2021). https://doi.org/10.21533/pen.v9i4.2444

23. Chauhan, S., Singh, M., Aggarwal, A.K.: Cluster head selection in heterogeneous wireless sensor network using a new evolutionary algorithm. Wirel. Pers. Commun. **119**(1), 585–616 (2021). https://doi.org/10.1007/s11277-021-08225-5

24. Patro, S.G.K., Sahu, K.K.: Normalization: a preprocessing stage. Int. Adv. Res. J. Sci. Eng. Technol. **2**(3), 20–22 (2015)

25. Tawfeeq, M.A., Abdullah, M.Z.: Prolonging WSNs lifetime in IoT applications based on consistent algorithm. TELKOMNIKA Telecommun. Comput. Electron. Control **19**(3), 829–837 (2021)

26. Sachan, P., Ahmad, F.: Comparative analysis of protocols for WSN applications. Int. J. Innov. Res. Sci. Eng. Technol. **8**(6), 7352–7356 (2019)

The Impact of Mobility Models on Ad-Hoc Networks: A Review

Alyaa Safaa$^{(\boxtimes)}$ (iD) and Suhad Faisal Behadili$^{(\boxtimes)}$ (iD)

Department of Computer Science, College of Science, University of Baghdad, Baghdad, Iraq
{Alaia.abd2201m,Suhad.f}@sc.uobaghdad.edu.iq

Abstract. Nowadays, the effectiveness of ad hoc networks proves their importance in many applications, such as smart cities, the military, entertainment, health care, etc. This type of network has the ability to work without infrastructure, access points, or central control. A large field of researches focuses on the effects of various mobility models (MMs) on the functionality of ad hoc networks. The present study discusses the impact of different MMs on the performance of ad hoc networks. Methods of classifying MMs were discussed, with details for each model. The current study discusses how these models can impact the performance of various network components. It is concluded that choosing the appropriate mobility model will help to ensure that the performance evaluation is realistic and that the results can be generalized to real-world scenarios. Also, the enhancement could be focused on the terms of the most known parameters, such as packet delivery ratio (PDR), throughput, energy consumption, End-to-End delay (E2ED), and jitter. The importance of choosing the simulation tool is also discussed. The conclusion is presented at the end.

Keywords: Mobility Models · Ad-Hoc networks · Mobility

1 Introduction

Ad-hoc networks are wireless networks without a centralized Wireless Access Point (WAP), allowing devices to connect to each other directly. When two or more devices communicate with one another, these networks are instantly generated. In this kind of network, connections can be made at any time; hence, there is no central point [1]. Wireless mesh networks, wireless sensor networks, mobile ad-hoc networks, flying ad-hoc networks, and vehicular ad-hoc networks are a few examples of the various types of ad hoc networks. Ad-hoc networks have nodes that can enter and leave the network at any moment. These nodes can be portable electronics like mobile phones, laptops, or personal digital assistants [2]. In mobile ad-hoc networks (MANETs), a number of wireless mobile nodes (MNs) work together in a network that is only momentarily built [3], with no pre-existing infrastructure or centralized control. Figure 1 shows a brief description of the characteristics and features of MANET; some of them open doors for widespread use in a diverse range of fields, including the military, entertainment, emergency health care, and smart cities [4]. This characteristic can pose challenges for

© The Author(s), under exclusive license to Springer Nature Switzerland AG 2024
A. M. Al-Bakry et al. (Eds.): NTICT 2023, CCIS 2096, pp. 362–377, 2024.
https://doi.org/10.1007/978-3-031-62814-6_26

ad hoc services, like energy constraints, security threats, Quality of Services (QoS), scalability, and hidden or exposed terminals, which opens doors for researchers to study and conduct research to address these challenges in order to understand the benefits of ad hoc services. Figure 1 and 2 give a brief overview of the characteristics and benefits, respectively, of the MANET.

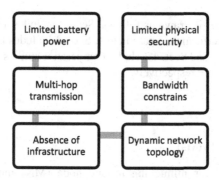

Fig. 1. Characteristics of MANET

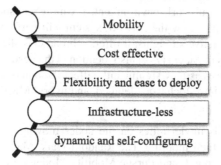

Fig. 2. Benefits of MANET

Mobility is the ability of nodes or devices to move within a network or system. It is a fundamental aspect of wireless networks and plays a crucial role in determining network efficiency and performance [5]. Ad-hoc networks are designed to support mobile devices and nodes that are constantly moving, allowing for dynamic and flexible communication between them. While mobility offers benefits in terms of flexibility and adaptability, it also introduces big challenges that need to be addressed in the design and implementation of ad hoc networks. Because it can cause frequent changes in the network topology and connectivity. This can lead to problems such as packet loss, increased latency, and decreased throughput. One of the methods used to address the challenges posed by mobility in ad hoc networks is mobility modeling. MMs are developed to capture and simulate the movement patterns of wireless nodes, allowing for a better understanding of the impact of node mobility on network efficiency. MMs are used to study and analyze the behavior of nodes in MANETs, including their movement patterns, speed, direction, and pause times [6]. By using MMs, researchers can simulate and analyze the impact of node mobility on factors such as packet loss, delay, and energy consumption in MANETs [2]. There is a trend towards developing MMs that capture human patterns of movement, both at an individual and group level, in order to improve the realism and applicability of these models [5]. Overall, mobility is a key factor in wireless networks, and MMs are developed to analyze and optimize network performance in the presence of node movement.

The rest of this paper is drawn as follows: Sect. 2 includes a background on MMs, simulation tools, and performance parameters; Sect. 3 includes the impact of MMs on the security of networks; and Sect. 4 includes comparative studies; Sects. 5 and 6 include discussion and conclusion, respectively.

2 Background

2.1 Mobility Models (MMs)

MMs are a technique for mimicking mobile node (MN) movement. Researchers classified MMs according to specific characteristics, as follows:

Classification Based on the Level of Realism. It is based on how well MMs anticipate the motions of MNs in a real-world environment. The more realistic the model, the better it will be able to anticipate how the network system will perform under actual conditions. Based on their relative strengths, it offers a broad classification of the currently used MMs. The authors in papers [7–9] also mentioned the need for more realistic models in network simulators, as they significantly affect the outcomes. This classification categorizes the models into stochastic, detailed, and hybrid MMs, as depicted in Fig. 3 [10].

Stochastic Models. This model relies on the nodes moving randomly and does not impose many limits on their mobility. The Random Walk, Manhattan Model, Random Direction, Random Waypoint, and Pursue Mobility Model are examples of stochastic models. These models can be applied in a variety of settings, including universities, hospitals, and pedestrian tracking in public areas. Each model has a similar type of movement pattern, and they all share some mobility aspects [8].

Detailed Models. Detailed models are highly customized and made with a particular scenario in mind. Lists of students and their exact placements on campuses are only a few examples of detailed models. Consider a student schedule for traveling to college by bus, car, or other means. All of these elements can be taken into account, and these MMs incorporate detailed information on how students might navigate around these various organizations in a particular setting [8].

Hybrid Models. Hybrid models primarily strive to stabilize the level of correctness of full MMs by allowing stochastic models to accommodate a variety of real-world conditions. As a result, the hybrid models are split into three groups based on a specific goal: group MMs, obstacle MMs, and trace-based MMs.

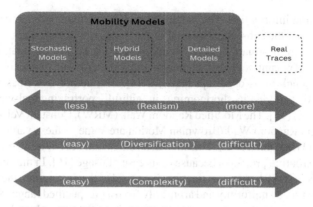

Fig. 3. Classification of MMs based on the level of realism

Classification Based on the Movement. Another classification was proposed by many authors, such as [4, 5, 11–13], which determined by the node movement, group, and entity MMs, implies that entity models imply random, geographic, and temporal dependency MMs. This classification is illustrated in Fig. 4 [4].

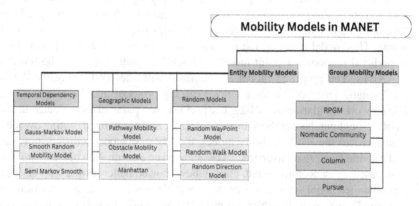

Fig. 4. A classification of MMs based on movement nature

Entity MMs. Models of entity mobility depict how specific MNs move independently of the other nodes.

1. Random models

   • Random Walk (RW): A MN can reach a new site by moving in any direction and speed, chosen uniformly from an interval, according to the RW. MNs cannot roam freely since they are limited to a small simulation zone during the time or distance range selected. This implies that each MN has limited transit time or space. A node arrives at the destination within time and distance constraints. A new direction and speed are predicted after that. New directions and speeds can be chosen without the existing information. The pattern of mobility is memoryless because, as was already said, the subsequent motion of the node does not depend on the previous node motion information. Thus, in addition to the lack of destination and pause intervals, RW cannot explain human mobility patterns due to unexpected motions such as quick turns or abrupt halts [4, 14]. The author describes variations of the RW that address its shortcomings in faithfully portraying real-world MANET scenarios in [15]. The Modified Random Walk (MRW), Constant Velocity Random Walk, and Random Walk/Brownian Motion are some of these variants.

- Random Waypoint (RWP): This model, an extension of RW, is the sole one used to simulate routing protocols because of its ease of usage [16]. In the simulated field, each MN in RWP has one destination. For travel to the destination, the MN chooses a constant speed uniformly and arbitrarily from a prespecified range. Nodes choose their own velocity and direction independently of the network nodes. The "pause time" (*Tpause*) is the period of time when the node stops after it reaches its destination. When *Tpause* = 0 persistent mobility occurs. After the pause ends, a node randomly selects and travels to another simulation field location. This repeats until the simulation ends. In the RWP paradigms, velocity and pause time control MN mobility. These factors affect ad-hoc network efficiency and stability. Small maximum velocity *Vmax* and long *Tpause* stabilise the ad-hoc network structure. Ad-hoc network topologies are highly dynamic when *Vmax* is high and *Tpause* is short. The RWP model, which includes distinct nodal speeds, can be employed in many mobility scenarios by adjusting these variables, especially *Vmax*. This model relies on the nodal speed estimation [4, 8, 15]. [17] presents a modified random waypoint mobility (MRWM) model based on the RWP for MANETs. The proposed model is compared with the RWP using the Ad-hoc On-Demand Distance Vector (AODV) routing protocol in a large wireless ad-hoc network simulation in (NS-2). The performance parameters evaluated include PDR, throughput, packet dropping ratio, E2ED, and normalized routing overhead. Compared to the RWP mobility model, the MRWM model reduced movement and provided a more realistic trace. [18] showed an examination of GM with RWP that, based on PDR, throughput, and latency, showed better results with GM with an unmanned aerial vehicles (UAV) network; NS-3 was used as a simulation tool.
- Random Direction (RD): MN in this model has the ability to walk near the border of the simulated area, where the target is chosen near this boundary. It was developed to fix a problem with the RW Model where the direction is randomly determined rather than the destination, and it was intended to reduce the density of the waves. Nodes are placed randomly in the simulation area assigned with angular direction in the range [0, 2π], and a uniform random velocity in a predefined range. When the node reaches the border of the simulated area, it pauses for a specific time, then a new movement angle and new velocity assign to the node, the process is repeated until the simulation ends [16, 19]. A more realistic rendition of the RD mobility model, which can produce non-homogeneous movement patterns, was provided by the authors in [20]. It was created by creating distinct, non-overlapping sections out of the simulated area. In [21], a 3D implementation of RD was proposed to improve the QoS in MANETs. The proposed model was compared with the other

3D MMs, RWP and Gauss Markov (GM), and it's been found that 3D-RD provided better throughput, delay, and PDR with the AODV routing protocol in MANETs.

2. Geographic models. Geographic models are used to simulate the movements of MNs, taking into account their geographical locations and movements [22].

- Pathway mobility model (PW): MNs move following a predetermined path on a map rather than at random in real life; therefore, their mobility is limited by geography. In the simulation field, the map is predefined [23]. In order to simulate the city map, the authors of [24] use a random graph. This graph can either be generated at random or precisely designed using a specific map of a real city. The city structures, intersections, and turns may be represented by the graph vertices, while the streets and motorways connecting any two junctions are its edges. Nodes are first positioned at random along the graph edges. The node then advances towards this destination, following the shortest path along the edges after a destination is randomly selected for each node. After arriving, the node stops for a set amount of time *Tpause* before selecting a new location for the following migration. This process repeated until the simulation ended [15]. By depicting global limitations on space, [24] developed a graphic-based mobility model and compared it to the RW for more realistic movements. The simulations showed that spatial constraints significantly affect ad-hoc performance.
- Obstacle Mobility model (OM): MN must alter its trajectory in order to avoid obstacles in its path. As a result, barriers affect MN movement behavior. Additionally, the obstructions have an effect on how radio waves propagate. For instance, the radio technology often could not carry the signal across obstacles in an indoor environment without significant attenuation. The radio is also susceptible to the radio shadowing effect when used outside. Consider both the impact of impediments on node mobility and radio propagation when incorporating them into MMs [23]. The authors in [25] proposed an OM model for modeling the nodes in terrains. Those are topographic maps of the actual world. It is based on a scenario where some obstacles are placed in the simulation field at random. These barriers are used to obstruct wireless signals as well as node movement. In order to simulate the location of buildings within a setting, such as a college campus, the barriers are positioned within a network area. The Voronoi diagram of obstacle vertices is used [26] to generate moving paths in addition to the obstacles. In [27], the authors addressed the challenges of describing the avoidance and curve movement patterns of the nodes in multivariate obstacle scenes by proposing a mobility model based on Hermite curves, filling the gap in curve MMs for complex scenarios. To create curve movement, the model mixes node movement with Hermite curves, and it adds perception parameters to avoid obstructions, creating a more realistic application landscape for MANET simulation.
- Manhattan Mobility model (MM): This model was developed in order to move the nodes in urban areas; it could be used in smart city infrastructures; there are numerous outdoor IoT applications [21]. It's based on a grid road topology where the streets are mapped vertically and horizontally. This model employs a probabilistic approach to determine the direction in which the nodes will travel at the intersection. The authors in [28] simulated Manhattan alongside other models in order

to provide more realistic movement patterns, and the model was used to simulate traffic events with a semaphore. Based on [29], with few restrictions, including probability for the changing direction, MM and RW are specific examples of the Correlated Random Walk (CRW)-based MMs. The velocity of the node at the previous time interval and the velocity of the node in front of it on the same street line serve as additional specifications and constraints for a node's velocity in the MM model. An enhanced MM model was proposed in [30] to enhance the performance of QoS. The proposed model provided an enhancement in the performance of packet overhead and packet loss as parameters that were used in the simulation environment. The same authors in [31] used the enhanced MM model and the Ant Colony Optimization (ACO) technique to come up with an ant-based bottleneck routing method for MANETs that would fix problems with routing overhead and delays. The results showed improvement in most of the known QoS parameters.

3. Temporal dependency models. Node mobility is influenced by prior motions and the physical principles of motion [23].

- Gauss-Markov Model (GM): GM was first presented for the modeling of personal communication system networks, which was later widely utilized for the simulation of ad-hoc networks for realistic behavior. The speed of a MN is correlated over time in this system; hence, speed and direction of a place in time t are determined using the speed and direction in t-1, so this model has the characteristic of temporal dependency. The Random GM Mobility Model eliminates the abrupt stops and abrupt turns that occur in the RW Mobility Model [18, 21]. In [32], an analysis was provided with variations in the number of MNs to test the performance metrics such as E2ED, PDR, and routing overhead in comparison to the Modified Gauss-Markov (MGM) and RW, it showed the dependency on the number of nodes in the performance of the parameters. A novel smooth Gauss-Semi-Markov mobility model was proposed in [33] for wireless networks. This model avoids all types of unrealistic node motions and assumes all realistic node movement statements. An Enhanced Gauss Markov (EGM) was proposed in [34] for Unmanned Aerial Ad hoc Networks (UAANETs) based on the existing Gauss Markov mobility model. It maintains uniform trajectories at the margins and lessens sudden halts and sharp twists in the boundaries. A critical analysis was performed by the author in [35] on the EGM, specifically focused on the impact of QoS parameters such as PDR, delay, and throughput. It looks at how varied distance and tuning parameter values affect the performance of routing protocols for both small and large groups of nodes.

- Smooth Random Mobility model (SRM): It is a memory-based model that creates fluid and realistic movement patterns by taking into account the physical rules of motion. In this model, MNs travel in a straight line for a predetermined amount of time at a steady speed. The maximum and minimum speeds of a node are limitations on its speed. At the conclusion of each time interval, node directions change randomly. The distribution of speed and direction changes correlates [36]. In [37] the authors proposed 3D Smooth Random Walk (3DSRW) mobility model to mimic the movement of Unmanned Aerial Vehicles in FANETs, the model provides better network performance in comparison with traditional 2D MMs in FANETs.

- Semi Markov Smooth (SMS): In a semi-Markov process, the time spent in each state is not exponentially distributed, but a transition matrix governs the state transitions. Thus, the SMS model is more realistic than memoryless models, where the duration of time in each state is exponentially dispersed. MNs move at a steady speed in a straight line for a predefined period of time. The maximum and minimum speeds of a node limit its speed. The travel direction of a node is changed at random at the end of each time interval. The distribution of direction changes and speed exhibits a correlation. Additionally, it is more flexible because it allows a variety of different mobility patterns to be generated by adjusting the model's parameters [38]. For Aeronautical Ad hoc Networks (AANET), a unique Semi-Markov Smooth mobility model (ISMS) was created in [39] to address the speed decay issue in the RWP model and the turning-state issue in the SMS model. This model divides the movement into seven states and guarantees a theoretically determined and simulated constant average speed.

Group MMs. Models of node movement that prefer to travel in groups and exhibit cooperative behavior are known as group-based MMs.

- Reference Point Group Mobility model (RPGM): This model is often used for group movement. It follows a logical center to simulate their movements. The reliability of a collection of nodes can be predicted by locating a reference point for every node. To find the true position of a node, a random movement vector is applied to the reference point location. The absolute positions of the reference points fluctuate according to the arbitrary mobility model. However, reference point positions within a group stay stable [4, 32, 40, 41]. Reference Point Group Mobility (RPGM) model design and performance study for MANETs were given in [42] The study discusses model performance with reactive routing protocols.
- Nomadic Community mobility model (NC): The NC model has been proposed to be an exception to the RPGM since each group member moves in the same direction towards a randomly chosen place inside a radius *r_max* circle centered on its reference point. This method gives each group a reference point and lets its members move around it while keeping it. The group members could move autonomously and according to a pre-simulation entity mobility model like RW [15, 43]. The authors in [44] used NC for bomb disposal in a building with robotic equipment and analyzed performance for military applications with two MMs, NC and RWP, using throughput, latency, and overhead. They found that routing protocols with RWP performed better than those with NC.
- Column Mobility model (CM): This model depicts MNs moving in a straight line with a fixed direction. An initial reference grid with a column of MNs is used to accomplish this strategy. Next, each MN is placed in the reference grid relative to its reference point. An entity mobility model lets the MN move arbitrarily around its reference point. Mimic human movement in a straight line. This method can mimic seeking and scanning, like the military robot detonating mines. The RPGM moves randomly, whereas this model moves in columns [4, 5, 23]. In contrast to NC, every

MN has its own reference point, while every group of MNs has a single reference point. CM motions are continuous, while NC motions are irregular [14]. A modified version of the CM mobility model was proposed in [15], where the nodes move parallel to the direction of movement.

- Pursue Mobility model (PM): The mobility model involves MNs pursuing a single MN or monitoring a target. It may show police following a fugitive. The pursued node can move according to any mobility model (such as RW or RWP), while the new position is calculated based on the old position, movement information, and the random vector, a random offset that can be used to adjust randomization for each node [4, 5, 23]. MNs move autonomously. All MNs attempt to reach the target node at the end of the movement time interval [14].

2.2 Simulation Tools

Many studies [41, 45, 46] provide a comprehensive survey of simulation tools for ad-hoc networks, offering researchers a valuable resource for selecting the most suitable tool for their specific needs. They highlight the features, capabilities, and limitations of each simulation tool, enabling researchers to make informed decisions when designing and evaluating algorithms and protocols. By presenting various simulation tools, the papers facilitate the advancement of research in ad-hoc networks by promoting the use of appropriate and reliable simulation environments. Finding the simulator that is most appropriate for the relevant job before performance evaluation is a crucial effort because each simulator offers distinct features in the areas of protocols and network models [45]. Table 1 lists the most popular simulators with the main features.

Table 1. Review for most popular simulators.

Simulator	Lang	Main Features
NS2	C++, OTcL	Open source, free, performance: Slow for large and sophisticated simulations, fast for small simulations
NS3	C++, Python	Open source, free, performance: fast for large and sophisticated simulations
OPNET	C++, Python API	Not open source, commercial, performance: Slow for large and sophisticated simulations, medium for small and straightforward simulations
OMNET	C++, NED, Python interface	Open source, free, performance: very fast for small simulation, fast for medium, and in sophisticated simulations slow

2.3 Performance Parameters

In ad-hoc networks, the performance parameters of MMs are quantitative measurements that are used to assess the realism and efficacy of MMs. Typically, simulations of ad-hoc networks are used to measure these properties. The most commonly used performance indicators are throughput, which is the speed at which data is successfully delivered through a network; PDR, the percentage of packets that are successfully delivered to their destination; E2ED, the amount of time it takes for a packet to go from where it started to where it is going; End-to-end latency fluctuation is known as jitter; moreover, Energy consumption is the quantity of energy used by network nodes. Performance metrics make it possible to assess the viability of various MMs for various applications and to compare and contrast them. For applications like real-time video streaming, for instance, a mobility model that achieves high throughput and low delay may be a good fit. For applications like sensor networks, a mobility model that uses less computing power and bandwidth might be more appropriate [46–48].

3 Security

The choice of MMs can have a significant impact on the security of the network architecture and protocols. It is known that mobility can present security challenges. Therefore, it is crucial to select MMs that are in line with the network security requirements and limitations. Researchers Give priority to models that provide a harmonious combination of predictability, resource utilization, and security [15, 49]. Improving QoS in MANET is crucial for secure and effective network communication as wireless networks become more prevalent in various areas [21].

4 Comparative Study

A comparison of recent papers that study the effectiveness of choosing MMs on the performance of ad-hoc networks using suitable simulation tools with a verification of environmental metrics by evaluating the performance parameters (Table 2).

Table 2. Studies according to performance metrics of ad-hoc network.

Year	Ref	MMs	Sim tools	QoS parameters	Results
2020	[50]	RWP in combination with GM (RGIM)	NS-2	PDR, jitter, throughput, and E2ED	In comparison with RWP, it showed better PDR and throughput, with less jitter and delay. Weakness: Low performance is reported at high node speeds
2020	[21]	RD-3D	NS-3	Throughput, delay, and PDR	In comparison with 3D for both RWP and GM, it showed better throughput, delay, and PDR. Weakness: it was limited with AODV
2020	[51]	RPGM, RWP, GM, MGM	NS-2	Throughput, E2ED, Drop packets, PDR, and NRL	Evaluation and analysis paper showed: RPGM performs the best in terms of routing protocol performance, specifically with the AODV routing protocol, when compared to other MMs such as RWP, GMM, and MGM. Weakness: was limited with AODV
2021	[9]	RWP, GM, RW-2D, RD-2D, constant velocity	NS-3	PDR, delay, and throughput	Analysis paper, GM has better results in PDR and delay than other models, whereas RWP and constant velocity showed better value in throughput among other models. Weakness: was limited with AODV
2022	[17]	MRWM	NS-2	Throughput, PDR, packet dropping rate E2ED, and NRO	More dynamic, which leads to more packet dropping, makes the model more realistic. Weakness: less throughput and PDR, which results in more NRO, and it Reduces mobility compared to the RWP
2022	[35]	MGM	NS-3	Throughput, delay, and PDR	Analysis paper showed that for a small set of nodes with both protocols, the minimum distance gives less delay. In large groups of nodes, with AODV, it gives high throughput and PDR using a large distance, but less delay with an average distance, whereas with DSDV, it gives better throughput, delay, and PDR using a maximum distance. Weakness: No flaws were noticed in the analysis

(continued)

Table 2. (*continued*)

Year	Ref	MMs	Sim tools	QoS parameters	Results
2022	[52]	MGM, Enhanced Modified Gauss-Markov (EMGM), RD-3D	NS-3	PDR, delay, and throughput	Performance analysis showed that EMGM performed better in PDR and delay in comparison with other models, whereas RD-3D showed better results in throughput. Weakness: No flaws were noticed in the analysis
2022	[53]	RWP, MM grid model	NS-2	Throughput, PDR, and Residual Energy	The DSR protocol shows highest throughput and PDR values, while the AODV protocol performs better with UDP/CBR traffic agents. The residual energy values for AODV and DSR are relative, but DSR has the highest value across all variations of node numbers. The highest residual energy value is observed in the variation of 100 nodes with RWP and TCP/FTP. Weakness: no flaws were noticed

5   Discussion

Many studies examine how MMs affect ad-hoc network operation. The performance of various routing protocols and other network elements can be assessed under various mobility situations by simulating node movement in an ad-hoc network using MMs. In ad-hoc network research, some of the most popular MMs are RWP, RD, MM model, OM model, GM model, and group mobility model. The use of these models depends on the field application. Some prominent articles evaluate existing parameters to increase network performance, while others improve the realistic to be more realistic in real life. In general, the mobility model should reflect the real-world situations where the ad-hoc network would be deployed; This will make the performance evaluation realistic and applicable to real-world situations.

6   Conclusion

The papers demonstrated that greater randomness in mobility led to increased unrealistic behavior. However, based on the recent studies, it can be concluded that random mobility models can be effectively utilized in FANETs due to the absence of obstacles in the majority of scenarios. Further, a combination of realistic and stochastic MMs (or any other two types) might be utilized to extract features from both and improve network performance with the realistic feature. The choice of a mobility model significantly influences the performance of ad-hoc networks. For more precise results, it is recommended to select models based on the performance difficulties associated with the necessary performance of an ad-hoc network.

References

1. Sultan, R.N., Flayyih, W.N., Ali, H.M.: Extending Wi-Fi direct single-group network to multi-group network based on android smartphone. Iraqi J. Sci. **64**(1), 419–438 (2023). https://doi.org/10.24996/ijs.2023.64.1.38
2. Agrawal, R., et al.: Classification and comparison of ad hoc networks: a review. Egyptian Inform. J. **24**(1), 1–25 (2023). https://doi.org/10.1016/j.eij.2022.10.004
3. Ismaeel, T.Z., Mohsen, D.R.: Estimation and improvement of routing protocol mobile Ad-Hoc network using fuzzy neural network. J. Eng. **22**(7), 142–163 (2016)
4. Abdullah, A.M.: Mobile Ad hoc networks: a survey of existing mobility models and routing protocols (2021)
5. Ribeiro, N., Mendes, P.: A survey on mobility models for wireless networks. (SITI Technical Report No. SITI-TR-11-01). COPELABS (SITI), Lusofona University (2011)
6. Alqaysi, M.K., Behadili, S.F.: A review of flow migration through mobile networks. Iraqi J. Sci. **63**(5), 2243–2261 (2022). https://doi.org/10.24996/ijs.2022.63.5.36
7. Issariyakul, T., Hossain, E.: Introduction to Network Simulator 2 (NS2), pp. 1–18. Springer, New York (2009). https://doi.org/10.1007/978-0-387-71760-9
8. Rathod, V.U., Gumaste, S.V.: Role of routing protocol in mobile Ad-Hoc network for performance of mobility models. In: 2023 IEEE 8th International Conference for Convergence in Technology (I2CT), pp. 1–6. IEEE (2023)
9. Khan, M.F., Das, I.: Analysis of various mobility models and their impact on QoS in MANET. In: Bansal, J.C., Paprzycki, M., Bianchini, M., Das, S. (eds.) Computationally Intelligent Systems and their Applications. SCI, vol. 950, pp. 131–141. Springer, Singapore (2021). https://doi.org/10.1007/978-981-16-0407-2_10
10. Sichitiu, M.L.: Mobility models for ad hoc networks. In: Misra, S., Woungang, I., Misra, S.C. (eds.) Guide to Wireless Ad Hoc Networks, pp. 237–254. Springer, London (2009). https://doi.org/10.1007/978-1-84800-328-6_10
11. Zheng, Q., Hong, X., Ray, S.: Recent advances in mobility modeling for mobile ad hoc network research. In: Proceedings of the 42nd Annual Southeast Regional Conference, pp. 70–75. Association for Computing Machinery (ACM). USA (2004)
12. Kim, B.S., Kim, K.H., Kim, K.I.: A survey on mobility support in wireless body area networks. Sensors (Switzerland) **17**(4), 797 (2017). https://doi.org/10.3390/s17040797
13. Wheeb, A., Al-Jamali, N.: Performance analysis of OLSR protocol in mobile Ad Hoc networks. Int. J. Interact. Mobile Technol. **16**(1), 106–119 (2022). https://doi.org/10.3991/IJIM.V16I01.26663
14. Safaei, B., et al.: Impacts of mobility models on RPL-based mobile IoT infrastructures: an evaluative comparison and survey. IEEE Access **8**, 167779–167829 (2020)
15. Roy, R.R.: Handbook of Mobile Ad Hoc Networks for Mobility Models, vol. 170. Springer, New York (2011). https://doi.org/10.1007/978-1-4419-6050-4
16. Alenazi, M.J., Abbas, S.O., Almowuena, S., Alsabaan, M.: RSSGM: recurrent self-similar Gauss–Markov mobility model. Electronics **9**(12), 2089 (2020)
17. Vyas, A., Suthar, M.: Development of modify random waypoint mobility model of routing protocol for the mobile ad-hoc network. Preprints (2022)
18. Naser, M., Wheeb, A.: Implementation of RWP and Gauss Markov mobility model for multi-UAV networks in search and rescue environment. Int. J. Interact. Mobile Technol. **16**(23), 125–137 (2022). https://doi.org/10.3991/ijim.v16i23.35559

19. Royer, E.M., Melliar-Smith, P.M., Moser, L.E.: An analysis of the optimum node density for ad hoc mobile networks. In: ICC 2001. IEEE International Conference on Communications, vol. 3, pp. 857–861. IEEE, Finland (2001)

20. Gloss, B., Scharf, M., Neubauer, D.: A more realistic random direction mobility model. Technical report, COST 290, 52 (2005)

21. Khan, M.F., Das, I.: Implementation of random direction-3D mobility model to achieve better QoS support in MANET. Int. J. Adv. Comput. Sci. Appl. 11(10), 195–203 (2020). https://doi.org/10.14569/IJACSA.2020.0111026

22. Camp, T., Boleng, J., Davies, V.: A survey of mobility models for ad hoc network research. Wirel. Commun. Mob. Comput. 2(5), 483–502 (2002)

23. Batabyal, S., Bhaumik, P.: Mobility models, traces and impact of mobility on opportunistic routing algorithms: a survey. IEEE Commun. Surv. Tutor. 17(3), 1679–1707 (2015). https://doi.org/10.1109/COMST.2015.2419819

24. Tian, J., Hahner, J., Becker, C., Stepanov, I., & Rothermel, K.: Graph-based mobility model for mobile ad hoc network simulation. In: Proceedings 35th Annual Simulation Symposium. SS 2002, pp. 337–344. IEEE, San Deigo (2002)

25. Jardosh, A., Belding-Royer, E. M., Almeroth, K. C., Suri, S.: Towards realistic mobility models for mobile ad hoc networks. In: Proceedings of the 9th Annual International Conference on Mobile Computing and Networking, pp. 217–229. Association for Computing Machinery, New York (2003). https://doi.org/10.1145/938985.939008

26. Ayawli, B.B.K., Mei, X., Shen, M., Appiah, A.Y., Kyeremeh, F.: Mobile robot path planning in dynamic environment using Voronoi diagram and computation geometry technique. Ieee Access 7, 86026–86040 (2019). https://doi.org/10.1109/ACCESS.2019.2925623

27. Shi, F., Wang, W., Zhang, Y.: A novel mobility model adapted to multi-obstacle scenarios: obstacle avoidance and smooth motion. Preprints (2023)

28. Martinez, F.J., Cano, J.C., Calafate, C.T., Manzoni, P.: Citymob: a mobility model pattern generator for VANETs. In: ICC Workshops-2008 IEEE International Conference on Communications Workshops, pp. 370–374. IEEE, Beijing (2008)

29. Bandyopadhyay, S., Coyle, E.J., Falck, T.: Stochastic properties of mobility models in mobile ad hoc networks. IEEE Trans. Mob. Comput. 6(11), 1218–1229 (2007). https://doi.org/10.1109/TMC.2007.1014

30. Kour, S., Ubhi, J.S.: Performance analysis of mobile nodes in mobile ad-hoc networks using enhanced manhattan mobility model. JSIR 78(02), 69–72 (2019)

31. Kour, S., Ubhi, J.S., Singh, M.: QoS improvement using enhanced Manhattan mobility model on proposed ant colony optimization technique in MANETs. JSIR 82(06), 616–628 (2023)

32. Nurwarsito, H., Putra, E.S.: Analysis of OLSR routing protocol performance based on Gauss-Markov mobility and random walk in mobile Ad-Hoc network (MANET). J. Inf. Technol. Comput. Sci. 7(3), 183–195 (2023)

33. Zhang, H., Zheng, B., Zhuo, K.: A novel smooth Gauss–Semi-Markov mobility model for mobile wireless networks. Frequenz 69(5–6), 245–262 (2015)

34. Biomo, J.D.M.M., Kunz, T., St-Hilaire, M.: An enhanced Gauss-Markov mobility model for simulations of unmanned aerial ad hoc networks. In: 2014 7th IFIP Wireless and Mobile Networking Conference (WMNC), pp. 1–8. IEEE, Vilamoura (2014)

35. Khan, M.F., Das, I.: Critical analysis of modified Gauss Markov mobility model using varying values of parameters to check the impact of QoS In MANET. J. Eng. Sci. Technol. 17(5), 3393–3409 (2022)

36. Bettstetter, C.: Smooth is better than sharp: a random mobility model for simulation of wireless networks. In: Proceedings of the 4th ACM International Workshop on Modeling, Analysis and Simulation of Wireless and Mobile Systems, New York, pp. 19–27 (2001)

37. Lin, N., Gao, F., Zhao, L., Al-Dubai, A., Tan, Z.: A 3D smooth random walk mobility model for FANETs. In: 2019 IEEE 21st International Conference on High Performance Computing and Communications; IEEE 17th International Conference on Smart City; IEEE 5th International Conference on Data Science and Systems, pp. 460–467. IEEE, Zhangjiajie (2019). https://doi.org/10.1109/HPCC/SmartCity/DSS.2019.00075

38. Roy, R.R.: Semi-Markov smooth mobility. In: Roy, R.R. (ed.) Handbook of Mobile Ad Hoc Networks for Mobility Models, pp. 345–377. Springer, Boston (2011). https://doi.org/10.1007/978-1-4419-6050-4_11

39. Li, J., Lei, L., Liu, W., Shen, Y., Zhu, G.: An improved semi-Markov smooth mobility model for aeronautical ad hoc networks. In: 2012 8th International Conference on Wireless Communications, Networking and Mobile Computing, pp. 1–4. IEEE, Shanghai (2012)

40. Sakthivel, T., Balaram, A.: The impact of mobility models on geographic routing in multi-hop wireless networks and extensions–a survey. Int. J. Comput. Netw. Appl. **8**(5), 634–670 (2021). https://doi.org/10.22247/ijcna/2021/209993

41. Mahiddin, N.A., Affandi, F.F.M., Mohamad, Z.: A review on mobility models in disaster area scenario. Int. J. Adv. Technol. Eng. Explor. **8**(80), 848 (2021). https://doi.org/10.19101/IJA TEE.2021.874084

42. Dorge, P.D., Meshram, S.L.: Design and performance analysis of reference point group mobility model for mobile ad hoc network. In: 2018 First International Conference on Secure Cyber Computing and Communication (ICSCCC), pp. 51–56. IEEE, Jalandhar (2018)

43. Sánchez, M., Manzoni, P.: ANEJOS: a Java based simulator for ad hoc networks. Futur. Gener. Comput. Syst. **17**(5), 573–583 (2001). https://doi.org/10.1007/978-981-19-2397-5_19

44. Usha, R., Premananda, B.S., Reddy, K.V.: Performance analysis of MANET routing protocols for military applications. In: 2017 International Conference on Intelligent Computing and Control Systems (ICICCS), pp. 1063–1068. IEEE, Madurai (2017)

45. Kang, S., Aldwairi, M., Kim, K.I.: A survey on network simulators in three-dimensional wireless ad hoc and sensor networks. Int. J. Distrib. Sensor Netw. **12**(10), (2016). https://doi.org/10.1177/1550147716664740

46. Wheeb, H., Nordin, R., Samah, A.A., Alsharif, M.H., Khan, M.A.: Topology-based routing protocols and mobility models for flying Ad Hoc networks: a contemporary review and future research directions. Drones **6**(1), 9 (2022)

47. Haile, H., Grinnemo, K.J., Ferlin, S., Hurtig, P., Brunstrom, A.: End-to-end congestion control approaches for high throughput and low delay in 4G/5G cellular networks. Comput. Netw. **186**, 107692 (2020). https://doi.org/10.1016/j.comnet.2020.107692

48. Mujahid, M.A.A., Bakar, K.B.A., Darwish, T.S., Zuhra, F., Ejaz, M.A., Sahar, G.: Emergency messages dissemination challenges through connected vehicles for efficient intelligent transportation systems: a review. Baghdad Sci. J. **17**(4), 1304 (2020). https://doi.org/10.21123/bsj.2020.17.4.1304

49. Naif, O., Mohammed, I.: WOAIP: wireless optimization algorithm for indoor placement based on binary particle swarm optimization (BPSO). Baghdad Sci. J. **19**(3), 605–618 (2022). https://doi.org/10.21123/BSJ.2022.19.3.0605

50. Kaur, P., Singh, A., Gill, S.S.: RGIM: an integrated approach to improve QoS in AODV, DSR and DSDV routing protocols for FANETS using the chain mobility model. Comput. J. **63**(10), 1500–1512 (2020). https://doi.org/10.1093/comjnl/bxaa040

51. Albu-Salih, A., Al–Abbas, G.: Performance evaluation of mobility models over UDP traffic pattern for MANET Using NS-2. Baghdad Sci. J. **18**(1), 0175–0175 (2021). https://doi.org/10.21123/bsj.2020.18.1.0175

52. Khan, M.F., Das, I.: Performance analysis of improved mobility models to check their impact on QoS in MANET. In: Shaw, R.N., Das, S., Piuri, V., Bianchini, M. (eds.) Advanced Computing and Intelligent Technologies, vol. 914, pp. 173–182. Springer, Singapore (2022). https://doi.org/10.1007/978-981-19-2980-9_15

53. Hasanah, Barokatun, Mahardika, Indra, Wicaksono, Himawan, Farid, Mifta Nur: Analysis of the effect of mobility model and traffic agent variations on routing protocol performance on mobile Ad-Hoc network. In: Yang, X.S., Sherratt, S., Dey, N., Joshi, A. (eds.) Proceedings of Seventh International Congress on Information and Communication Technology: ICICT 2022, vol. 4, pp. 191–200. Springer, Singapore (2023). https://doi.org/10.1007/978-981-19-2397-5_19

Authentication System Based on Zero-Knowledge Proof Employing the Rabin Cryptosystem and a Secret Sharing Schema

Sajjad Mohammed Shlaka[✉] and Hala Bahjat Abdul Wahab

Department of Computer Science, University of Technology, Baghdad, Iraq
cs.21.17@grad.uotechnology.edu.iq,
Hala.B.AbdulWahab@uotechnology.edu.iq

Abstract. Passwords play a significant role in the authentication process for web applications. These applications are widely employed to deliver a diverse range of crucial services, making their security a top priority. Due to their regular usage and essential nature, passwords are very vulnerable to theft or unauthorized access through the process of guesswork. This paper presents a groundbreaking authentication system aimed at tackling the security issues commonly linked to traditional password-based authentication in web applications. The system leverages a combination of zero-knowledge proof (ZKP), the Rabin cryptosystem, and a secret sharing schema to bolster security and safeguard user privacy. During the registration phase, user passwords are fragmented into shares, distributed across multiple databases, and subsequently reassembled during authentication as an additional layer of security. ZKP facilitates password verification without the need to transmit sensitive information, while the Rabin cryptosystem adds an additional layer of complexity to key generation. The proposed approach offers heightened security and mitigates the risk of data breaches. NIST testing has confirmed the randomness of the generated keys, while time testing has demonstrated their efficient performance. The results underscore the system's effectiveness in delivering secure and efficient web application authentication while ensuring the protection of user data.

Keywords: zero-knowledge proof · Rabin cryptosystem · secret sharing schema · authentication · password · security

1 Introduction

The World Wide Web has been a tremendous success. Many web apps have been developed today in a variety of industries, including banking and finance, e-commerce, education, the government, and entertainment. As a result, most companies and organizations today have websites to cater to their clients' and customers' needs and desires [1, 2].

Many of these customers need to have mail accounts that require authentication before signing in and accessing resources. However, relatively few people are aware that a lot of those companies and organizations employ the insecure PAP (Password Authentication Protocol) to authenticate users.

© The Author(s), under exclusive license to Springer Nature Switzerland AG 2024
A. M. Al-Bakry et al. (Eds.): NTICT 2023, CCIS 2096, pp. 378–394, 2024.
https://doi.org/10.1007/978-3-031-62814-6_27

In the case of PAP, even though the password and the corresponding username are stored on the server in a hashed format, which makes them less vulnerable to attacks, the username-password pair used by PAP travels clear on the wire, making it vulnerable to attacks like eavesdropping and packet sniffing that will easily reveal the sensitive data to the intruder [3]. As a result, maintaining the security of websites has become a crucial component in gaining customers' confidence [4].

Despite years of extensive studies into alternative authentication technologies, passwords still dominate the authentication notion. This is mostly attributable to an unmatched mix of security, deployability, and usability [5].

The objective of organization-set password policies is to enhance the security of passwords used by specifying various requirements during the selection and usage process. These policies aim to protect user account data and prevent unauthorized users from engaging in malicious activities. This could lead to network or service-wide compromise and potential harm to the company, raising liability concerns. By enforcing strong passwords that are difficult to guess, password policies can mitigate the risk of compromised accounts due to dictionary attacks [6].

Authentication is a communication procedure that gives the recipient confidence in the veracity of incoming communications. The network services offered by the remote server allow for the network-wide sharing of resources across the hosts. On the one hand, the remote server must be able to authenticate users before offering such services; otherwise, an adversary may login using the identity of a legitimate user and get access to services. On the other side, the user should be able to verify the remote server before adopting the services it offers, or a hacker may create a false server [7]. Finding a balance between simplicity and security, as shown in Fig. 1, is vital for authentication [8].

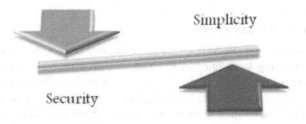

Fig. 1. Simplicity vs Security [8].

These systems presume that the user is the only one who is aware of the private information (password), that he keeps secret [3]. The systems suffered from significant security challenges as they involved transmitting sensitive information over public networks if it was assigned to a particular location. This made them highly vulnerable to attacks by malicious individuals and potentially resulted in data loss. There is a need for a cryptographic challenge-response authentication system [9], like a firewall [10].

1.1 Preliminaries

In this section, we will discuss the theoretical background of the used techniques.

1.1.1 Zero-Knowledge Proof

Zero-knowledge protocols (ZKP) provide the execution of fundamental cryptographic operations, such as authentication, identification, and key exchange, without revealing any confidential information during the process of active communication. One advantage of these protocols is that they have low computational requirements, making them ideal for implementation in embedded systems and smart card applications. In the context of zero-knowledge proofs, the prover (referred to as A) and the verifier (referred to as B) partake in an interactive protocol to ascertain the possession of a credential by the prover, such as a credit card number, while ensuring that the precise value of the credential remains undisclosed to the verifier. A zero-knowledge proof is necessary for authentication in this scenario to fulfill the following criteria [6, 11]:

- **Completeness:** where a truthful verifier must always be able to prove the validity of the statement to an impartial verifier.
- **Soundness:** where it is impossible to manipulate the result to make an incorrect statement appear accurate to the verifier with only a small chance of success.
- **Zero-knowledge:** where the verifier knows only that the statement is true if it is valid, with no information being released on the specifics of the document.

The Feige-Fiat-Shamir identification system was created in 1988 by Uriel Feige, Amos Fiat, and Adi Shamir as a parallel zero-knowledge proof in cryptography. It enables Peggy (the prover) to demonstrate to Victor (the prover) that she has access to sensitive information without giving Victor that information, like previous zero-knowledge proofs. The quantity of communications between Peggy and Victor is nevertheless constrained by the Feige-Fiat-Shamir identification method, which makes use of parallel verification and modular arithmetic [12].

1.1.2 The Rabin Cryptosystem

The Rabin cryptosystem (RCS) is a type of public-key cryptography that relies on the difficulty of factoring large integers to ensure the security of its encryption. Unlike the RSA cryptosystem, which is also based on factoring, the Rabin cryptosystem has the advantage that the underlying problem has been proven to be as hard as integer factorization. This means that breaking the encryption requires an attacker to factor in large integers, which is currently believed to be a computationally difficult problem.

One disadvantage of the Rabin cryptosystem is that each output of the Rabin function can be generated by any of four possible inputs. This implies that in order to determine which of the four potential inputs is the actual plaintext, the decryption process must be made more difficult. As long as the keys are kept a secret, this does not materially affect the system's security.

The Rabin cryptosystem, introduced by Michael O. Rabin in January 1979, marked a significant milestone in cryptography. It represented the first asymmetric cryptosystem, demonstrating that the computational difficulty of recovering the complete plaintext from the ciphertext was equivalent to the challenging problem of factoring. Over the years, the Rabin cryptosystem has found applications in diverse fields, including digital signatures, key exchange protocols, and ensuring secure communications [13, 14].

1.1.3 Secret Sharing Schema

The secret sharing (SS) schema is a crucial cryptographic technique that enables a group of participants to share a secret in a way that prevents any one person from having complete knowledge of the secret but allows a minimum number of members to cooperate to reconstruct the secret. This offers a mechanism to prevent private data from being abused or maliciously altered by a single user. It may be utilized in a number of functions, including voting, access control, and auctions.

In 1979, Shamir put forward the initial theory of (k, n) secret sharing, which divides a secret into n bits of data, S1, S2, and Sn. At least k pieces of data must be merged in order to reveal the secret S. Based on this idea, it depicts a secret S that is kept by n different persons, each of which can only possess one fragment of it (represented as S1–Sn). The secret can only be discovered if k or more individuals cooperate [15, 16].

1.2 Rising Use of Web Applications

Web-based apps are now being used more often. This is mainly because web applications have a number of advantages over conventional software.

- Compatibility with the devices, web browsers, and operating systems.
- The capability of carrying out tasks that were previously carried out by host-based software.
- Development of new web applications, languages, and processes has given way to the development of new dynamic applications (i.e., AJAX).
- Easy to use and look more presentable and attractive.
- Does not need additional hardware or software (installation, etc.) configuration.
- Data security exists because the information is kept on a centralized server rather than multiple PCs.
- A custom-built web application is unquestionably less expensive than ready-made ones and offers more productivity and less maintenance.

Given this, we must concentrate on enhancing web application security while preserving its usability in terms of both development and implementation [17].

1.2.1 Security Issues Relating to Web Applications

For web-based apps, there are many weak and harmful viruses. This covers both web-specific (like cross-site scripting) and general (like password sniffing) attacks, all of which put the user at risk of experiencing identity theft. This indicates that the security vulnerability on the network is a much more significant concern. While some solutions may be investigated or discovered, they will not be adequate. One potential limitation of SSL in ensuring the complete security of an application arises from the emergence of novel concepts, such as cloud computing and the intricate nature of attack vectors [18].

1.2.2 Authentication on the Web

An online application that asks a user for a login and password is considered the traditional method of authenticating. The server then responds to the request after receiving the login information and password (see Fig. 2). An HTTPS protocol is used if security is required. Even though the password is stored in a hashed and salted format, it is still delivered over the Internet, where servers can read it. As an illustration, a person often accesses a website whose certificate the user's browser is unable to verify. While the certificate's verification failed, the browser often allows the user to accept the certificate. Also, in order to fully utilize HTTPS technology, the user must purchase and install a certificate issued by an established certificate authority on their browser. The installation of a certificate on a web browser occurs rarely and is vulnerable to several potential threats [19].

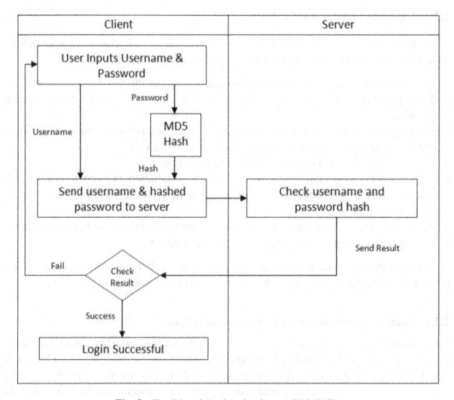

Fig. 2. Traditional Authentication on Web [17].

1.2.3 Problem of Sending Over Password Hashes

As shown in Fig. 3, the concern with the transmission of password hashes stems from the fact that the transmitted information alone is sufficient for a hacker to impersonate an authorised user. In the context of conventional authentication, the user transmits their username and a hashed representation of their password. In the event that a hacker successfully intercepts the username and hashed password, they have the ability to access the user's account at any given moment. Nevertheless, a more urgent concern regarding the transmission of password hashes pertains to the vulnerability associated with disclosing a password in its original, unencrypted form. An instance of concern arises when an individual employs an identical password across many services, resulting in adverse consequences. The hacker has the capability to obtain the user's password hash through the process of network sniffing [17].

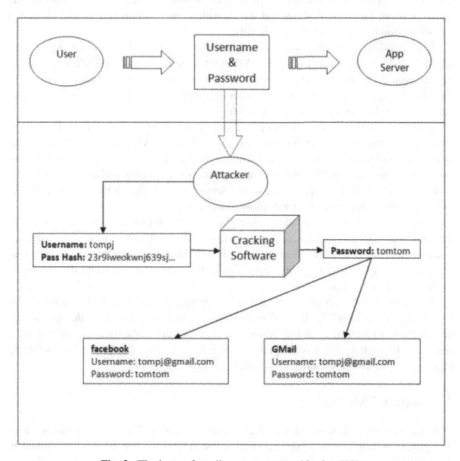

Fig. 3. The issue of sending over password hashes [17].

2 Related Works

This section will discuss some of the related works that employed zero-knowledge proof with different techniques for authentication systems.

In [19], authors proposed zero-knowledge proof authentication (ZKP) based on isomorphic graphs for safe transactions on the web. In their system, the user's web browser performed the majority of calculations without disclosing the password or login information. Instead, it generates random isomorphic graphs and permutation functions depending on the user's login and password, which may then be exchanged. Their approach necessitates trusting the server.

In [17], the authors proposed using Python-based scripts on the web to implement zero-knowledge authentication with zero knowledge (ZKA_wzk) to offer higher levels of security by proving the password of the user without sending over the actual password. The author implemented the algorithm in three phases: initialization, registration, and authentication. The authors think that the algorithm can be improved, so she suggested a modification by editing Python script files.

In [3], authors proposed two protocols based on the idea of zero-knowledge proof: ZK-PAP and ZK-PAP with PKE. ZK-PAP enables the user to prove his/her possession of the password to the authentication server without disclosing the actual password. On the other hand, ZK-PAP with PKE enhances the protocol security and facilitates mutual authentication between the client and server.

In [18], the authors proposed using the zero-knowledge protocol with the RSA cryptography algorithm for authentication on the web (Z-RSA) to let the server set the confidence and security levels for authentication. The system consists of two phases: the registration phase and the authentication phase for registering and authenticating the user by demonstrating his password to the server without disclosing it. Compared to other systems of authentication, the algorithm is complex.

In [12], the authors proposed a secure SSL protocol with zero-knowledge proof for authenticating users to prevent hackers from using fake certificates and masquerading as clients or servers. The system used the Feige-Fiat-Shamir identification scheme, wherein one party proves knowing of the certificate to the other party without disclosing its contents, and vice versa.

In [6], the authors proposed an authentication system based on ZKP, the novel 6D-Hyper chaotic, and RSA for the web to prevent passwords from being stolen or guessed due to their frequent use. Initialization, registration, and authentication are the three stages of the system that work together to solidly construct the system by producing high-quality random numbers and storing them in a chaotic list.

3 Proposed Method

Web applications still need a design that increases security. This paper proposes a new authentication solution to solve one of the most important Top 10 risks explained in the OWASP report [6] by using the ZKP protocol with RCS and SS schema. ZKP is a strong password authentication protocol that allows the user to demonstrate knowledge of the password associated with their account without revealing the actual password. This confirms that the user is the legitimate account holder.

Our contribution to this paper will be as follows:

- A zero-knowledge proof (ZKP) protocol is used to build a secure authentication system. This protocol is responsible for creating security challenges and responses for exchanging the password through a communication channel between the clients and the website without disclosing any secret information during active communication.
- Using the Rabin cryptosystem makes the system more complicated and unpredictable by generating a pair of large keys as prime numbers, which makes it hard for an attacker to break them down into their integer factorization.
- Secret sharing splits the password into shares; the system keeps its share for the description process and distributes the other claims on different databases, so it's hard for the attacker to know the whole password, which makes the verification process between the user and the system more secure.

In our proposal, whether someone understands the formula of the process (the attacker) or not (the user), ZKP, RCS, and SS schema offer enhanced security and privacy features compared to traditional authentication systems by providing an additional level of protection to the protocol, allowing no password hashes to be stored on a single database, and reducing user interaction with the system. The system is divided into two phases: the registration phase and the authentication phase. The three methods work in the system as follows:

3.1 Registration Phase

During the registration phase, shown in Fig. 4, the user provides their username and password to the system. The system generates a key pair. The system proves knowledge of the password without revealing it. The password is split into shares and distributed to different databases. Algorithm (1) describes the full process. The following is a step-by-step procedure for the registration process.

Algorithm (1): The registration process

Input: User's username (U_username), User's password (U_password).

Output: User registration completed with enhanced security measures.

Steps:

1. Generate a key pair (n, p, q) using the Rabin cryptosystem for the user.

- n: Public key

- p, q: Private key

2. Securely store the user's private key (p, q).

3. Associate the public key (n) with the user's account.

4. Perform a zero-knowledge proof protocol with the user to validate their knowledge of the password (U_password) without revealing it.

- This ensures that the user possesses the correct password without transmitting it in plaintext.

5. Store the proofs generated during the zero-knowledge proof process in the database.

6. Split the user's password (U_password) into shares (s1, s2, sn) using a secret sharing scheme.

- s1, s2, sn: Shares of the user's password

7. Distribute the password shares (s1, s2, sn) to different databases within the system.

End of registration.

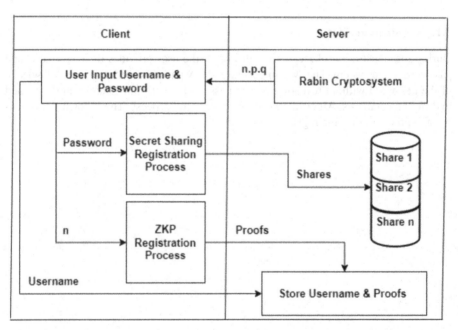

Fig. 4. The registration phase.

3.2 Authentication Phase

In the authentication phase, shown in Fig. 5, the user provides their username and password. The system validates the username and generates a challenge for the user. The user signs the challenge. The system requests password shares from the databases. If the passwords match, the user will login successfully. Algorithm (2) describes the full process. The following is a step-by-step procedure for the authentication process.

Algorithm (2): The authentication process
Input: User's username (U_username), User's password (U_password).
Output: User authentication completed with enhanced security measures.
Steps:
1. User requests the login page and provides their username (U_username) and password (U_password).
2. The system validates the provided username (U_username).
3. The system generates a challenge (C) as an additional layer of security.
4. The system sends the challenge (C) to the user.
5. User performs a zero-knowledge proof to validate their knowledge of the password (U_password) without revealing it.
6. User sends the proof to the system.
7. The system verifies the proof.
8. User uses their private key (p, q) to sign the challenge (C) provided by the system.
9. User sends the signature to the system.
10. The system verifies the signature using the user's public key (n).
11. The system requests password shares (s1, s2, sn) from the databases holding them.
12. The shares (s1, s2, sn) are combined to reconstruct the original password.
13. The reconstructed password is compared with the user's provided password share.
14. If the passwords match, the user is granted access and can login successfully.
End of authentication.

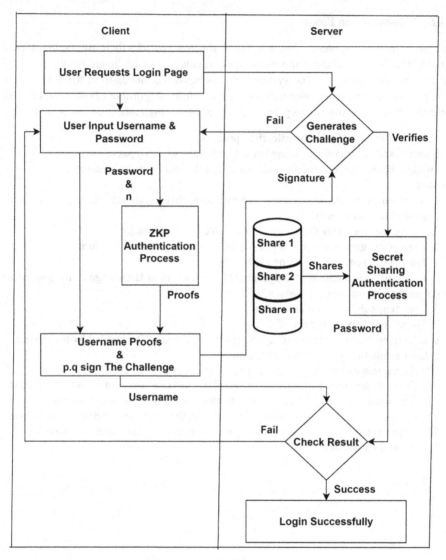

Fig. 5. The authentication phase.

4 Experimental Results

This proposed framework is specifically designed for e-banking applications that are accessed through web-based platforms. The system ensures the security and authenticity of the framework's most critical step, which pertains to user login or registration. Subsequently, a proposed protocol suggests employing a zero-knowledge proof (ZKP) protocol combined with an RCS algorithm and SS schema. This combination serves to safeguard passwords that are susceptible to being stolen or exploited, thereby ensuring their non-disclosure and secure usage.

In this system, the secret sharing schema operates by first transforming the password into a digital ASCII value and subsequently splitting it into shares. These shares are then distributed across multiple secure locations within a database. The system establishes a threshold value (k) that specifies the minimum number of shares required to reconstruct the password. For instance, if the threshold is set to $k = 3$, then any grouping of 3 or more shares enables the reconstruction of the original password.

The results of the proposed research demonstrated the viability of the method employed as well as the accomplishment of the safeguards set up using the CIA [20], which is a crucial foundation for protecting websites and apps that everyone on the internet utilizes. The system outcomes demonstrated an improved user experience while concurrently upholding security measures.

4.1 Rabin Key Generation Results

In this subsection, we present the results of the Rabin key generation process, which is a crucial component of our proposed authentication system. The Rabin key pairs, consisting of the public key (n) and private key (p, q), were generated to enhance the security of our authentication system. We generated a set of samples for evaluation purposes, each of which includes the public key (n) and private key (p, q). Table 1 below explains the results:

Table 1. Samples of the Rabin keys.

No.	public	p	q
1	3d294a6c886f071004eca54623199fb80b48721f25c8aa40e7a21a9b279a5b61	8f7d38e8b3c242f9ac363347755ea891	6d1e484a280166b92cefc2158d3d6dd1
2	483e8936cfdb4ccb8c5d98909402ff9be8d2df6a2ae9980c2a363545b00cdb43	ecbf6b97a7430a397646618120f07783	4e1e81c2238414c3df245d49b5ad0141
3	574be72e548293794032375bb8a6aab19e4f0e4e0c97a22546241a4d1933b667	695e30f847b115f7666e690c83a44b2b	d4180e15d20f6a2eac3b82bf34b333b5
4	2fbdc5b346af18082ee50bbd0f55a43fdd6a1c25087961a55619779d29bf55c9	4ae951aa2e7529b15a4fee71eea8d191	a326524123afd50c0b96cb29040fa4b9
5	152bf1c9ce393e4d1e5f30bae556a3244b829da1c650f6dfd91f1508b7778e75	386110397b0185e488f73b364cd2049d	60224342b67f9b3ed99e36238edb4db9

The sample of Rabin key pairs mentioned above exemplified the effective generation of cryptographic keys through the utilization of the Rabin cryptosystem. These keys are vital to our authentication system, contributing to its overall security and robustness.

4.2 NIST Testing

All encryption algorithms strive to provide outcomes that are as unexpected and random as possible. The National Institute of Standards and Technology (NIST) of the United States Department of Commerce is used to assess randomness [6]. Our system creates a key pair for the user consisting of the public key (n) and the private key (p, q). So, we tested all the keys. Table 2 and Table 3 display the testing outcomes.

Our key pairs passed 12 out of 16 NIST randomness tests, confirming the system's ability to generate secure public keys.

Similarly, the private key pairs passed 12 out of 16 NIST randomness tests, demonstrating the system's ability to generate secure private keys.

Table 2. NIST testing results of public key (n).

Index	Test	P-value	Results
1	Frequency Test (Monobit)	0.080	Pass
2	Frequency Test within a Block	0.209	Pass
3	Run Test	0.655	Pass
4	Longest Run of Ones in a Block	0.327	Pass
5	Binary Matrix Rank Test	−1.0	Fail
6	Discrete Fourier Transform (Spectral) Test	0.135	Pass
7	Non-Overlapping Template Matching Test	0.999	Pass
8	Overlapping Template Matching Test	nan	Fail
9	Maurer's Universal Statistical test	−1.0	Fail
10	Linear Complexity Test	−1.0	Fail
11	Serial test	0.498	Pass
12	Approximate Entropy Test	1.0	Pass
13	Cummulative Sums (Forward) Test	0.105	Pass
14	Cummulative Sums (Reverse) Test	0.121	Pass
15	Random Excursions Test:	2.25	Pass
16	Random Excursions Variant Test	5	Pass

Table 3. NIST testing results of private key (p, q).

Index	Test	private key p		private key p	
		P-value	Results	P-value	Results
1	Frequency Test (Monobit)	0.723	Pass	0.859	Pass
2	Frequency Test within a Block	0.723	Pass	0.859	Pass
3	Run Test	0.587	Pass	0.593	Pass
4	Longest Run of Ones in a Block	0.908	Pass	0.436	Pass
5	Binary Matrix Rank Test	−1.0	Fail	−1.0	Fail
6	Discrete Fourier Transform (Spectral) Test	0.871	Pass	0.330	Pass
7	Non-Overlapping Template Matching Test	0.999	Pass	0.999	Pass
8	Overlapping Template Matching Test	nan	Fail	nan	Fail
9	Maurer's Universal Statistical test	−1.0	Fail	−1.0	Fail
10	Linear Complexity Test	−1.0	Fail	−1.0	Fail
11	Serial test	0.498	Pass	0.498	Pass
12	Approximate Entropy Test	1.0	Pass	1.0	Pass
13	Cummulative Sums (Forward) Test	0.818	Pass	0.574	Pass
14	Cummulative Sums (Reverse) Test	0.984	Pass	0.737	Pass
15	Random Excursions Test:	3.400	Pass	15.0	Pass
16	Random Excursions Variant Test	7	Pass	4	Pass

4.3 Time Testing

In the context of our proposed system, prioritizing the reduction of time-related costs and expediting the verification process enhances overall performance, resulting in greater client satisfaction and heightened security. The following Table 4 illustrates the average time consumption for each of the implemented techniques:

Table 4. Time testing results of our system.

The technique	Type of time	Average time
Zero-knowledge proof	Challenge time	(1.23838383) s
The Rabin cryptosystem	Key generation time	(0.06778860) s
Secret sharing schema	Split and reconstructed time	(0.00343704) s

Our system demonstrates efficient performance, with minimal time consumption for key generation, challenge-response, and secret sharing operations.

4.4 Security of Our Model

- ZKP comprises fundamental principles. Furthermore, our model could function as a security system and incorporate additional cybersecurity concepts, as explained below:
- **Zero-knowledge:** data exchanged during authentication over the network holds no value for attackers. The attackers do not know what is going on.
- **Confidentiality:** password exchanging through communication between the clients and the system without disclosing any sensitive information.
- **Integrity:** only users who pre-register their information within the system can login; our system knows who is authorized and unauthorized to access and modify.
- **Availability:** our system ensures high availability and protects against denial-of-service attacks. Secret sharing distributes the passwords to different databases to protect them, and it does not have harmful consequences if one of the databases fails.
- **Authenticity:** our system ensures authenticity because it employs one of the most known authentication techniques: zero-knowledge proof.

4.5 Comparison of Our Model with Other Alternative Authentication Systems

Security in the registration phase is crucial, guarding user data and system integrity. It establishes trust and repels threats from the outset. Therefore, we worked to enhance protection in this phase. Table 5 shows where our model sets itself apart from others.

Table 5. Registration phase security comparison.

Reference	Technique	layers of protection
[19]	Zero-knowledge proof	No additional layer of protection is implemented during the registration phase
[17]	Zero-knowledge proof	No additional layer of protection is implemented during the registration phase
[6]	Zero-knowledge proof and novel 6D chaotic system	No additional layer of protection is implemented during the registration phase
[12]	SSL handshake protocol with zero-knowledge proof	No additional layer of protection is implemented during the registration phase
[18]	Zero knowledge protocol with RSA cryptography	No additional layer of protection is implemented during the registration phase
Our model		An additional layer of protection is implemented during the registration phase, where the system splits the password into shares, distributing it across multiple databases. These shares can then be reconstructed during the final verification process

5 Conclusion

This paper presents a robust framework tailored for web applications, primarily focusing on enhancing user authentication and bolstering security. The framework's core innovation involves the integration of a zero-knowledge proof (ZKP) protocol with an RCS algorithm and SS schema for the first time. This special combination successfully protects user passwords from theft and abuse, protecting their confidentiality and secure usage.

Our study's extensive findings support the proposed strategy's efficacy when combined with the CIA (confidentiality, integrity, and availability) guiding principles. Notably, the NIST testing outcomes demonstrate the strong random properties of the produced key pairs, with 12 out of 16 tests yielding positive results. This indicates that the framework can create safe cryptographic keys, a critical component of any secure online application.

Additionally, time-testing findings show that the suggested method maximizes efficiency by reducing the amount of time needed for various activities. As a result of the system, the authentication procedure is quick and responsive, improving the user experience while ensuring security.

References

1. Ziemer, S.: An architecture for Web applications. Essay in DIF 8914 Distributed Information Systems (2002)
2. Althobaiti, M.M., Mayhew, P.: Security and usability of authenticating process of online banking: user experience study. In: 2014 International Carnahan Conference on Security Technology (ICCST), pp. 1–6. IEEE (2014)
3. Datta, N.: Zero knowledge password authentication protocol. In: Patnaik, S., Tripathy, P., Naik, S. (eds.) New Paradigms in Internet Computing. Advances in Intelligent Systems and Computing, vol. 203, pp. 71–79. Springer, Heidelberg (2013). https://doi.org/10.1007/978-3-642-35461-8_7
4. Hussien, F.T.A., Rahma, A.M.S., bdul Wahab, H.BA.: Structureal deasign of secure E-commerce websites employing multi-agent system. J. Al-Qadisiyah for Comput. Sci. Math. **14**(3), 88 (2022)
5. Ruoti, S., Andersen, J., Seamons, K.: Strengthening password-based authentication. In: Twelfth Symposium on Usable Privacy and Security (SOUPS 2016), Denver, Colorado (2016)
6. Mohammed, S.J., Mehdi, S.A.: Web application authentication using ZKP and novel 6D chaotic system. Ind. J. Electr. Eng. Comput. Sci. **20**(3), 1522–1529 (2020)
7. Yang, D., Yang, B.: A new password authentication scheme using fuzzy extractor with smart card. In: 2009 International Conference on Computational Intelligence and Security, vol. 2, pp. 278–282. IEEE (2009)
8. Kalayeh, M.R.G., Nik, M.H., Kordestani, H.: Using template-based passwords for authentication in e-banking. In: 7th International Conference on e-Commerce in Developing Countries: with focus on e-Security, pp. 1–9. IEEE, Kish Island, Iran (2013)
9. Idrus, S.Z.S., Cherrier, E., Rosenberger, C., Schwartzmann, J.J.: A review on authentication methods. Aust. J. Basic Appl. Sci. **7**(5), 95–107 (2013)
10. Alaa, A.H., Hashem, S.H.: A proposed firewall security method against different types of attacks. IRAQI J. Comput. Commun. Control Syst. Eng. **5**(1) (2005).
11. Tariq, A.: SMSCC: smarter and more secure credit card using neural networks in zero knowledge protocol. Al-Rafidain University College For Sciences, pp. 227-243 (2014). ISSN 16816870
12. Zaw, T.M., Thant, M., Bezzateev, S.V.: User authentication in SSL handshake protocol with zero-knowledge proof. In: 2018 Wave Electronics and its Application in Information and Telecommunication Systems (WECONF), pp. 1–8. IEEE (2018)
13. Srivastava, A.K., Mathur, A.: The Rabin cryptosystem & analysis in measure of Chinese Reminder Theorem. Int. J. Sci. Res. Publ. **3**(6), 1–4 (2013)
14. GeeksforGeeks Homepage. https://www.geeksforgeeks.org/rabin-cryptosystem-with-implementation/. Accessed 5 Oct 2023
15. Tso, R., Liu, Z.Y., Hsiao, J.H.: Distributed E-voting and E-bidding systems based on smart contract. Electronics **8**(4), 422 (2019)
16. Geng, C., Wang, J.: A multi-secret sharing scheme with combiner identification authentication. In: 2020 International Conference on Computer Communication and Network Security (CCNS), pp. 92–98. IEEE (2020)
17. Jun, B.L.J.: Implementing Zero-Knowledge Authentication with Zero Knowledge (ZKA wzk). Python Papers Monograph, 2 (2010)
18. Mainanwal, V., Gupta, M., Upadhayay, S.K.: Zero knowledge protocol with RSA cryptography algorithm for authentication in web browser login system (Z-RSA). In: 2015 Fifth International Conference on Communication Systems and Network Technologies, pp. 776–780. IEEE (2015)

19. Grzonkowski, S., Zaremba, W., Zaremba, M., & McDaniel, B.: Extending web applications with a lightweight zero knowledge proof authentication. In: Proceedings of the 5th International Conference on Soft Computing as Transdisciplinary Science and Technology, pp. 65–70, Cergy-Pontoise, France (2008)
20. Bhattacharjya, A.: A holistic study on the use of blockchain technology in CPS and IoT architectures maintaining the CIA triad in data communication. Int. J. Appl. Math. Comput. Sci. **32**(3) (2022)

A New Approach to Design KM1 Encryption Algorithm for Alarm in IoT

Mohammed D. Taha[✉] and Khalid A. Hussein[✉]

Computer Sciences Department College of Education, Mustansiriyah University, Baghdad, Iraq
muhammed84@uomustansiriyah.edu.iq, dr.khalid.ali68@gmail.com

Abstract. The research introduces a hybrid encryption system for parallel computing environments, focusing on safeguarding IoT image data for fire incident detection and response. The system uses a six-dimensional chaotic system to generate secure encryption keys, integrated into the RC6 encryption framework. Additionally, the study introduces eight new S-boxes and eight P-layers derived from the initial chaotic state parameters. The encryption process consists of eight rounds of RC6 encryption, incorporating these S-boxes and P-layers to produce a securely encrypted text suitable for protecting sensitive data. The primary objective of this system is to establish a secure channel for transmitting fire incident images to designated personnel via the Telegram platform, ultimately reducing response times and enhancing situational awareness. The parallel processing environment enhances image encoding speed, enabling real-time transmission during fire incidents. The system achieves a Peak Signal-to-Noise Ratio (PSNR) of 7.748 decibels, an Average Unified Change Intensity (UACI) of 33.4842, and a Number of Pixel Change Rate (NPCR) of 99.6082. The encryption speed for the Baboon image registers at 11.558 s, representing an approximate improvement of 1155.8% in performance. These elevated levels of security and computational efficiency underscore the potential applicability of the proposed algorithm in real-world scenarios, where secure encryption of diverse data types is of paramount importance.

Keywords: parallel computing · Chaotic · Hybrid algorithm · image encryption · S-Box · P-Layer

1 Introduction

Alarm systems are essential in both residential and commercial environments, protecting property, assets, and lives [1, 2]. The Internet of Things (IoT) has revolutionized the network, enabling seamless interconnection between devices [3]. However, the exponential rise in data traffic has increased the need for data privacy and security [4]. Encryption, a venerable cryptographic technique, is a potential solution for mitigating security challenges in the IoT domain [5]. However, the computationally intensive nature of encryption can engender processing delays, particularly in real-time applications [6].To address these challenges, lightweight encryption algorithms have been developed, such

A. M. Al-Bakry et al. (Eds.): NTICT 2023, CCIS 2096, pp. 395–410, 2024.
https://doi.org/10.1007/978-3-031-62814-6_28

as the RC6 [7] and PRESENT [8] algorithms. These algorithms are suited for deployment in resource-constrained systems and offer resistance against attacks. This research aims to propose a pioneering lightweight encryption algorithm that operates within a parallel processing milieu, catering to resource-constrained IoT devices. The proposed algorithm will accelerate the processing of voluminous data streams, ensuring the integrity of sensitive data in commercial environments.

2 Related Work

The work focuses on building efficient methods for data encryption and decryption using chaotic systems, RC6 and PRESENT, and other techniques. To perform various encoding and decoding processes, many algorithms are used. Mohammed et al. [9] This paper proposes an improved S-box and p-layer to address the problem that the primary PRESENT S-box and P-layer have an anti-fixed point. It also gives a brief description of the workings of the PRESENT algorithm. Ten new S-boxes and ten new P-layers are generated at random for the PRESENT algorithms by utilizing 6D chaotic systems. The security research has finally been finished, and the findings show that the chaos S-box and P-layer are excellent for securing sensitive data because they can withstand differential attacks and linear assaults better. Zaid M. et al. [10].The suggested algorithm has been altered using a chaotic system's five-dimensional equations as a basis. It is advised to use the newly designed lightweight algorithm in conjunction with a novel chaotic system to maintain good performance while increasing complexity. A text file with a size of 28 kb is used to calculate the execution time for the updated algorithm and the PRESENT cipher. The encryption operation took 42.45923 s to execute using the PRESENT cipher, but the modified technique took 1.89193 s to execute. Mai Helmy, and et al. [11].Research proposes a new encryption technique using 3-D Rubik's cube for 3D encoding photos. The method uses RC6 as the initial encryption method and combines permutation and diffusion techniques. The system is tested using a wireless OFDM system, achieving a PSNR of 45.3784 dB and 0.0043, and a time encryption of 1,012.1734 s. were obtained from the PSNR and correlation tests used to assess the quality of the decrypted images. Artan Berisha and Hektor Kastrati in [12]. The algorithm was put into practice both sequentially and concurrently, and it was evaluated for words of different lengths by the following values were used during testing: Round = 10, n. characters = 80000000, RC6 execution time = 26.113 s, RC6Par execution time = 10.158 s, ratio = 0.39, speedup = 2.54 s It was discovered that while the parallel version is quicker, it also uses up more resources.

3 Research Methods

3.1 Chaotic System

Chaos, additionally referred to as the "butterfly effect, is a phenomenon that results from the difficult aperiodic behaviors of deterministic systems and exhibits remarkable sensitivity to minute changes in initial conditions. In the framework of Shannon's confusion and diffusion principles, this characteristic of chaos has been used in cryptography.

Chaotic events have demonstrated potential as a source of pseudo-random information security by utilizing their mixing property and great sensitivity to tiny alterations. Due to their deterministic character, chaotic maps nonlinear dynamical systems that display chaotic behavior have proven to be especially helpful in the field of cryptography. These chaotic system characteristics have been used by researchers to increase the security of a number of cryptographic applications, including block and stream ciphers and picture encryption algorithms [5, 13, 14] [9]. It exhibits hyper chaotic behavior in the six-dimensional hyper chaotic system. It has the mathematical form of Eq. (1) [15].

$$\dot{x} = -a * x + b * y + c * w - dv$$

$$\dot{y} = e * x - f * x * z - g * e^v$$

$$\dot{z} = -h * z + x * y + i * v \tag{1}$$

$$\dot{w} = -w - y * z - g * v$$

$$\dot{v} = x + j * y - i * z$$

$$\dot{u} = k * x - L * u - j * z * w$$

The hyper chaotic behavior of the system given in Eq. (1) is comprised of six states: x, y, z, w, v, u, and Positive constants make up the following parameters: a, b, c, d, e, f, g, h, i, j, k, and l [15]. Six-dimensional chaotic systems offer enhanced security in cryptography due to their complexity, entropy, and dynamics. They provide a larger key space, making it difficult for adversaries to predict. However, their computational complexity makes them less suitable for low-resource devices and cryptographic schemes relying on chaotic systems.

3.2 Encryption Algorithm

RC6 Algorithm
The RC6 is a symmetric block cypher developed from the RC5, offering improvements to meet AES competition demands [16, 17]. It increases the number of working registers, strengthens resistance to linear and differential attacks, and approaches the avalanche effect more quickly. RC6 offers key sizes of 128, 192, and 256 bits and a block size of 128 bits. It uses a different multiplication operation, RC6, to make the rotation operation reliant on each bit in a word. The algorithm consists of key expansion, encryption, and decryption. The RC6 algorithm, shorthand for w = 32, b = 16, and r = 20, uses w/r/b for parameter values and consists of three parts: key expansion, encryption, and decryption [16, 18]. The RC6 algorithm is computationally efficient, flexible, and secure, withstanding years of cryptanalysis. However, its performance impact may be slower due to multiple rounds and complex operations, and it uses the same key for each block of encrypted text.

PRESENT Algorithm

Asymmetric-key encryption is a secure cryptographic method that uses 64-bit data blocks and either 80-bit or 128-bit key lengths. It employs a substitution-permutation network (SPN) operation, replacing input values with permutation steps. The algorithm uses a 4-bit S-box and a fixed table for protection. It also employs a key whitening technique, merging plaintext and key before encryption, making it more resistant to attacks [8]. In summary, the PRESENT algorithm offers advantages such as lightweight design, low power consumption, and security strength, but it also has limitations, including a small block size and potential susceptibility to certain types of attacks. The choice of whether to use Present depends on the specific requirements and constraints of the intended application [8, 9].

4 Proposed System

This section provides an in-depth exploration of the design and implementation phases of an alarm system that leverages lightweight encryption algorithms. The alarm system comprises unique algorithm designed to execute encryption and decryption processes in both serial and parallel models. The selected cryptographic primitives include the 6-D chaotic system, coupled with the RC6, and Present algorithms. These cryptographic components are employed to devise novel hybrid algorithm. The hybrid algorithm integrates six-dimensional chaotic system s with the RC6 and Present algorithms. These hybrid algorithms are instrumental in the real-time transmission of images when a fire incident occurs. The images are securely transmitted via Telegram to the primary security guard responsible for safeguarding the building. The guard can subsequently decrypt and view the transmitted images to ascertain the fire's location, The proposed algorithm is designed to ensure secure transactions in various sectors, including financial, healthcare, IoT, and educational institutions. It protects sensitive information, ensures confidentiality and integrity of financial data, and facilitates secure communication between cloud services, logistics, supply chain management, and research data security. The proposed KM1 a robust algorithm in different computing environments is one that demonstrates resilience, adaptability, and consistent performance across various platforms, operating systems, and usage scenarios. The design and implementation of the algorithm play a crucial role in achieving these characteristics. as visually depicted (see Fig. 1).

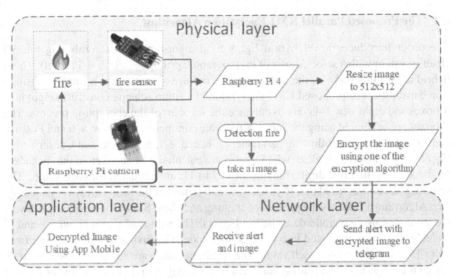

Fig. 1. Main diagram of the proposed alarm in IoT system

4.1 Design Proposed Alarm in IoT System

To designing a proposed IoT-based alarm system with a three-layer architecture involves creating a system that's:

Physical Layer: The physical layer of a system refers to the lowest level of the network stack, where physical devices and Connections are managed. In the context of a fire detection and monitoring system using a Raspberry Pi 4 and a camera, the physical layer involves the hardware components and their interactions.

Roseberry pi 4: Roseberry Pi 4 represents a central processing unit or microcontroller at the heart of the alarm system. It serves as the brain of the system, orchestrating the interactions between the various components and executing the necessary algorithms for fire detection and image capture.

Fire Sensor: The fire sensor is a critical sensor device designed specifically for detecting the presence of fires or elevated temperatures. It operates on the principle of monitoring environmental conditions for anomalies associated with fires.

Webcam: The webcam is an optical device equipped with a camera and imaging capabilities. It is employed to capture visual data in the form of images or videos when a fire event is detected.

Network Layer: IoT (Internet of Things) devices can be integrated with network layers, including. Cellular networks like 4.5G.LTE-Advanced) by using modems and communication modules.

Application Layer: Integrating IoT with the application layer using Telegram as a communication channel is a powerful way to receive real-time notifications and alerts, and control your IoT devices remotely.

4.2   The Proposed Parallel KM1 Encryption Algorithm

The paper introduces novel hybrid lightweight algorithms, each combining the 6D chaotic system with a selected pair of cryptographic primitives (RC6 + Present). These hybrid algorithms serve as the cornerstone of the fire alarm system's image transmission capability. The proposed KM1 encryption algorithm scheme constructs eight new S-boxes and eight new P-layers to enhance the security of the encryption process. The S-boxes are generated using the input initial state parameters x, y, z, w, v, u and Positive constants make up the following parameters a, b, c, d, e, f, g, h, i, j, k, and l, described in Eq. (1). The values of xi, yi, zi, wi, vi, and ui are calculated and converted into hexadecimal code. The first five digits, ranging from 7 to 11, are extracted for each value,. The resulting values are stored in a 16x8 array, which is used as the S-box array for encoding. (see Algorithm 1). Shows the process of creating an S-box. Similarly, eight new P-layers are generated using chaotic described in Eq. (1). The values of xi, yi, zi, wi, vi, and ui are calculated and converted into hexadecimal code. The first five digits, ranging from 11 to 15, are extracted for each value. The resulting values are stored in a 64 × 8 array, which is used as the P-layer array for encoding (see Algorithm 2). Shows the process of creating a P-layer. The encryption process uses the RC6 algorithm, S-box, and P-layer, with eight rounds of encryption. In each round, the data is divided into 16-byte blocks, which are further divided into four blocks of 4-byte size stored in A, B, C, and D. A and D are merged into eight bytes and entered into the first S-box. The resulting values are divided into A and D of four bytes each. Similarly, B and C are merged into eight bytes and entered into the first S-box. The resulting values are divided into B and C of four bytes each. A and B are merged and entered into the first P-layer. The resulting values are divided into A and B of four bytes each. Similarly, D and C are merged and entered into the first P-layer. The resulting values are divided into D and C of four bytes each. The resulting values are entered into the first round of the RC6 algorithm to obtain an encrypted and secure block. This process is repeated over eight rounds to obtain the encrypted (A, B, C, and D) block. Finally, the above process is repeated for all plaintext blocks to obtain the cipher text, which is stored in a text file. The resulting cipher text is highly secure and suitable for protecting sensitive data. See in algorthim1 depicts the block diagram of the encoding process utilizing the proposed KM1 encryption algorithm (see Fig. 2).

Algorithm 1: Generating S-Box using 6D chaotic system.

Input: initial values for the 6D chaotic
Output: New S-Box 16
Begin
Step1: read initial conditions.
Step2: Set index= 0, string array S-Box [16]
Step3: x,y,z,u,v,w = Generation 6D Chaotic
Step 4: Convert (x,y,z,u,v,w) to hex and get only five digits from index (7 to 11) and then the remainder of the partition is taken to 16 and then saved to S
Step5: Set i= 0
Step6: If S[i] Contents in S-Box then   i = i + 1
        else
            S-Box[index] = S[i], index = index + 1
        End if
Step7: While (index <16)
        i = i + 1
        if (i < 16) go to Step 6
         else go to Step 3
        end if
        end while
End

Algorithm 2: Generating P-Layer using 6D chaotic system

Input: initial values for the 6D chaotic
Output: New P-Layer [64]
Begin
Step1: read initial conditions.
Step2: Set index= 0, string array P-Layer [64]
Step3: x,y,z,u,v,w = Generation 6D Chaotic
Step 4: Convert (x,y,z,u,v,w) to hex and get only five digits from index (11 to 15) and then the remainder of the partition is taken to 64 and then saved to S
Step5: Set i= 0
Step6: If S[i] Contents in P-Layer then   i = i + 1
        else
            P-Layer [index] = S[i], index = index + 1
         End if
Step7: While (index <64)
        i = i + 1
        if (i < 16) go to Step 6
         else go to Step 3
        end if; end while
End

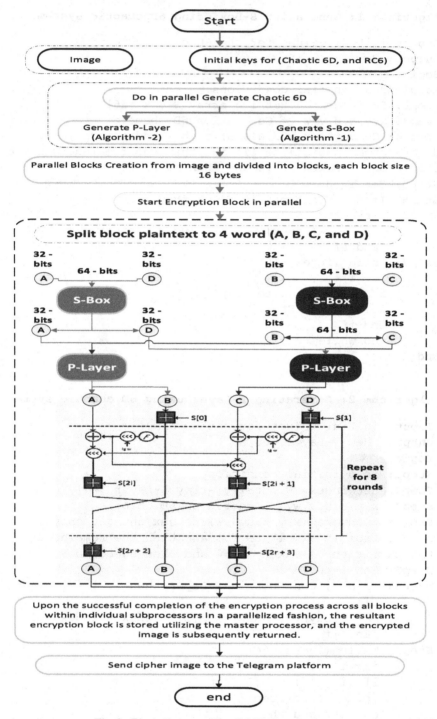

Fig. 2. Block diagram of Parallel KM1 method steps.

5   Results and Discussion

The experimental evaluation of the proposed approach was conducted on a computational system with a 3.30 GHz CPU, 16 GB of RAM, and utilizing Python programming language on the 64-bit Windows 11 Home operating system. The test images used in the experiments were selected from a dataset, including Baboon, Bird, and Lena gray images with color depths of up to 24 bits per pixel for Baboon and Bird, and 8 bits per pixel for Lena. While the proposed system was tested on ten images from the dataset, these images exhibited a color depth of 24 bits per pixel. (see Fig. 3).

| Baboon | Lena | Bird | image from web |
| (512 * 512) | (512 * 512) | (620 * 413) | camera (512 * 512) |

Fig. 3. A set of experimental images

The effectiveness of the current cryptosystem is assessed using the following standards:

5.1   Key Space

Key space is the range of possible keys that the cryptographic algorithm generates. Increasing the dimensionality of the key space makes cryptographic assaults more complex and strengthens security measures by expanding the pool of possible keys. A careful evaluation of the critical area requires consideration of the necessary security measures in relation to the projected threat environment. The suggested KM1 algorithm makes use of a set of 16 keys, which include 10 parameters and 6 initial conditions as the confidential key. The precision of each key ranges from 10^{16}. As a result, for a hyper chaotic system, the size of the keys space reaches $(10^{16})12 = 10^{192}$, or roughly 2^{638}. By contrast, the key space of RC6 with 2^{256} exceeds the 2^{128} threshold, coming in at about 2^{894}. This large key space magnitude strengthens the encryption technique by providing a significant defense against brute force attacks.

5.2   Avalanche Criteria (AC)

A block cipher's performance must be evaluated using the Avalanche Criterion (AC). The effect of a small change in plaintext on the cipher text is assessed by the AC to determine whether there is a lack of correlation between the input bits and output sequence. The optimal value for the AC value, which spans from 0 to 1, according to Eq. (2), is 0.5[6].

The results of the AC test for the proposed cryptosystem are reported the following Table 1.

$$C = \frac{\text{Number of Flipped Bits in Cipher Text}}{\text{Number of All Bits in Cipher Text}} \qquad (2)$$

Table 1. Avalanche test for Proposed KM1 parallel Algorithm

Image	AC
Baboon	0.5001
Bird	0.50024
Lena	0.500
Image from web camera	0.5102

5.3 NIST Randomness Test

The suggested encryption algorithm's resilience was examined by applying the NIST test, which is an accepted method to assess the randomness of randomized bit sequences, to the cipher text it produced. The outcomes of the 16 distinct NIST tests, along with the P-values for each test. The test is considered to have been successful if the P-value is greater than 0.01 and the sequence is shown to be random [19]. The following Table 2 demonstrates that the suggested technique produced a random cipher for the Baboon picture encryption after passing all 16 of the NIST tests.

5.4 Entropy

The randomness of the data in an image is measured by entropy. The data within an image is said to be more disordered when the entropy value is high [20]. Equation (3) can be used to calculate entropy, with a target entropy value of 8 or near to it. The equation is provided by, specifically:

$$Entropy = \sum_i P(S_i)\log_2\left(\frac{1}{P(S_i)}\right) \qquad (3)$$

In this case, P(Si) denotes the likelihood that pixel Si (i = 0 to 255) would appear in an image. The entropy value, which is close to 8, the following Table 3. A number that is nearer 8 indicates a lesser likelihood of unintentional information disclosure. The suggested plan performs better than earlier research.

Table 2. NIST Test for Proposed KM1

Type of Test	P-Value	Status	Type of Test	P-Value	Status
Frequency Test (Monobit)	0.62696	Ok	Maurer's Universal Statistical	0.56697	Ok
Frequency Test within a Block	0.25652	Ok	Linear Complexity	0.02272	Ok
Run Test	0.69523	Ok	Serial test	0.0638	Ok
Longest Run of Ones in a Block	0.43326	Ok	Approximate Entropy	0.74404	Ok
Binary Matrix Rank	0.71919	Ok	Cumulative Sums (Forward)	0.74396	Ok
Discrete Fourier Transform (Spectral)	0.57563	Ok	Cumulative Sums (Reverse)	0.89059	Ok
Non-Overlapping Template Matching	0.11091	Ok	Random Excursions (+1)	0.42370	Ok
Overlapping Template Matching	0.53420	Ok	Random Excursions Variant (+1)	0.05623	Ok

Table 3. The results of the information entropy are presented in comparison to previous studies

Image	Entropy
Baboon	7.9995
Bird	7.9985
Lena	7.999
Image from web camera	7.998

5.5 Correlation Analysis

The efficacy of an encryption method is examined in this study by looking at the analysis of correlations between the original and encrypted photographs. The suggested encryption technique outperforms existing techniques and keeps correlation values low. A mathematical method is used to generate correlation values, which are then observed to be close to zero using Eq. (4), showing that there is no meaningful association among nearby uninteresting pixels in any of the encrypted images. The proposed encryption method is successful in preserving low correlation values, according to the results [21]. The following Table 4.

$$\text{Correlation} = \sum \left(\frac{(i - \mu i)(j - \mu j)}{\sigma_i \sigma_j} \right) \qquad (4)$$

Table 4. The outcomes of the correlation coefficients

Image	correlation
Baboon	0.0018
Bird	0.000067
Lena	0.00083
Image from web camera	0.001093

5.6 PSNR and MSE Analysis

Based on the Mean Squared Error (MSE) and Peak Signal to Noise Ratio (PSNR) values of the original and encrypted photos [22], which are shown in Table 5 for the Baboon, Peppers, and Bird images, the encryption technique' quality is assessed. Indicators of a better encryption outcome include a higher MSE value and a lower PSNR value, respectively. The decrypted image should also have an MSE value of 0 and a PSNR value of infinity, it is important to note [23].

Table 5. MSE and PSNR values were calculated for both encrypted and decrypted images.

image	Encrypt		Decrypt	
	MSE	*PSNR*	*MSE*	*PSNR*
Baboon	10913.15	7.7513	0	∞
Bird	10917.430	7.748	0	∞
Lena	10920.65	7.749	0	∞
Image from web camera	10916.65	7.747	0	∞

5.7 Differential Attack

Using two metrics—NPCR and UACI—calculated using Eqs. (5) and (6) for MH picture size[19, 22], the suggested image encryption technique was tested against differential assaults. The results, which are shown in Table 6, demonstrate that the suggested technique performs better than earlier research and is extremely sensitive to a change of one pixel.

$$UACI = \frac{1}{W * H} \left[\sum_{i,j} \frac{|C_1(i,j) - C_2(i,j)|}{255} \right] * 100\,\% \qquad (5)$$

$$D(i,j) = \begin{cases} 1, C_1(i,j) \neq C_2(i,j) \\ 0, C_1(i,j) = C_2(i,j) \end{cases} \qquad (6)$$

The terms "original image encryption" and "original image encryption after a single pixel change" are denoted by the variables "C1" and "C2," respectively.

Table 6. NPCR and UAC values with comparison to related works.

Image	NPCR	UACI
Baboon	99.6085	33.4285
Bird	99.6082	33.4842
Lena	99. 651	33.4626
Image from web camera	99. 6317	33.4705

5.8 Analysis Proposed KM1 Algorithm Based on Running Execution Time.

Table 7. Shows the sequential and parallel execution times for the proposed encryption and decryption methods on different kinds of files with various sizes. According to the results, the parallel method is more efficient than the serial method at executing tasks [24, 25].

Table 7. Execution Time for sequential and parallel KM1 algorithm on different kinds of files

File Type	Size (bytes)	Sequential KM1(sec)		Parallel PKM1(sec)	
		Encryption	Decryption	Encryption	Decryption
Docx	57838	0.6402	0.6333	0.41647	0.3546
Pptx	104386	1.0485	0.9253	0.5942	0.5795
Jpg	176120	1.1294	1.1019	0.6937	0.6814
Png	188790	2.2201	2.0214	0.8479	0.8094
Bmp	588020	10.1387	10.0970	4.8001	4.6916
Xlsx	1081797	47.7686	44.8829	14.0172	13.1664
Pdf	1363583	76.3565	73.7135	18.2781	17.3185
Mp4	1950351	143.9913	141.9913	30.6611	29. 8534
Txt	2128609	147.6129	145.9173	32.3768	31. 0638

5.9 Speed up and Efficiency.

Speed up represents the ratios between sequential and parallel runtimes [23] as "Eq. (7)".

$$\text{Speed up} = T_S/T_P \tag{7}$$

Parallel efficiency denotes ratios between speed-up values and processor counts [12] [21] as "Eq. (8)".

$$\text{Efficiency} = \text{speed up}/p \tag{8}$$

where p is the number of processors, T_S is the run time for the sequential implementation, and T_P is the run time for the parallel variant. The following Table 8.

Table 8. Execution Time, Speed up and Efficiency for KM1 parallel KM1(PKM1) algorithm.

Image	Sequential KM1(S)		Parallel KM1(S)		Speed up		Efficiency	
	Encrypt	Decrypt	Encrypt	Decrypt	Encrypt	Decrypt	Encrypt	Decrypt
Baboon	3.93	3.78	0.34	0.31	11.558	12.193	2.8895	3.04825
Bird	3.55	3.36	0.24	0.34	14.791	9.8823	3.6427	2.4705
Lena	3.98	3.83	0.34	0.36	11.705	8.906	2.9262	2.226
Image from web camera	3.99	3.93	0.37	0.35	10.783	11.228	2.6959	2.807

5.10 Comparative Analysis of the Proposed Image Encryption Method with Previous Studies

Table 9 Undertake a comparative examination of the Performance Evaluation Metrics derived from the proposed algorithms against those derived from preceding studies. The tabulated data indicates a superiority of outcomes in favor of the suggested methods in comparison to their antecedents.

Table 9. Comparing the proposed for encrypting baboon images with previous studies

Methodology	Entropy	PSNR	NPCR	UACI	Correlation Coefficient
[26]	-	-	99.6	31.9	0.063
[11]	-	10	-	-	0.0043
Proposed KM1	7.9995	7.7513	99.5981	33.4315	0.0018
Proposed PKM1	7.9995	7.7397	99.60029	33.5043	-0.00066

6 Conclusion

The proposed KM1 encryption methodology is designed for resource-constrained environments, providing robust safe-guards for sensitive data. It uses a hybrid cryptographic framework, combining Chaotic system 6D, PRESENT (S-Box, P-Layer) and RC6 algorithms, executing both sequential and in parallel. The algorithm's efficacy has been proven through comprehensive evaluations, showing superior computational efficiency. The parallel encryption method achieves execution times of 0.34 s and 0.31 s for baboon image encryption and decryption, compared to 3.93 s and 3.78 s for serial methods. The integration of Telegram for transmission of notifications and encrypted images has expedited incident response, streamlined responsibilities and enhanced operational efficiency.

Reference:s

1. Aziz, A., Singh, K.: Lightweight security scheme for Internet of Things. Wireless Pers. Commun. **104**(2), 577–593 (2019). https://doi.org/10.1007/s11277-018-6035-4
2. Daoud, R.W., Al-Khashab, Y.M.B.I.: Design and simulation of smart control system for internet traffic distribution on servers by using fuzzy logic system. Al-Kitab J. Pure Sci. **2**(1) (2018)
3. Jasim, O.A., Hussein, K.A.: A hyper-chaotic system and adaptive substitution box (S-Box) for image encryption. In: 2021 International Conference on Advanced Computer Applications (ACA), pp. 144–149 (2021)
4. Jafer, A.S., Hussein, K.A., Naif, J.R.: Secure wireless sensor network authentication by uti-lized modified SHA3 hash algorithm. In: 2022 International Conference on Data Science and Intelligent Computing (ICDSIC), pp. 20–24 (2022)
5. Hoomod, H.K., Naif, J.R., Ahmed, I.S.: A new intelligent hybrid encryption algorithm for IoT data based on modified PRESENT-Speck and novel 5D chaotic system. Period. Eng. Nat. Sci. **8**(4), 2333–2345 (2020)
6. Hussein, K.A., Mahmood, S.A., Abbass, M.A.: A new permutation-substitution scheme based on Henon chaotic map for image encryption. In: 2019 2nd Scientific Conference of Computer Sciences, pp. 63–68 (2019)
7. Rivest, R.L., Robshaw, M.J., Sidney, R., Yin, Y.L.: The RC6TM block cipher. In: First Advanced Encryption Standard (AES) Conference, p. 16 (1998)
8. Bogdanov, A., et al.: PRESENT: an ultra-lightweight block cipher. In: Cryptographic Hard-ware and Embedded Systems-CHES 2007: 9th International Workshop, Vienna, Austria, 10–13 September 2007, Proceedings, vol. 9, pp. 450–466 (2007)
9. Taha, M.D., Hussein, K.A.: Generation S-box and P-layer For PRESENT algorithm based on 6D hyper chaotic system. Al-Kitab J. Pure Sci. **7**(1), 48–56 (2023)
10. Kubba, Z.M.J., Hoomod, H.K.: Modified PRESENT encryption algorithm based on new 5D Chaotic system. In: Conference Series: Materials Science and Engineering, vol. 928, no. 3 (2020). https://doi.org/10.1088/1757-899X/928/3/032023
11. Helmy, M., El-Rabaie, E.S.M., Eldokany, I.M., El-Samie, F.E.A.: 3-D image encryption based on Rubik's cube and RC6 algorithm. 3D Res. **8**, 1–12 (2017)
12. Berisha, A., Kastrati, H.: Parallel implementation of RC6 algorithm. J. Comput. Sci. Technol. Stud. **3**(2), 1–9 (2021)
13. Wang, H. Zheng, H., Hu, B., Tang, H.: Improved lightweight encryption algorithm based on optimized S-box. In: Proceedings - 2013 International Conference on Computational and Information Sciences, ICCIS 2013, pp. 734–737 (2013). https://doi.org/10.1109/ICCIS.2013.198
14. Tang, Z., Cui, J., Zhong, H., Yu, M.: A random PRESENT encryption algorithm based on dynamic s-box. Int. J. Secur. Appl. **10**(3), 383–392 (2016). https://doi.org/10.14257/ijsia.2016.10.3.33
15. Mehdi, S.A., Ali, Z.L.: A new six-dimensional hyper-chaotic system. In: 2019 International Engineering Conference (IEC), pp. 211–215 (2019)
16. Rivest, R., Robshaw, M.J.B., Sidney, R., Yin, Y.L.: The RC6 Block Cipher. First Adv. Encryption … (1998). http://citeseerx.ist.psu.edu/viewdoc/summary?doi=10.1.1.35.7355
17. Ganavi, M., Prabhudeva, S., Nayak, S.N.: A secure image encryption and embedding approach using MRSA and RC6 with DCT transformation. Int. J. Comput. Netw. Appl. **9**(3), 262–278 (2022). https://doi.org/10.22247/ijcna/2022/212553
18. Faisal, Z., Abdul Ameer, E.H.: Encryption image by using RC6 and hybrid chaotic map. Webology **17**(2), 189–199 (2020). https://doi.org/10.14704/WEB/V17I2/WEB17024

19. Rashid, A.A., Hussein, K.A.: Image encryption algorithm based on the density and 6D logistic map. Int. J. Electr. Comput. Eng. **13**(2), 1903–1913 (2023). https://doi.org/10.11591/ijece.v13i2.pp1903-1913

20. Salman, R.S., Farhan, A.K., Shakir, A.: Creation of S-Box based one-dimensional chaotic logistic map: colour image encryption approach. Int. J. Intell. Eng. Syst. **15**(5), 378–389 (2022). https://doi.org/10.22266/ijies2022.1031.33

21. Rashed, A.A., Hussein, K.A.: A lightweight image encryption algorithm based on elliptic curves and chaotic in parallel. In: 2022 3rd Information Technology to Enhance e-learning and Other Application (IT-ELA) (2022). https://doi.org/10.1109/IT-ELA57378.2022.10107924

22. Jasem, N.N., Mehdi, S.A.: Multiple random keys for image encryption based on sensitivity of a new 6D chaotic system. Int. J. Intell. Eng. Syst. **16**(5), 576–585 (2023). https://doi.org/10.22266/ijies2023.1031.49

23. Salman, M.R., Farhan, A.K.: Color image encryption depend on DNA operation and chaotic system. In: 2019 First International Conference of Computer and Applied Sciences (CAS), pp. 267–272 (2019)

24. Hussein, K.A., Kareem, T.B.: Proposed parallel algorithms to encryption image based on hybrid enhancement RC5 and RSA. In: Proceedings of 5th International Engineering Conference IEC 2019, pp. 101–106 (2019). https://doi.org/10.1109/IEC47844.2019.8950593

25. Rashid, A.A., Hussein, K.A.: A lightweight image encryption algorithm based on elliptic curves and a 5D logistic map. Iraqi J. Sci., 5985–6000 (2023)

26. Aguila, C.B.B., Sison, A.M., Medina, R.P.: Enhanced RC6 permutation-diffusion operation for image encryption. In: Proceedings of the 2018 International Conference on Data Science and Information Technology, pp. 64–68 (2018)

Author Index

A. M. Al-Bakry et al. (Eds.): NTICT 2023, CCIS 2096, pp. 411–412, 2024.
https://doi.org/10.1007/978-3-031-62814-6

Printed in the United States
by Baker & Taylor Publisher Services